Applied Econometrics

Massimiliano Marcellino

BUP

An Introduction

EGEA S.p.A.
Via Salasco, 5 - 20136 Milan, Italy
Phone + 39 02 5836.5751 - Fax +39 02 5836.5753
egea.edizioni@unibocconi.it - www.egeaeditore.it

First Edition: September 2016

ISBN Domestic Edition 978-88-99902-04-9
ISBN International Edition 978-88-85486-50-8
ISBN Pdf edition 978-88-85486-51-5

Print: Digital Print Service, Segrate (Milan)

Index

Applied Econometrics. An Introduction

Massimiliano Marcellino, Bocconi University

Econometrics deals with the quantitative study of economic relations and requires combining statistics with economics. The emphasis can be more on the analysis and development of econometric methods or on their application. In the second case, we talk about applied econometrics.

Applied econometrics requires a basic knowledge of mathematics and statistics, such as elementary functional analysis (continuity, derivatives, optimization etc.), possibly matrix calculus, the notion of random variable, expected value and moments, cumulative distribution and density functions. These skills are further developed and combined with economic theory studied in courses such as economics, monetary economics or political economy to lead to the formulation of empirical models of economic reality.

A further fundamental component of applied econometrics is the practical implementation of the theoretical notions — empirical economic analyses using the appropriate econometric software.

The goal of this book is to facilitate both teaching of applied econometrics, particularly in undergraduate and Master courses, and learning by students and, more generally, by those concerned with a formal measurement of economic events. It is not an easy task because it requires combining statistics, economics, and computer science in the right proportions.

Statistics is needed for a correct formulation of the problem and interpretation of the results, but an excess of formalization may discourage students. For this reason, the statistical content of this book is rigorous but limited to what is strictly necessary for a proper application of the methods. All theoretical concepts are then illustrated empirically, with examples that use either simulated data, in order to have a more immediate and controlled feedback, or actual data on economic variables. The software used is EViews, usually available in academic computer rooms or otherwise at an affordable price.

Chapter 1 contains a brief introduction to the problems faced by econometrics and to different types of economic (or, more generally, social) data that can be analyzed with econometric techniques. Chapter 2 presents the linear regression model, which often provides a good representation of the relationships between economic variables, and the estimators of the parameters of the model and their properties. Chapter 3 focuses on techniques to test hypotheses about the parameters of the linear regression model. Chapter 4 assesses the effects of violations of the assumptions underlying the linear model and develops extended versions of the model that require less restrictive conditions of applicability. Chapter 5 examines the consequences of unmodelled changes in the model parameters and possible remedies. Chapter 6 explicitly considers the case of stochastic explanatory variables. Chapter 7 proposes an introduction to dynamic models. Chapter 8 discusses models for panel data, which have both a longitudinal and a temporal dimension. Chapter 9 deals with models for binary variables, such as those resulting from questionnaires or other types of qualitative analyses.

Each chapter begins with the necessary theoretical background, continues with the practical applications based on simulated and real data using EViews, and concludes with a summary of the main concepts developed in the chapter and with both theoretical and applied exercises as a way to test and improve learning. The solutions of the exercises are available on http://mybook.egeaonline.it.

The material contained in the book is the evolution of lecture notes prepared during several years of academic courses in Econometrics and Applied Econometrics, mostly taught at Bocconi University. These notes were based on several textbooks, including Greene (1991), Pindyck and Rubinfeld (1998), Hill, Griffiths and Judge (2001a, 2001b), Judge, Hill, Griffiths, Lutkepohl and Lee (1988) and Spanos (1986).

Although these texts remain a good reference for further study, with more recent publications such as Wooldridge (2003) and Stock and Watson (2003), the contents of this book are an original synthesis, designed explicitly for undergraduate and Master courses, enriched by further statistical insights, many practical examples, and an extensive number of exercises.

This book is a translated and slightly extended version of the second edition of the book *Introduzione all'Econometria Applicata*. I am grateful for comments from several students and colleagues, with special thanks to Novella Maugeri and Claudia Foroni, though I retain responsibility for any remaining errors.

References

Greene, W.H., (1991), *Econometric Analysis*, Macmillan.
Hill, R.C., Griffiths, W.E., Judge, G.G. (2001a), *Undergraduate Econometrics*, 2nd edition, Wiley 2001
Hill, R.C., Griffiths, W.E., Judge, G.G. (2001b), *Using E-views for Undergraduate Econometrics*, 2nd edition, Wiley.
Judge, G.G., Hill, R.C., Griffiths, W.E., Lutkepohl, H., Lee, T.C. (1988), *An Introduction to the Theory and Practice of Econometrics*, Wiley.
Pindyck, R.S., Rubinfeld, D.L., (1998), *Econometric Models and Economic Forecasts*, McGraw-Hill.
Spanos, A. (1986), *Statistical Foundations of Econometric Modelling*, Cambridge University Press
Stock, J.H., Watson, M.W. (2003), *Introduction to Econometrics*, Addison-Wesley.
Wooldridge, J.M. (2003), *Introductory Econometrics*, 2nd edition, Thomson.

Additional resources are available online via MyBook:
http://mybook.egeaonline.it

1. Introduction

1.1 What is econometrics?

Econometrics deals with the quantitative study of economic relations. It is a tool to interpret reality in the light of economic theory, using statistical techniques.
The starting point is always a question that requires a quantitative answer. This can be for example the determination of which fraction of disposable income is consumed and which spared, of the effect of the increase in the price of a good or asset on the quantity demanded, of what happens to a stock index if the Central Bank raises the interest rate, of the effect of an increase in advertising expenditure on the sales of a product, of the extent to which public investment can increase economic growth, of the amount by which real wages vary when productivity increases, of the effect of a reduction in the rate of pollution on health spending, of the increase in aggregate investments if business taxes are reduced, of the interest rate on deposits or loans offered by a bank to different types of customers, of the effect of increased research and development spending on the number of registered patents, of the changes in the likelihood of losing the job determined by the level of education, of the effect of terrorist attacks on the growth rate of an economy, of the impact of a change in oil prices on the price of gasoline, of the extent of the reduction in exports when the exchange rate appreciates, …
Virtually every economic question that requires a quantitative response may be the subject of an econometric study.

1.2 Elements of an econometric study

Once you have identified the problem you want to tackle, you must assess whether and at what level of accuracy economic theory provides guidance for its solution.

For example, if you want to determine which fraction of disposable income is consumed and which spared, you can resort to the microeconomic theory of optimizing consumers or to Keynesian macroeconomic theory, depending on whether the interest focuses on a specific consumer or the entire community. Assuming you want to study the entire collectivity, Keynesian theory links the consumption level to that of disposable income:

$$C = C_0 + c * Y_d , \tag{1.2.1}$$

where C indicates aggregate consumption, C_0 autonomous consumption, Y_d disposable income and c the marginal propensity to consume. Moreover, it must be $C_0 > 0$ and $0 \leq c \leq 1$. The proposed relationship between consumption and income is therefore very precise but to calculate savings as

$$S = Y_d - C = (1-c) * Y_d - C_0 , \tag{1.2.2}$$

we need to know the precise values of c and C_0.
Also, the answer to the original question is based on the assumption hat the Keynesian theory is correct, but there is no consensus about this at the theoretical level. For example, the alternative life cycle theory suggests that consumption depends not only on income but also on wealth.
Therefore, we need to provide a value for the parameters c and C_0, but also to check whether the Keynesian theory is correct or not. Providing a value for c and C_0, econometrics allows *a more accurate description of the economic reality.* Veryfing whether consumption also depends on wealth, econometrics allows you to *test hypotheses about the validity or otherwise of an economic theory.*
For other questions of interest – for example, to evaluate the effects of a reduction in the rate of pollution on health spending –, there is no specific economic theory. In these cases, the role of econometrics is, therefore, all the more important because it allows you to *derive empirical regularities* from the analysis of the economic reality, which can then *provide an opportunity to develop an appropriate economic theory.*
Having established an economic theory of reference, or lack thereof, the next step is to develop an *econometric model.* This typically requires specifying a relationship between the expected value of the variable of interest and potential explanatory variables, as suggested by economic theory or empirical observation, with additional assumptions about the difference between the actual and the expected value of the variable.
Continuing with the example of consumption, and taking the Keynesian theory as valid, an econometric model is expressible as:

$$E(C) = C_0 + c * Y_d \quad (1.2.3)$$

$$C - E(C) \sim N(0, \sigma_C^2) \quad (1.2.4)$$

where $E(C)$ indicates the expected value of consumption and N the Normal (Gaussian) density. Combining (1.2.3) and (1.2.4), we get:

$$C \sim N(C_0 + c * Y_d, \sigma_C^2) \quad (1.2.5)$$

In the example of pollution, although economic theory does not help us, we can assume that the expected value of health spending (*SS*) is still linearly related to the pollution level (*LI*),

$$E(SS) = a + b * LI, \quad (1.2.6)$$

and that deviations of SS from its expected value satisfy:

$$SS - E(SS) \sim N(0, \sigma_{SS}^2). \quad (1.2.7)$$

Hence, the econometric model becomes:

$$SS \sim N(a + b * LI, \sigma_{LI}^2). \quad (1.2.8)$$

The problem is now better defined: the parameters that we want to estimate, or on whose size we want to conduct tests, are those of the expected value of a variable that has a certain Normal distribution, in the case of the examples. Statistical theory for parameter estimation and hypothesis testing are then helpful, as we shall see in detail in the following chapters.
In order to apply statistical theory to the econometric model, it is necessary to collect a sample of data that provides relevant information about the parameters of the model. Continuing the previous example, to determine which fraction of disposable income is consumed and which spared — to estimate the parameters C_0 and c in (1.2.5), we need data on aggregate consumption, on disposable income and, possibly, on wealth. Similarly, in order to assess the effect of a reduction in the rate of pollution on health spending, which is to estimate the parameters a and b in (1.2.8), we need measurements of the rate of pollution and data on health care expenditures. We will see in the next section that the data can be of many different types, and we will therefore also need to select the most appropriate one for the application of interest.

Assuming at this point that we have estimated the model parameters, it is necessary to interpret them correctly. With reference, for example, to equation (1.2.3), the parameter C_0 indicates that if the disposable income is 0, then the expected consumption will be equal to C_0. If instead there is a marginal change in disposable income, then there will be a corresponding variation in the expected value of consumption given by $c * \Delta Y_d$. We will return in more detail to the interpretation of the results in the next chapters, but it is good to clarify here three concepts.

First, we do not know the model parameters, rather we estimate them using statistical procedures and so there is a *more or less broad uncertainty around their values* that must be borne in mind when interpreting the results. For example, if the estimated value for the marginal propensity to consume, *c*, is 0.8, with a confidence interval at 90% for *c* of [0.4, 1.2], then we need to keep in mind that the effects of a change in disposable income on consumption are very uncertain. This is particularly important when the results of the econometric study are used to support economic policy decisions. For example, on the basis of the estimated coefficient *b* in (1.2.6) you may decide to allocate more funds to reduce pollution to obtain savings on health spending. However, if there is substantial uncertainty regarding the value of *b*, then the positive effects on health spending are very uncertain, while the costs to reduce pollution are certain.

Second, you might think to use the model to assess the effects on the variable of interest of changes of any size in the explanatory variables, and not just their marginal variations as previously indicated. For example, you may want to estimate the effects of a 10% increase in disposable income on aggregate consumption to gauge whether a substantial tax reduction will help to revive the economy. The problem is that when the explanatory variable changes substantially even the model parameters could change. For example, if the massive increase in disposable income is associated with a tax reduction, consumers might decide to reduce the marginal propensity to consume and increase the savings rate to cope with possible future tax increases needed to rebalance the fiscal budget. This interpretation problem was noted by Bob Lucas in the 1970s and is known as the *Lucas Critique.*

Third, the causal interpretation of the results of an econometric model is sustainable only when the model is based on economic theory. For example, although there is disagreement on other details, all theories of consumption agree that an increase in income leads to an increase in consumption. Therefore, an econometric model in which the estimated marginal propensity to consume is positive and of meaningful size can be used to support the statement that income causes consumption.

If instead the econometric model is not based on a valid and shared economic theory, the fact that a variable *x* has an estimated coefficient significantly different from 0 to explain a variable y does not imply that x causes y. This is because the estimators of the parameters of the model, as we shall see in detail in the next chapter, rely fundamentally on the statistical notion of covariance, which cannot be used to substantiate causal relationships. For example, if y causes x, but the econometric model assumes by mistake that y depends on x, then x will typically have a significant estimated coefficient, but from this we cannot infer that x causes y. Similarly, when the relationship between x and y is not based on economic theory, the link between x and y might be spurious and due for example to a third variable z is not included in the model.

1.3 Data

In order to apply statistical techniques to the econometric model, it is necessary to have a sample of data. In experimental sciences, such as physics or biology, data are the result of experiments and are therefore available in virtually unlimited quantities (apart from a cost factor). Instead, in economics the data are usually not the result of experiments but choices and outcomes of real actions of economic actors. For example, we cannot ask ourselves what Mr. Smith's consumption would be if he had an income of EUR 1,000,000, we can only observe what Mr. Smith's consumption is given his actual level of income.

The problem of not being able to repeat the experiment with Mr. Smith is partly compensated by having data on many individuals (or units, more generally). There are for example polls where a vast group of people are asked to indicate their income, consumption and a host of other potentially interesting characteristics for the econometric analysis. Data with only one observation for a large number of units are referred to as longitudinal or *cross-section*.

Another partial solution to the non-availability of experimental data in econometrics is the possibility to observe the features and choices of the same individual or unit over an extended period of time. For example, if we are interested in studying the relationship between aggregate consumption and disposable income, we have a very long sample of observations on these two variables. Data consisting of a number of observations over time for the same unit are defined *time series*.

As a third possibility, we could have a sample with both a longitudinal and a temporal dimension. Such data are defined *panels*. For example, we could observe the consumption and income of several individuals for several months, or aggregate income and consumption of different countries for several years.

Note that when we use cross-section, time series or panel data we make an implicit assumption of homogeneity of the econometric model across units and/or over time. For example, it is assumed that several individuals have the same marginal propensity to consume, or that this remains unchanged over time for the same individual, or that both of these propositions are valid in the case of panel data. This assumption is required to have a sufficiently large sample of observations to ensure that statistical procedures (e.g., parameter estimation or hypothesis testing) give reliable results.
Finally, it is worth noting that the examples considered so far include *continuous variables*, such as consumption, income, the rate of pollution or health spending. There are also cases where the variable of interest or those that are used to explain it are instead of *discrete type*. For example, we might be interested in assessing what determines whether an economy is in recession or expansion, by associating the value 1 to the phases of expansion and the value 0 to those of recession, so that the variable of interest is binary or dichotomous. Or we might want to explain what determines the choice of a group of individuals to enroll in the University or not, to grant a mortgage or not, or to buy a certain product or not. Even in these cases the choices we want to explain can be represented by a binary variable. We will see in Chapter 9 that when the variable of interest is binary (or, more generally, discrete), it is necessary to adopt different econometric models from those used in the case of continuous dependent variables.

1.4 The descriptive analysis

A very important component of an econometric study is the descriptive analysis, which precedes and often simplifies the specification of a formal model.
This requires you to produce and analyze simple descriptive statistics for the variables under consideration, such as the mean, the variance or the correlation, possibly for the whole sample and subsamples. A graph of the behavior of the variables is also always recommended.
These simple steps can already provide useful guidance on which variables are potentially more important to explain that of interest, on the stability of the relation in the sample and/or the presence of abnormal observations, different from most others, that could distort the results of the analysis.
It is also important to consider whether and to what extent the available data represent a good approximation for the theoretical variables to which they relate. While there are typically no problems for variables such as interest rates or the number of employees, for other notions such as the potential output or

the natural rate of unemployment or the expectation of a future variable, the matching with the available data is much less clear. We will see that these cases require special treatment in order to avoid possible distortions in the results.
Finally, it is also useful to include in the descriptive analysis a discussion of the institutional context, although this cannot be fully formalized in the model. For example, the presence of exceptional events such as wars or substantial increases in the price of raw materials, but also temporary changes in legislation on value added tax, can have a significant impact on the relationship between income and consumption. We will see how to, at least partially, take into account these events in the formulation of the econometric model.

1.5 Some examples

Various computer softwares are available to conduct econometric analysis. Among them we chose EViews for its popularity, ease of use, flexibility and the availability of an online clear and complete user manual (examples and exercises have been carried out with version 7 but run with version 8 as well).
EViews workfiles use the extension ".wf1" and all those related to the examples and exercises in this book can be downloaded from the webpage www.igier.unibocconi.it/marcellino.
Each workfile has a predetermined data type (undated to be used for cross-sectional data, dated for time series data, or panel) and sample size. The data type and sample size should be indicated when creating a new file or automatically loaded when opening an existing file. Each workfile can then contain different types of objects, such as the "series" objects that contain data on the variables to be considered in the study.
By opening as an example the workfile "example_cons_chap1.wf1", we see from Figure 1.1 that it contains the following items.

- Two so-called default objects, used by the program for internal processing: a vector of coefficients *c*, denoted by the β icon, and the series "resid", with the  icon typically associated with data series.
- The series labeled CONS, YD, WEALTH and DEFLATOR, containing, respectively, data on nominal private consumption, disposable income and wealth, all at the aggregate level, and the GDP deflator for Italy. Data are sampled at a quarterly frequency and expressed in millions of euros, with the first observation relating to the first quarter of 1990 and the last one to the first quarter of 2012.
- Other series and objects that we will define shortly.

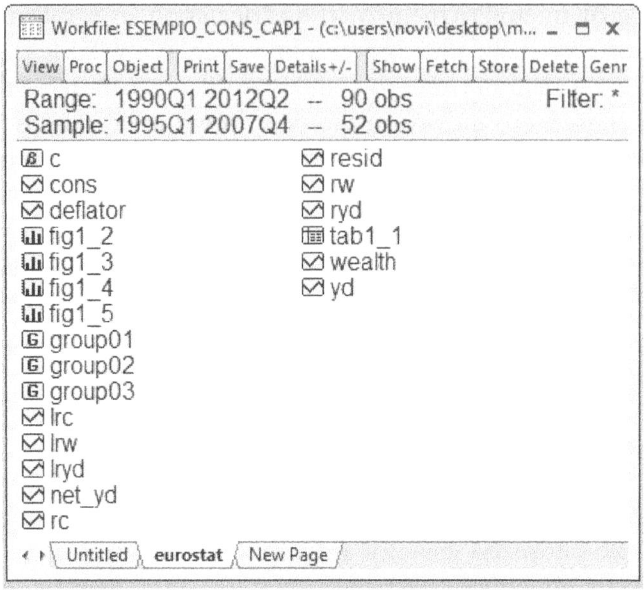

Figure 1.1: An example of an EViews workfile

The four variables CONS, YD, WEALTH and DEFLATOR, reporting national accounts data, can be used for a basic econometric study of the aggregate consumption function. For example, we could compare the simple Keynesian theory, where consumption depends only on current disposable income, with the life cycle theory of consumption, where wealth also plays an important role.

In this first example it is convenient to start the sample in 1995 and end it in 2007, excluding from the analysis the complex years of the financial crisis. To work with a subsample, we simply write in the command line of EViews:

```
smpl 1995 2007
```

indicating, after the command "*smpl*" the start date of the subsample (1995) and then the end date (2007). It is convenient to also introduce the commands "@first", to indicate the first observation in the sample (e.g., smpl @first 2000), "@last" for the last observation in the sample (e.g., smpl 1995 @last) and @all (e.g., smpl @all is equivalent to smpl 1990 2012).

It is always appropriate to start an econometric analysis with a graph of the temporal (and/or cross-sectional) evolution of the variables under consideration. To do this, you can highlight the four series by clicking with the left mouse button on their name (and holding down the Ctrl key), then click with the right mouse button and select the "open group" option from the scroll down menu, click on "View", and select the options "graph" and then "line & symbol", leaving all the other options at their default values. The result is shown in Figure 1.2:

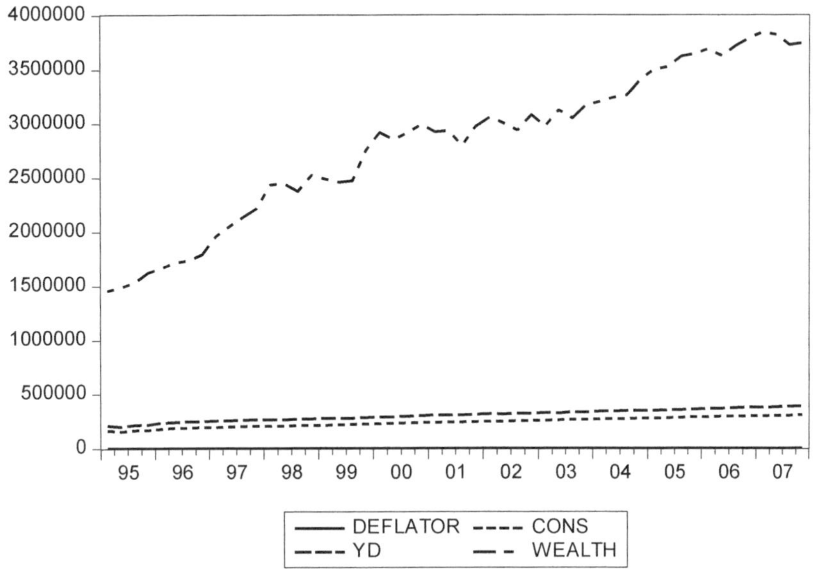

Figure 1.2: An example of a graph of the variables

Many other types of charts are available in the menu and the characteristics of the figure can be changed by clicking on it with the right mouse button and selecting "options". For example, in Figure 1.2 the GDP deflator is hardly distinguishable due to the different measuring scale of the variables. From the menu "Axes and scaling", we can click on the "+" sign and get some submenus such as "scaling". Choosing "normalized data" and the box "left axis scaling method", EViews subtracts to each variable its sample mean and divides the resulting variable by its standard deviation. The result is shown in Figure 1.3.

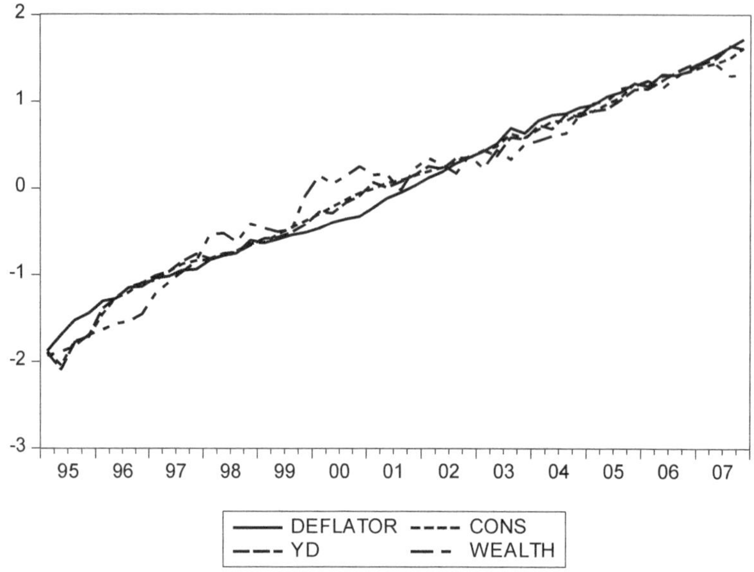

Figure 1.3: An example of a graph of standardized variables

After being "frozen" (by clicking the button "freeze"), each figure can be saved in the EViews workfile by assigning it a name (from the menu "name") or as a separate file (by clicking with the right mouse button and choosing "save graph to disk"), to be then imported directly into other computer software.

The variables that we have considered so far are expressed in nominal terms while economic theory typically refers to real variables. We can create real variables by dividing the nominal variables by the GDP deflator. We use the commands:

```
series rc = cons/deflator
series rw = wealth/deflator
series ryd = yd/deflator
```

and then present a graph of the real variables in Figure 1.4.

Comparing Figures 1.2 and 1.4, we note that the pattern of the three real variables is quite similar to that of their nominal counterparts, with an increasing trend for real wealth at least throughout the first half of the sample, and a fairly constant relationship between real consumption and real disposable income.

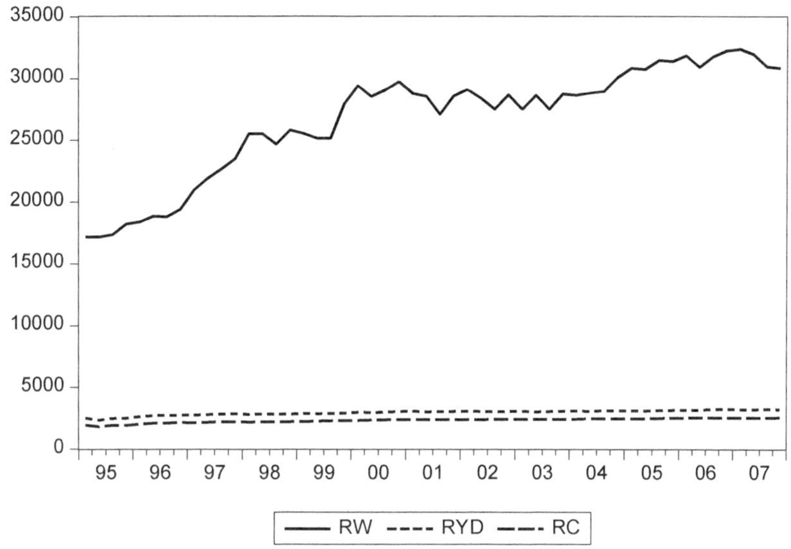

Figure 1.4: An example of a graph of real variables

The logarithmic (log) transformation is also common in *econometrics*, when the variables take non-negative values, as in the case of income, consumption and wealth. The range of data values can be so reduced and made more consistent across variables, allowing also a decrease in the volatility of the variables and making more apparent linear relations across them. The log transformation is also monotonic and strictly increasing, so that it leaves unaffected both variable trends and the position of the minimum and maximum values, if any. We will see though that, being the logarithmic transformation nonlinear, we should pay attention to the interpretation of the parameters in the model for the transformed variables.

To run a logarithmic transformation with EViews, we can use the following commands:

```
series lrc = log (rc)
series lrw = log (rw)
series lryd = log (ryd)
```

and the resulting variables are graphed in Figure 1.5.

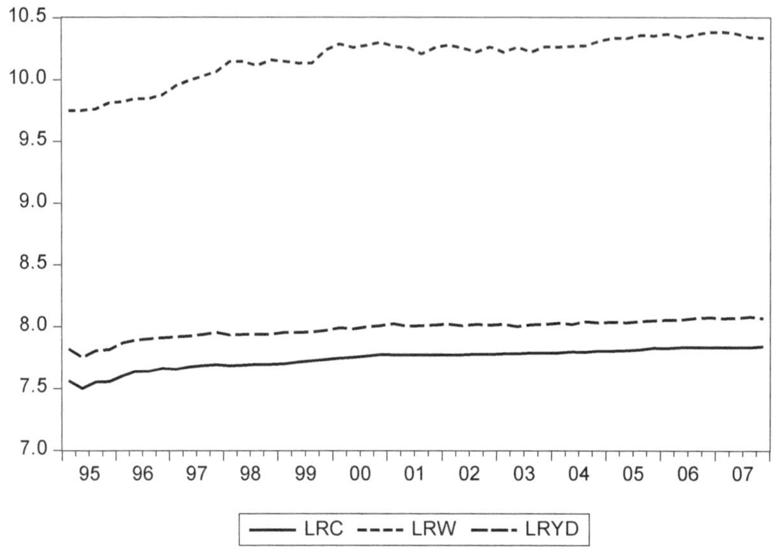

Figure 1.5: An example of a graph of log transformed variables

In Figure 1.5, there seems to be a high and positive correlation between the logs of real consumption and real income. Therefore, there is also a high and positive correlation between the two un-transformed variables (since the log transformation is monotonic). The growing trend of wealth could also generate a positive correlation with consumption. To calculate the three correlations, we select the three variables LRYD, LRC and LRW and from the Group object we click on "View", "Covariance analysis", and check the box "Correlations". The result, shown in Table 1.1, confirms our expectations. Interestingly, the positive correlation between consumption and wealth is in line with the theory of the life cycle, although the latter refers to the consumption-wealth link when also conditioning on disposable income, while the correlations in Table 1.1 are unconditional.

	LRC	LRW	LRYD
LRC	1.000000	0.962472	0.994394
LRW	0.962472	1.000000	0.951921
LRYD	0.994394	0.951921	1.000000

Table 1.1: Correlations between the logs of real variables

As a second example, the workfile named "example_regional_chap1.wf1" contains data on total real wages (W), employment (E) and per worker productivity (PR) in all the 20 Italian regions, for the period 1995-2009. It is therefore a *panel* type dataset, where the longitudinal dimension is represented by the regions. If we focus on data for a single year for all the 20 regions we obtain a *cross-sectional* dataset, whereas if we consider data from 1995 to 2009 for a given region we have a *time series*, as in the first example. In both cases, however, the overall sample size would be somewhat reduced – hence the usefulness of taking advantage of both the temporal and the longitudinal dimensions.
We may construct graphs of the variables even for panels, but the large number of series to be considered suggests using other tools. For example, we can build a table reporting the temporal average of the three series for each region. This can be done manually, selecting for each variable the option "Descriptive Statistics" in the menu "View", or by writing a simple EViews program, which is a series of commands that tell EViews to compute the quantities of interest.
Writing programs may seem more time consuming than using the EViews menus but it can actually save a lot of time when the same or similar actions or analyses have to be repeated several times, for example for different countries or when periodically updating the data.
After clicking on "File", "New", "Program", the list of commands required by EViews to generate the table of interest is reported below, where comment lines are preceded by the symbol " ' " and are not read by EViews when running the program.

smpl 2007 95 ' specifies the sample amplitude (in this example too, we leave aside the years of crisis)
! q = 1 ' defines a counter variable whose value is equal to 1
table (20.4) medie ' defines a table object with 20 rows and 4 columns, called "medie"
for %s LOM PIEM VAL FRI FRI LIG EMIL TOS UMB MAR LAZ ABR MOL PUG BAS CAL CAM SIC SAR TREN **' defines a "loop" on all regions**
series {%s}ow={%s}w/{%s}e ' generates wages per employee
!q=!q+1 ' updates the counter
medie (!q,1) =%s ' replaces the !q,1 element of the table with the region's name %s
medie(!q,2)=@mean({%s}e) ' replaces the !q,2 element of the table with the average of E for region %s
medie(!q,3)=@mean({%s}pr) ' replaces the !q,3 element of the table with the average of PR for region %s
medie(!q,4)=@mean({%s}ow) ' replaces the !q,4 element of the table with the average of OW for region %s
next ' indicates the end of the loop (i.e., the code switches to the next region)
!t = 1 ' sets another counter
for %u EMPL PROD WAGEPE ' defines another loop
!t = !t+1 ' updates the counter
medie (1,!t) =%u ' defines the header of each column in the table

```
next ' indicates the end of the loop
smpl @all ' reconsiders the whole sample
```

The result is a new object in the workfile, a table named "medie", shown in Table 1.2. The table points out sizable differences across regions in the total number of employees, caused by different surfaces and population, but also by differences in the average productivity of labor, partly reflected in the wages per worker.

	EMPL	PROD	WAGEPE
LOM	4313.023	59.54618	23.51026
PIEM	1894.546	53.07935	21.49163
VAL	56.66923	56.96148	21.49799
VEN	2132.685	52.59079	20.96652
FRI	549.7846	50.18644	22.18595
LIG	638.2231	53.55893	21.20674
EMIL	1980.431	53.29711	21.29569
TOS	1579.077	51.70392	20.35534
UMB	355.4769	47.61522	19.30201
MAR	668.8538	46.68140	19.01356
LAZ	2250.577	59.01183	24.65786
ABR	487.8462	46.63591	19.21301
MOL	114.8846	42.99700	17.88512
CAM	1756.277	44.90474	19.18502
PUG	1268.785	43.84201	19.19500
BAS	204.0077	42.47083	18.87000
CAL	611.9462	43.35827	18.61562
SIC	1443.938	47.27515	20.19538
SAR	579.2923	45.83074	19.32625
TREN	454.2385	54.04006	22.26839

Table 1.2: Average values of the variables for each region

So far in our examples we focused on sample averages and correlations. In the online guide of EViews, available under the "Help" button, "EViews Help Topics" under the heading "Descriptive Statistics" you can find a list of available commands to compute other descriptive statistics, with explanations and examples.

The data of the two previous examples were already in the form of an EViews workfile. However, you will typically download data from large databases, such as those of Eurostat or the OECD, as text files or Excel spreadsheets. Assuming that you have data in Excel, for example in the file "example_regional_chap1.xls", you can import them into an EViews workfile by using the following procedure (in a similar way, you can import data into other formats).

- Create the workfile by clicking on the main window of EViews on "File", "New", "Workfile". You get the window reported in Figure 1.6.
- From the "Frequency" menu, for this example you should select "Annual"; in the "Start date" field, type 1995, the earliest date for which observations are available in the Excel file; in the "End date" field, type 2009. By clicking OK you get the EViews workfile (containing the default objects only).
- Click on the main window of EViews on "File", "Import", "Import from file". From the resulting menu, browse your folders until you find the "example_regional_chap1.xls" file. The Excel file should be closed and the data to be imported must be in the first worksheet.
- Click on the file, getting the window in Figure 1.7.

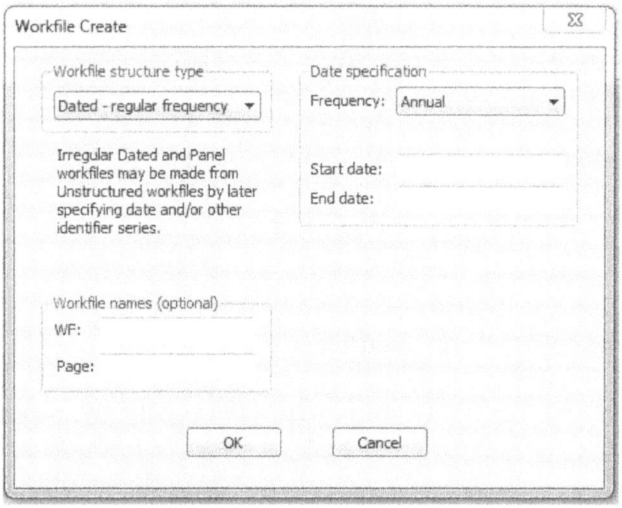

Figure 1.6: Creating a new Workfile

By clicking on "Next", you reach a similar screen where you can tell the program how many rows match the header file. In our case, it is a single line, so we just write "1" in the box "Header lines" and click the "Finish" button. At this point, the data are imported into the workfile, and series are created with the name specified in the Excel sheet.

You can follow a similar procedure to export data from Excel to EViews. In addition, EViews interacts with external programs to also allow the export of output. For example, by clicking with the right mouse button on a graph of EViews and choosing "Copy" and the option "EMF – enhanced metafile", you copy the chart to the Clipboard – ready to be pasted, for example, in MS Word.

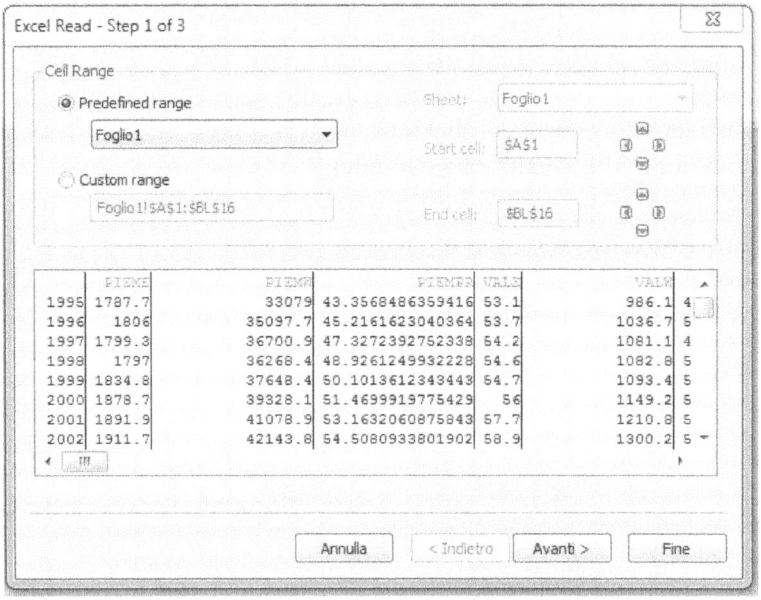

Figure 1.7: EViews window to import data

The main concepts of this chapter

In this chapter we saw that econometrics deals with the quantitative study of economic relations. econometrics males it possible to more accurately describe the economic reality, allowing you to test hypotheses about the validity or otherwise of an economic theory. When there is no specific economic theory, the role of econometrics is even more important, because it allows you to derive empirical regularities from the analysis of economic data, which can then provide an opportunity to develop an appropriate economic theory.

The results of econometric analysis should be interpreted with caution. We do not know the parameters of the model, we estimate them using statistical procedures, and so there is a more or less broad uncertainty around their values, which must be borne in mind when interpreting the results. The problem is that when the explanatory variable changes substantially, the model parameters could change too. This interpretation problem was spotted by Bob Lucas in the 1970s and is known as the Lucas Critique.

In addition, if the econometric model is not based on a valid and broadly accepted economic theory, the fact that a variable x has an estimated coefficient significantly different from 0 to explain a variable y does not imply that x causes y.

We have then considered different types of data sets, which require different techniques of analysis. Data sets composed of a single data point for a number of units are referred to as *cross-sections*. Data sets consisting of many temporal observations for the same variable are known as *time series*. Data sets with both a longitudinal and temporal dimension are defined as *panels*. In addition, variables can be continuous, as consumption, income, the rate of pollution or health spending, or discrete, as in the case of binary variables.

2. The Linear Regression Model

2.1 Definitions and notation

Let us assume we want to explain the behavior of an economic variable Y according to that of another variable X. For example, we ask ourselves what is the relationship between consumption and income, or investment and profits, or the demanded quantity of a given good and its price. Let us assume that this relationship is linear,

$$Y = \beta_1 + \beta_2 X \tag{2.1.1}$$

Often, we are interested in analyzing the relationship between Y and X not only in one period or for one unit (e.g., a family or a company) but for a set of periods or units. Assuming there are T periods or units of interest, we have

$$Y_i = \beta_1 + \beta_2 X_i, \qquad i = 1,\dots,T \tag{2.1.2}$$

Note that we are assuming that the relationship does not change at different times or for different units: the parameters β_1 and β_2 do not depend on the index i.

From the statistical point of view, we think of Y_i as a random variable, whose T realizations y_i, $i = 1, \dots, T$, we can observe. Similarly, we suppose we have available T corresponding observations of X_i, x_i, $i = 1, \dots, T$. If the model suggested in (2.1.2) is correct, it should be

$$y_i = \beta_1 + \beta_2 x_i, \qquad i = 1,\dots,T \tag{2.1.3}$$

In practice, due to the randomness of Y, the relationship (2.1.2) will not hold exactly. Each Y_i will be only explained up to an error term E_i, so that

$$Y_i = \beta_1 + \beta_2 X_i + E_i, \qquad i = 1,...,T \tag{2.1.4}$$

and

$$y_i = \beta_1 + \beta_2 x_i + e_i, \qquad i = 1,...,T \tag{2.1.5}$$

Since $E_i = Y_i - \beta_1 - \beta_2 X_i$, E_i is also a random variable. Moreover, if the relationship in (2.1.2) is correct, we can require that the expected value of E_i is 0 for all i, i.e.,

$$E(E_i) = 0 \,. \tag{2.1.6}$$

Alternatively, we may think of E_i as a random variable that contributes to explaining the variable Y. In this case, the randomness of the Y variable results from that of the error E_i. The two interpretations of the error term, as the unexplained component of Y or as an additional explanatory variable, are not empirically distinguishable but fortunately, as we shall see, do not require a different treatment.

It is also important to clarify the nature of the explanatory variable X. This can be deterministic or, in turn, stochastic. In practice, in economic applications, the X variable is almost always stochastic. In the examples in Chapter 1, X is private income or profits or the price of an asset, whose observed values can be thought of as realizations of a random variable. However, from the statistical point of view, it is easier to analyze the regression model by assuming that X is deterministic. Moreover, as we shall see in Chapter 6, many of the results obtained in the case of deterministic X are also valid in the case of stochastic X, under a proper set of additional assumptions. For this reason, we assume for the moment that X is deterministic, i.e., $X_i = x_i$.

When the variable X is deterministic, from (2.1.4), (2.1.5) and (2.1.6) it follows that:

$$E(Y_i) = \beta_1 + \beta_2 x_i \,, \tag{2.1.7}$$

and also

$$y_i = E(Y_i) + e_i \,. \tag{2.1.8}$$

Therefore, the X variable helps explain the expected value of the Y variable, while the difference between the realization y_i and its expected value corresponds to the realization of the error term, e_i.

In terms of notation, we use

- Y, y → dependent variable
- X, x → independent variable or regressor or explanatory variable
- E → error
- e → residual, realization of the error
- $E(Y_i)$ → systematic component, expected value of Y
- $\beta_1 + \beta_2 x_i$ → regression function

The model can be represented graphically as in Figure 2.1, which contains some realizations of Y and X, the corresponding expected values for Y, and the resulting residuals.

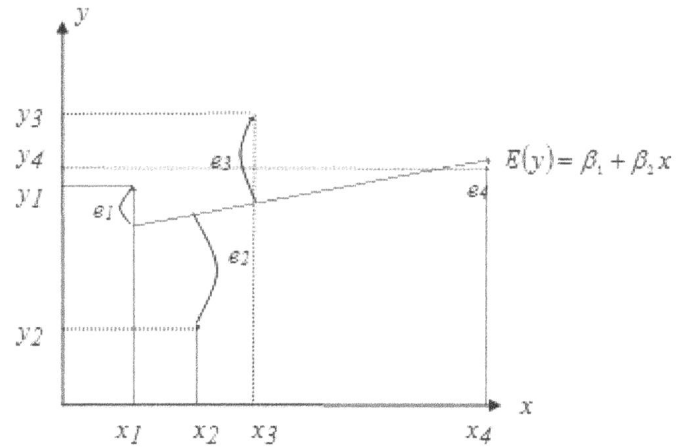

Figure 2.1: The regression function

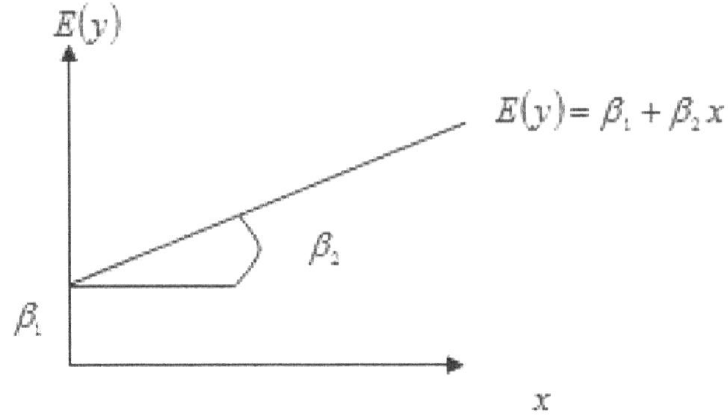

Figure 2.2: Interpreting the regression function parameters

Figure 2.2 helps interpret the parameters of the regression function as

$$\beta_2 = \frac{\partial E(y|X)}{\partial X} \qquad \text{and, only if } x=0,\ E(y)=\beta_1 \tag{2.1.9}$$

The *intercept* β_1 measures the expected value of the dependent variable when the explanatory variable is 0. The *slope* β_2 indicates how the expected value of Y changes when there is a marginal variation of the regressor.
Finally, in Figures 2.3 and 2.4 we report in 2D and 3D the entire density function of Y for given values of X, $f(y)$.

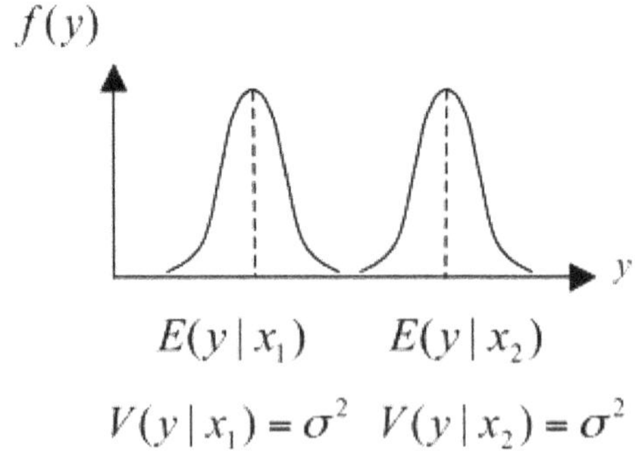

Figure 2.3: Density function of Y for given values of X

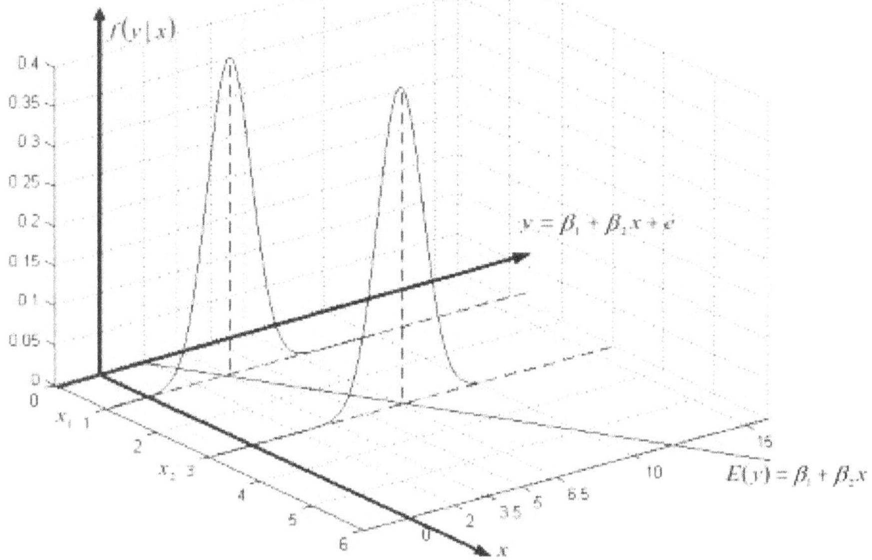

Figure 2.4: Joint representation of x, y and density function of Y

2.2 The assumptions of the linear regression model

In the previous Section, we assumed the existence of a linear relationship between the dependent variable Y_i and the explanatory variable X_i, whose parameters are constants across units, whereby:

1. $E(Y_i|X_i) = \beta_1 + \beta_2 x_i$ (linearity and stability of parameters).

We have also assumed that the X_i are deterministic, so that:

2. $E(Y_i|X_i) = E(Y_i)$ (deterministic regressors).

Let us also suppose:

3. $V(Y_i) = \sigma^2$ (homoscedasticity),

namely, the variance of Y is constant, it does not depend on the value of the index i.

4. Cov $(Y_i, Y_j) = 0$, $i \neq j$, (no correlation),

namely, Y_i and Y_j are uncorrelated for each $i \neq j$.

5. There exist i and j such that $x_i \neq x_j$.

The latter hypothesis is used because, in order to measure the change in y when x changes, we need at least two different observations on x.

Since the regressor is deterministic by assumption, Y and E only differ by a deterministic term, $\beta_1 + \beta_2 X$, so that $E = Y - \beta_1 - \beta_2 X$. We can then reformulate the assumptions about the linear regressor model based on the error term as:

1. $E(E_i \mid X_i) = E(Y_i - \beta_1 - \beta_2 X_i \mid X_i) = E(Y_i \mid X_i) - \beta_1 - \beta_2 X_i = 0$ (linearity and stability of parameters).
2. $E(E_i \mid X_i) = E(E_i)$ (deterministic regressors).
3. $V(E_i) = V(Y_i - \beta_1 - \beta_2 X_i) = V(Y_i) = \sigma^2$ (homoscedasticity).
4. Cov $(e_i, e_j) = \text{Cov}(y_i, y_j) = 0$, $i \neq j$, (no correlation)
5. There exist i and j such that $x_i \neq x_j$.

The two formulations of the hypotheses are exactly equivalent and the choice will depend on the context. Note, however, an important difference between the dependent variable Y and the error term E. The random variable Y is observable, its realization is y. Instead the error E is not observable, since:

$$e = y - E(Y) = y - \beta_1 - \beta_2 x,$$

y and x are observable, but the parameters β_1 and β_2 are unknown. For this reason, it may be preferable to formulate the hypotheses in terms of the properties of Y instead of E, but in practice, as we said, the two approaches are equivalent.
A further assumption, required to discuss some topics related to the linear regression model, relates to the distribution of the dependent variable. In particular, we have:

6. $Y_i \sim N(\beta_1 + \beta_2 X_i, \sigma^2)$ or $E_i \sim N(0, \sigma^2)$ (Normality).

The assumption of Normality implies that, if the parameters β_1, β_2 and σ^2 are known, it is possible to easily calculate the probability that Y takes values in a given interval of interest. For example, we could calculate the probability that consumption growth is in the range 2%-4%, or that aggregate investment growth is negative, or that the demand for a given good grows more than 5%.
The assumptions from (1) to (5) are often referred to as *weak OLS* hypotheses.
The assumptions from (1) to (6) are often referred to as *strong OLS* hypotheses.

2.3 The role of the error term

In the first Section we said that one reason for including the error term in the linear model is the randomness of the dependent variable, which allows explaining its expected value but not its actual realizations.
Other reasons to include an error term in the relationship between *Y* and *X* are:

(a) **Measurement errors in variables**. For example, if aggregate private consumption depends on the unobservable permanent income (*X*) rather than on the observable current income (*X**), and if the relationship between the two notions of income is $X_i^* = X_i + V_i$, where V_i is the measurement error, then, although the relationship between *Y* and *X* is deterministic as in (2.1.1), we have

$$y_i = \beta_1 + \beta_2 x^*_i - \beta_2 v_i = \beta_1 + \beta_2 x^*_i + e_i \quad (2.3.1)$$

We will return on the effects of measurement error in Chapter 6.

(b) **Heterogeneity.** Let us suppose that we want to analyze the consumption behavior (*Y*) of two families with different marginal propensities to consume.

$$Y_1 = \alpha + \beta_1 x_1 \text{ and } Y_2 = \alpha + \beta_2 x_2 .$$

By mistake, we assume that the propensities to consume are the same and equal to β. Then, the models become:

$$y_1 = \alpha + \beta x_1 + (\beta_1 - \beta)x_1 = \alpha + \beta x_1 + e_1 ,$$
$$y_2 = \alpha + \beta x_2 + (\beta_2 - \beta)x_2 = \alpha + \beta x_2 + e_2 ,$$

and the errors e_1 and e_2 are caused by the heterogeneity of the coefficients. We will analyze in more details the consequences of this form of error in Chapter 5.

(c) **Omitted variables**. Continuing the example about aggregate consumption (*Y*), if income (*X*) is not the only significant explanatory variable for *Y* but, for example, wealth (*Z*) also matters, then the error term reflects the consequences of the omission of *Z* from the model for *Y*. We will return to this topic in Chapter 3.

(d) **Specification error.** If the relationship between consumption and income is not linear but governed by a generic function, such as *Y* =

g(X), then the error in the linear model captures the approximation error we make when using a liner relationship to model the dependence of *Y* on *X* ($\beta_1 + \beta_2 X$) instead of the proper nonlinear function (*g(X)*). We will assess the consequences of errors of this type in Chapter 4.

Regardless of the source of the error term, the relevant question is whether the weak or strong OLS assumptions introduced in Section 2.2 are met. In this chapter we suppose that this is the case, while in Chapter 4 we will assess the consequences of the violation of the OLS assumptions and the possible remedies.

2.4 OLS estimators of the regression function parameters

The β_1 and β_2 parameters of the linear regression function are not known and must be estimated. For this, we use the assumption of stable parameters over time or across different sample units, so that we can use the available *T* observations on *Y* and *X*, drawn in Figure 2.5, to gather information on β_1 and β_2.

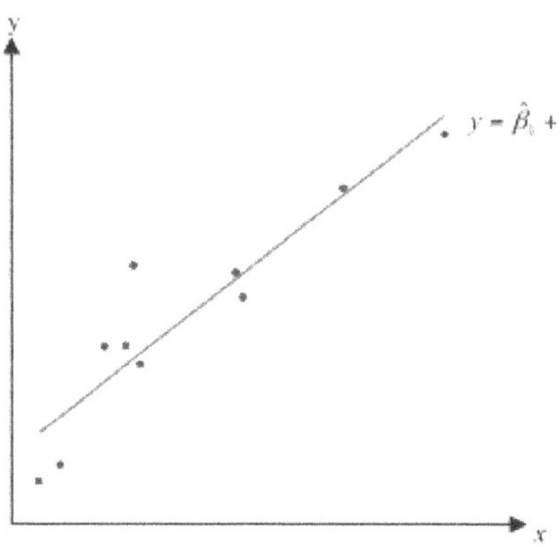

Figure 2.5: Observations on y and x and the regression line

We would like to choose β_1 and β_2 in order to «draw» a line that passes as close as possible to all points in Figure 2.5. Recalling that the difference between y_i and $\beta_1 + \beta_2 x_i$ is the residual e_i, we then determine the estimators of β_1 and β_2 in order to minimize the sum of squares of residuals, i.e.

$$\min_{\beta_1\beta_2} S = \sum_{j=1}^{T} e_j^2 \qquad (2.4.1)$$

namely,

$$\min_{\beta_1,\beta_2} S = \sum_{t=1}^{T} (Y_t - \beta_1 - \beta_2 X_t)^2 \qquad (2.4.2)$$

The minimizers of this optimization problem are the *Ordinary Least Squares (OLS) estimators* of β_1 and β_2, indicated with $\hat{\beta}_1$ and $\hat{\beta}_2$.

To begin with, we obtain the first-order conditions by differentiating the objective function in (2.4.2) with respect to β_1 and β_2. We get:

$$\partial S / \partial \beta_1 = \sum_{t=1}^{T} 2(Y_t - \beta_1 - \beta_2 X_t)(-1) = 2T\beta_1 - 2\sum_{t=1}^{T} Y_t + 2\beta_2 \sum_{t=1}^{T} X_t \qquad (2.4.3)$$

$$\partial S / \partial \beta_2 = \sum_{t=1}^{T} 2(Y_t - \beta_1 - \beta_2 X_t)(-X_t) = -2\sum_{t=1}^{T} X_t Y_t + 2\beta_1 \sum_{t=1}^{T} X_t + 2\beta_2 \sum_{t=1}^{T} X_t^2 \qquad (2.4.4)$$

By equating to 0 the first-order conditions, we have:

$$T\hat{\beta}_1 + \hat{\beta}_2 \sum_{t=1}^{T} X_t = \sum_{t=1}^{T} Y_t \qquad (2.4.5)$$

$$\hat{\beta}_1 \sum_{t=1}^{T} X_t + \hat{\beta}_2 \sum_{t=1}^{T} X_t^2 = \sum_{t=1}^{T} X_t Y_t \qquad (2.4.6)$$

By multiplying equation (2.4.5) by $\sum_{t=1}^{T} X_t$, and (2.4.6) by T, and subtracting (2.4.6) to (2.4.5), we obtain:

$$\hat{\beta}_2 \sum_{t=1}^{T} X_t \sum_{t=1}^{T} X_t - \hat{\beta}_2 T \sum_{t=1}^{T} X_t^2 = \sum_{t=1}^{T} Y_t \sum_{t=1}^{T} X_t - T \sum_{t=1}^{T} X_t Y_t,$$

from which it follows that:

$$\hat{\beta}_2 = \frac{T\sum_{t=1}^{T} X_t Y_t - \sum_{t=1}^{T} X_t \sum_{t=1}^{T} Y_t}{T\sum_{t=1}^{T} X_t^2 - \left(\sum_{t=1}^{T} X_t\right)^2} \tag{2.4.7}$$

Finally, from (2.4.5) we derive $\hat{\beta}_1$ as

$$\hat{\beta}_1 = \frac{\sum_{t=1}^{T} Y_t}{T} - \hat{\beta}_2 \frac{\sum_{t=1}^{T} X_t}{T} \tag{2.4.8}$$

After having obtained the formulae of the OLS estimators, let us make a few comments on them.

$\hat{\beta}_1$ and $\hat{\beta}_2$ are estimators, they depend on the random variable Y and therefore are themselves random variables. If we replace the random variable Y by its realizations, the estimators become estimates (observed values), which for simplicity we still define as $\hat{\beta}_1$ and $\hat{\beta}_2$:

$$\hat{\beta}_2 = \frac{T\sum_{t=1}^{T} x_t y_t - \sum_{t=1}^{T} x_t \sum_{t=1}^{T} y_t}{T\sum_{t=1}^{T} x_t^2 - \left(\sum_{t=1}^{T} x_t\right)^2}, \quad \hat{\beta}_1 = \frac{\sum_{t=1}^{T} y_t}{T} - \hat{\beta}_2 \frac{\sum_{t=1}^{T} x_t}{T} = \bar{y} - \hat{\beta}_2 \bar{x} \tag{2.4.9}$$

where $\bar{x}$ indicates the sample mean.

If the X values are all the same, from (2.4.7) $\hat{\beta}_2$ is not defined. For this reason, we have requested that at least two X values are different in hypothesis (5) of Section 2.2.

If we used $\sum_{j=1}^{T} e_j$ as objective function instead of $\sum_{j=1}^{T} e_j^2$, we could get a compensation of positive and negative errors. To avoid this issue, we could take $\sum_{j=1}^{T} \left| e_j \right|$ as the loss function to be minimized, but the derivation of the estimators would be more complicated since the function is not differentiable in 0.

Similarly, if we want to give more weight to large mistakes we could use $\sum_{j=1}^{T} e_j^4$ as loss function, or $\sum_{j=1}^{T} e_j^2$ when $e>0$ and $b\sum_{j=1}^{T} e_j^2$ when $e<0$ if positive and negative errors have different importance for us. These cases substantially complicate the derivation of the parameter estimators, so we will stick to the assumption of a quadratic loss function, $S = \sum_{j=1}^{T} e_j^2$.

In the sequel, for the sake of simplicity and when the difference is clear, we will no longer use a different notation for random variables (e.g., *Y* or *E*) and their realizations (*y* or *e*).

2.5 The linear model and the OLS estimators in vector notation

We now derive the same OLS estimators of β_1 e β_2 but using the vector notation. The latter is a useful tool for the analysis of more complex models. To start with, we write the linear model (2.1.5) in expanded form as

$$\begin{aligned} y_1 &= \beta_1 + \beta_2 x_1 + e_1 \\ y_2 &= \beta_1 + \beta_2 x_2 + e_2 \\ &\cdots \\ y_T &= \beta_1 + \beta_2 x_T + e_T \end{aligned} \tag{2.5.1}$$

or:

$$\begin{bmatrix} y_1 \\ y_2 \\ \vdots \\ y_T \end{bmatrix} = \begin{bmatrix} 1 \\ 1 \\ \vdots \\ 1 \end{bmatrix} \beta_1 + \begin{bmatrix} x_1 \\ x_2 \\ \vdots \\ x_T \end{bmatrix} \beta_2 + \begin{bmatrix} e_1 \\ e_2 \\ \vdots \\ e_T \end{bmatrix} \tag{2.5.2}$$

By grouping the elements of the equations in (2.5.2), we get

$$\underset{T\times 1}{y} = \underset{T\times 1}{v} \quad \underset{1\times 1}{\beta_1} + \underset{T\times 1}{X} \quad \underset{1\times 1}{\beta_2} + \underset{T\times 1}{e} \tag{2.5.3}$$

or,

$$\underset{T\times 1}{y} = \underset{T\times 2}{W} \underset{2\times 1}{\beta} + \underset{T\times 1}{e} \tag{2.5.4}$$

where:

$$W = (v \vdots X) \text{ and } \beta = (\beta_1, \beta_2)'$$

We can also formulate the hypothesis (3) and (4) of homoscedasticity and no correlation as:

$$E[\underset{T\times 1}{(y-W\beta)} \underset{1\times T}{(y-W\beta)'}] = E(\underset{T\times 1}{e} \underset{1\times T}{e'}) = \sigma^2 I_T = \sigma^2 \begin{bmatrix} 1 & & & \\ & 1 & 0 & \\ & 0 & \ddots & \\ & & & 1 \end{bmatrix} \tag{2.5.5}$$

$$S = \sum_{t=1}^{T} e_t^2 = \underset{1\times T}{e'} \underset{T\times 1}{e} = [e_1 \quad e_2 \quad \dots \quad e_T] \begin{bmatrix} e_1 \\ e_2 \\ \vdots \\ e_T \end{bmatrix}$$

Let us now reconsider the loss function we want to minimize, defined in (2.4.1): We can re-write it as:

$$S = e'e = (y-W\beta)'(y-W\beta) = \underset{1\times T}{y'} \underset{T\times 1}{y} - \underset{1\times T}{y'} \underset{(T\times 2)(2\times 1)}{W\beta} - \underset{(1\times 2)(2\times T)(T\times 1)}{\beta' W' y} + \underset{1\times 1}{\beta' W' W \beta},$$

or, since all the terms are scalars,

$$S = y'y - 2\beta' W' y + \beta' W' W \beta \, . \tag{2.5.6}$$

The first-order conditions become:

$$\partial S / \partial \beta = \frac{\partial}{\partial \beta}(y'y - 2\beta' W' y + \beta' W' W \beta) = -2W'y + 2W'W\beta = 0 . \tag{2.5.7}$$

It follows that the OLS estimators in vector notation are:

$$\underset{2\times 1}{\hat{\beta}} = \underset{2\times 2}{(W'W)^{-1}} \underset{2\times 1}{W'y} \tag{2.5.8}$$

As a check that the estimators in (2.5.8) are identical to those in (2.4.9), let us rewrite the former as

$$\underset{2\times 1}{\hat{\beta}} = (W'W)^{-1}W'y = [\underset{2\times T}{(v,X)'} \; \underset{T\times 2}{(v,X)}]^{-1} [\underset{2\times T}{(v,X)'}] \underset{2\times 1}{y} =$$

$$= \left[\begin{pmatrix} v' \\ X' \end{pmatrix} (v,X)\right]^{-1} \begin{bmatrix} v' \\ X' \end{bmatrix} y = \begin{bmatrix} v'v & v'X \\ X'v & X'X \end{bmatrix}^{-1} \begin{bmatrix} v'y \\ X'y \end{bmatrix}$$

By substituting v and X with their expressions, we obtain:

$$\hat{\beta} = \begin{bmatrix} T & \sum_{t=1}^{T} x_t \\ \sum_{t=1}^{T} x_t & \sum_{t=1}^{T} x_t^2 \end{bmatrix}^{-1} \begin{bmatrix} \sum_{t=1}^{T} y_t \\ \sum_{t=1}^{T} x_t y_t \end{bmatrix} = \frac{1}{T\left(\sum_{t=1}^{T} x_t^2\right) - \left(\sum_{t=1}^{T} x_t\right)^2} \begin{bmatrix} \sum_{t=1}^{T} x_t^2 & -\sum_{t=1}^{T} x_t \\ -\sum_{t=1}^{T} x_t & T \end{bmatrix} \begin{bmatrix} \sum_{t=1}^{T} y_t \\ \sum_{t=1}^{T} x_t y_t \end{bmatrix} = \begin{bmatrix} \hat{\beta}_1 \\ \hat{\beta}_2 \end{bmatrix}$$

For example, it is:

$$\hat{\beta}_2 = \frac{T\sum_{t=1}^{T} x_t y_t - \sum_{t=1}^{T} x_t \sum_{t=1}^{T} y_t}{T\left(\sum_{t=1}^{T} x_t^2\right) - \left(\sum_{t=1}^{T} x_t\right)^2},$$

which is exactly the same as the formula presented in equation (2.4.9).
Finally, note that the second-order conditions of the optimization problem are

$$\partial^2 S \big/ \partial\beta\partial\beta' = W'W \tag{2.5.9}$$

Since the matrix $W'W$ is positive definite, the OLS estimators are indeed the minimizers of the sum of squared regression residuals.

2.6 Properties of OLS estimators

Having obtained the OLS estimators for the regression function parameters, we now:

1. analyze their properties;
2. compare them with other possible estimators and show whether and in what sense the OLS estimators are better.

2.6.1. Unbiasedness

Let us work under the weak OLS assumptions (1-5 in Section 2.2). Note that assumption 6, Normality of Y or E, is not required. We have seen above that, using the vector notation, the model and parameter estimators are:

$$\underset{T\times 1}{y} = \underset{T\times 2}{W}\ \underset{2\times 1}{\beta} + \underset{T\times 1}{e} \quad (2.6.1)$$

$$\underset{2\times 1}{\hat{\beta}} = \underset{2\times 2}{(W'W)^{-1}}\ \underset{2\times 1}{W'y} \quad (2.6.2)$$

By substituting for y in (2.6.2) its expression in (2.6.1), we get:

$$\hat{\beta} = (W'W)^{-1}W'(W\beta + e) = \beta + (W'W)^{-1}W'e.$$

Therefore, it is:

$$E(\hat{\beta}) = E(\beta + (W'W)^{-1}W'e) = \beta + (W'W)^{-1}W'E(e) = \beta \quad (2.6.3)$$

since $E(e)=0$ by assumption (1).

The property $E(\hat{\beta}) = \beta$ is labeled *unbiasedness*. Intuitively, unbiasedness means that, if we could obtain several samples of $Y_1, X_1, ..., Y_T, X_T$ and for each sample we estimated the parameters of the linear model by OLS, then the average across all the samples of the OLS estimators $\hat{\beta}$ would coincide with the true value of the parameters, β. Unfortunately, in practice in economics we generally have one sample, and therefore we cannot use the property of unbiasedness to say that for our specific sample $\hat{\beta}$ is close to β.

In other words, the OLS estimation method ensures that, if the experiment were repeated many times, on average, the estimators would give the correct answer

about the parameter values. But with a single sample we have no information on whether the estimated value of the parameters is close to the true value.
Note also that, if $E(e) \neq 0$, then $E(\hat{\beta}) \neq \beta$. The expected value of the error term can be non-0 if, for example, we omit relevant explanatory variables. Therefore, the model specification is also relevant to determine the properties of the parameter estimators.

2.6.2 The variance of the OLS estimators
Defining by *var (Z)* the variance (or variance-covariance matrix) of *Z*, we have:

$$\underset{2\times 1}{\text{var}}(\hat{\beta}) = \begin{pmatrix} \text{var}(\hat{\beta}_1) & \text{cov}(\hat{\beta}_1, \hat{\beta}_2) \\ \text{cov}(\hat{\beta}_1, \hat{\beta}_2) & \text{var}(\hat{\beta}_2) \end{pmatrix} = E(\hat{\beta} - \beta)(\hat{\beta} - \beta)' = (W'W)^{-1} W' E(ee') W (W'W)^{-1} =$$
$$= (W'W)^{-1} W' \sigma^2 I W (W'W)^{-1} = \sigma^2 (W'W)^{-1} W'W (W'W)^{-1} = \sigma^2 (W'W)^{-1} \quad (2.6.4)$$

where we have used the assumptions (3) and (4) of homoscedasticity and no error correlation that, along with E(e)=0, imply that $E(ee')=var(e) = \sigma^2 I_T$.
From the formula above, since

$$\sigma^2 (W'W)^{-1} = \sigma^2 \begin{bmatrix} T & \sum_{t=1}^{T} x_t \\ \sum_{t=1}^{T} x_t & \sum_{t=1}^{T} x_t^2 \end{bmatrix}^{-1},$$

expanding the inverse matrix in square brackets, we get

$$\text{var}(\hat{\beta}_1) = \sigma^2 \left[\frac{\sum_{t=1}^{T} x_t^2}{T \sum_{t=1}^{T} (x_t - \bar{x})^2} \right] \quad (2.6.5)$$

$$\text{var}(\hat{\beta}_2) = \left[\frac{\sigma^2}{\sum_{t=1}^{T} (x_t - \bar{x})^2} \right] \quad (2.6.6)$$

$$\operatorname{cov}\left(\hat{\beta}_1, \hat{\beta}_2\right) = \sigma^2 \left[\frac{-\bar{x}}{\sum_{t=1}^{T} (x_t - \bar{x})^2} \right] \tag{2.6.7}$$

The greater the variance of $\hat{\beta}_1$ and $\hat{\beta}_2$, the greater the uncertainty about the estimated values. Hence, we would like the variance of the OLS estimators to be low. This happens when:

- $\sum_{t=1}^{T} (x_t - \bar{x})^2$ is large, meaning that there is a lot of variation in the independent variable and so a lot of information. This is the case in the graph in the right panel of Figure 2.6, while in the left one $\sum_{t=1}^{T} (x_t - \bar{x})^2$ is relatively small and therefore we can expect a greater variance of $\hat{\beta}_2$;
- σ^2 is small, namely, the variability of the error term is small (and also that of y since $\operatorname{var}(y) = \operatorname{var}(e)$);
- the sample size, T, increases, since in the formula for var($\hat{\beta}_2$) in (2.6.6) we add positive terms to the denominator, such as $(x_t - \bar{x})^2$, while in the expression for var($\hat{\beta}_1$) in (2.6.5) we divide by T.

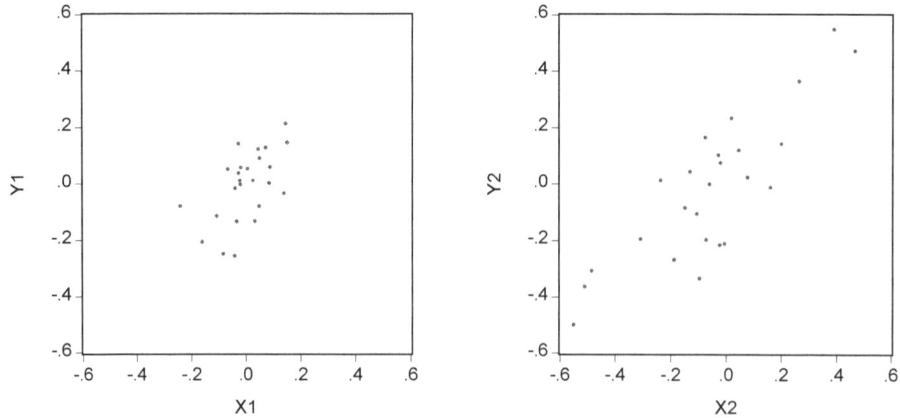

Figure 2.6: Samples with different dispersion

2.6.3 Consistency

If the sample size diverges, i.e. $T \to \infty$, then $\text{var}(\hat{\beta}_1) \to 0$ and $\text{var}(\hat{\beta}_2) \to 0$. Hence, the distribution of $\hat{\beta}_1, \hat{\beta}_2$ degenerates on their expected values, namely, $\hat{\beta}_1, \hat{\beta}_2$ converge (in probability) towards the true values β_1 and β_2, and the uncertainty disappears. This property is called *consistency* and is very important because it tells us that, if the sample that we have is very large, then the estimated parameter values are close to the actual values (always maintaining hypotheses (1)-(5)).

Unbiasedness, $E(\hat{\beta}) = \beta$, requires many samples to have a precise estimate. Consistency requires a single sample, but a very large one.

In summary, we have seen so far that the OLS estimators of the parameters β of the linear regression function, $\hat{\beta} = (W'W)^{-1} W'y$, are:

- linear, i.e. linear function of the random variable y;
- random, since they depend on y, and with an associated sample density function;
- unbiased ($E(\hat{\beta}) = \beta$);
- with variance given by $\sigma^2 (W'W)^{-1}$;
- consistent.

2.6.4 Efficiency of OLS estimators

If two estimators are both unbiased, we prefer the more efficient one, i.e., the estimator with the lowest variance, since it contains sharper information on the parameters. We use this simple criterion to compare alternative estimators.

Gauss-Markov Theorem

Under the OLS assumptions (1)-(5), the OLS estimator $\hat{\beta} = (W'W)^{-1} W'y$ is the Best Linear Unbiased Estimator (B.L.U.E.), i.e., it has the lowest variance (or highest precision) in the set of the linear and unbiased estimators of β. Note that this property does not require the hypothesis (6) of Normality. Note also that the theorem refers to estimators, not to estimates.

The proof is rather simple if we consider single parameter estimators, for example, if we want to show that $(\hat{\beta}_2) < \text{var}(\tilde{\beta}_2)$, where $\tilde{\beta}_2$ is a generic alternative linear and unbiased estimator of β_2. But we want to look at β_1 and

β_2 jointly. This is more complex because we have to compare the variance-covariance matrices of the alternative estimators.
To tackle this problem, we consider a generic linear combination of parameters,

$$\underset{1\times 2}{a'}\ \underset{2\times 1}{\beta} = a_1\beta_1 + a_2\beta_2. \tag{2.6.8}$$

For example, $a' = (1,0)\beta = \beta_1$, $a' = (0,1)\beta = \beta_2$. The OLS estimator of $a'\beta$ is $a'\hat{\beta}$. We then say that $\hat{\beta}$ is better than $\tilde{\beta}$, where $\tilde{\beta}$ is another linear and unbiased estimator of β, if $\operatorname{var}(a'\hat{\beta}) \leq \operatorname{var}(a'\tilde{\beta})$ for any a, namely,

$$a' \operatorname{var}(\hat{\beta})\ a \leq a' \operatorname{var}(\tilde{\beta})\ a, \tag{2.6.9}$$

for any a. This is equivalent to:

$$\underset{1\times 2}{a'}\ [\ \underset{2\times 2}{\operatorname{var}(\hat{\beta})} - \underset{2\times 2}{\operatorname{var}(\tilde{\beta})}\ ]\ \underset{2\times 1}{a} \leq 0, \tag{2.6.10}$$

that is, $\operatorname{var}(\hat{\beta}) - \operatorname{var}(\tilde{\beta})$ is negative semi-definite.
A generic linear estimator $\tilde{\beta}$ can be written as

$$\tilde{\beta} = [\ \underset{2\times T}{(W'W)^{-1}W'} + \underset{2\times T}{C}\ ]\ \underset{T\times 1}{y} \tag{2.6.11}$$

where C is a $2\times T$ matrix that must satisfy the constraints we will see shortly.
Substituting y with its expression we have

$$\tilde{\beta} = [(W'W)^{-1}W' + C]\ [W\beta + e] = \beta + (W'W)^{-1}W'e + CW\beta + Ce,$$

and, therefore,

$$E(\tilde{\beta}) = \beta + 0 + CW\beta + 0.$$

Since $\tilde{\beta}$ must be unbiased, it must be $CW\beta = 0$, which requires $CW = 0$. Hence, C must be such that $CW = 0$ in order for $\tilde{\beta}$ to be unbiased. Therefore,

$$\tilde{\beta} - \beta = (W'W)^{-1} W'e + Ce .$$

With this condition, the variance covariance matrix of $\tilde{\beta}$ is:

$$\begin{aligned}
\text{var}(\tilde{\beta} - \beta) &= E(\tilde{\beta} - \beta)(\tilde{\beta} - \beta)' = E\left[(W'W)^{-1}W'e + Ce\right]\left[e'C' + e'W(W'W)^{-1}\right] = \\
&= E\left[(W'W)^{-1}W'ee'C'\right] + E\left[(W'W)^{-1}W'ee'W(W'W)^{-1}\right] + E\left[Cee'C'\right] + E\left[Cee'W(W'W)^{-1}\right] = \\
&= \sigma^2 (W'W)^{-1}\underbrace{W'C'}_{=\ 0} + \sigma^2 (W'W)^{-1} + \sigma^2 CC' + \sigma^2 \underbrace{CW}_{=\ 0}(W'W)^{-1} = \sigma^2 (W'W)^{-1} + \sigma^2 CC' = \text{var}(\hat{\beta} - \beta) + \sigma^2 CC'
\end{aligned}$$

Hence, $\text{var}(\hat{\beta}) - \text{var}(\tilde{\beta}) = -\sigma^2 CC'$ and $\text{var}(\hat{\beta}) - \text{var}(\tilde{\beta})$ is indeed negative semi-definite.

Finally, note that we could get the OLS estimator as that B.L.U.E. (instead of the minimizer of the sum of squared residuals). Narrowing the search for parameter estimators to the set of linear and unbiased ones, i.e. of the type (2.6.11) with $CW=0$, and requiring that with minimum variance, we just get the OLS estimator ($C=0$ in (2.6.11)).

2.6.5 The distribution of the OLS estimators

The results we have seen so far do not require Normality of the dependent variable or error. However, if $Y \sim N(W\beta, \sigma^2)$, then it is also $\hat{\beta} \sim N$. This is so because $\hat{\beta}$ is a linear function of Y (or of E). In particular, given the results we have derived above, it is

$$\hat{\beta} \sim \text{N}\left(\beta, \sigma^2 (W'W)^{-1}\right) \tag{2.6.12}$$

If E or Y are not Normal, the OLS estimator still has a Normal distribution (after an appropriate standardization) when the sample size, T, is very large. The proof requires the use of the Central Limit Theorem that, given the weak OLS assumptions, ensures that

$$\left(1/\sqrt{T}\right)W'e \xrightarrow{d} N(0, \sigma^2 \Sigma_{WW}), \quad (W'W/T) \longrightarrow \Sigma_{WW} \tag{2.6.13}$$

From

$$\hat{\beta} = (W'W)^{-1}W'Y = \beta + (W'W)^{-1}W'e ,$$

it follows that

$$\sqrt{T}(\hat{\beta}-\beta)=\left(\frac{W'W}{T}\right)^{-1}\frac{1}{\sqrt{T}}W'e \longrightarrow \Sigma_{WW}^{-1}N(0,\sigma^2\Sigma_{WW})=N(0,\sigma^2\Sigma_{WW}^{-1}) \quad (2.6.14)$$

It is difficult to give a precise value for T such that the distribution of $\hat{\beta}$ is well approximated by a Normal distribution, but usually $T>50$ is enough.

2.6.6. Orthogonality between OLS fitted values and residuals

Using the OLS estimators, we can decompose the dependent variable y into the explained component or fitted value, $\hat{y}$, and the residual component or residual term, $\hat{e}$, where:

$$y = W\hat{\beta}+\hat{e} = \underset{T\times 1}{\hat{y}} + \underset{T\times 1}{\hat{e}} \quad (2.6.15)$$

Another relevant property of $\hat{\beta}_{OLS}$ is that $\hat{y}'\hat{e}=0$, namely, the explained and residual components are orthogonal (uncorrelated). In other words, the residuals cannot be used to improve the fitted values, i.e., the part of y we can explain. To prove this result, we note that:

$$\hat{y}'\hat{e}=\hat{\beta}'W'\hat{e}=\hat{\beta}'W'[I-W(W'W)^{-1}W']\ e=\hat{\beta}'[W'-W']\ e=0\ .$$

The OLS estimator $\hat{\beta}$ is the only linear unbiased estimator with this property. To show that this is the case, let us consider an alternative generic linear estimator:

$$\hat{\beta}=[\ (W'W)^{-1}W'+\underset{k\times T}{C}\ ]\ y$$

As in the proof of the Gauss-Markov Theorem,

$$\tilde{\beta}=[\ (W'W)^{-1}W'+C\ ]\ [W\beta+e]=\beta+(W'W)^{-1}W'e+CW\beta+Ce$$

so that $E(\hat{\beta})=\beta$ if and only if $CW=0$.

The residual component is:

$$\tilde{e}=y-W\tilde{\beta}=y-W[(W'W)^{-1}W'+C]y=[I-W(W'W)^{-1}W'+WC]e$$

so that

$$\tilde{\beta}'W'\tilde{e} = \left[\ e'C'+\beta'+e'W(W'W)^{-1}\ \right]W'\left[\ I-W(W'W)^{-1}W'-WC\ \right]e$$

$$=\left[\ e'C'+\beta'+e'W(W'W)^{-1}\ \right]\left[\ -W'WCe\ \right]=$$

$$=-e'C'W'WCe-\beta'(W'W)Ce-e'WCe=-(e'C'W'+\beta'W'+e')WCe$$

Consequently, it must be $C=0$ to have $\tilde{\beta}'W'\tilde{e}=0$, and in that case $\tilde{\beta}=\hat{\beta}_{OLS}$

.

2.7 An estimator of the error variance

A natural estimator for the variance of the error term σ^2 is $\sum_{t=1}^{T} e_t^2 \big/ T$, but we have seen that the errors e_t are unobservable since they depend on the unknown parameters β, $y-W\beta=e$. However, we could replace β with $\hat{\beta}$ and use the residuals:

$$\hat{e}=y-W\hat{\beta}=y-W(W'W)^{-1}W'y= \qquad (2.7.1)$$

$$=\underset{T\times T}{\left[\ I\right.}-\underset{T\times 2}{W}\ \underset{2\times 2}{(W'W)^{-1}}\ \underset{2\times T}{\left.W'\ \right]}\ \underset{T\times 1}{y} \qquad (2.7.2)$$

Let us have a closer look at the matrix in (2.7.2) since it is often used in subsequent derivations. Let us introduce the notation:

$$\underset{T\times T}{M}=\underset{T\times T}{\left[\ I\right.}-\underset{T\times 2}{W}\ \underset{2\times 2}{(W'W)^{-1}}\ \underset{2\times T}{\left.W'\ \right]} \qquad (2.7.3)$$

so that $\hat{e}=My$. The M matrix is *symmetric*, $M=M'$, and *idempotent*, $M\cdot M'=MM=M$. In fact,

$$\left[\ I-W(W'W)^{-1}W'\ \right]\left[\ I-W(W'W)^{-1}W'\ \right]=\left[\ I-W(W'W)^{-1}W'-W(W'W)^{-1}W'+W(W'W)^{-1}W'\ \right]=M$$

Using $y = W\beta + e$ in (2.7.2), we get:

$$\hat{e} = [I - W(W'W)^{-1}W']\ [W\beta + e] = [W\beta + e - W\beta - W(W'W)^{-1}W'e] = \underset{T\times T}{M}e \quad (2.7.4)$$

which provides the link between the residuals $\hat{e}$ and the errors e.

To estimate σ^2 we could then start from $\sum_{t=1}^{T}\hat{e}_t^2 = \hat{e}'\hat{e}$. From $\hat{e}'\hat{e} = e'M'Me = e'Me$, it follows that

$$E(\hat{e}'\hat{e}) = E(\underset{1\times T}{e'}\ \underset{T\times T}{M}\ \underset{T\times 1}{e})$$

Since this quantity is a scalar, we can also re-write as

$$E(\hat{e}'\hat{e}) = E[tr(e'Me)].$$

When the trace operator is applied to the product of matrices, it is $tr(ABC) = tr(CAB) = tr(BCA)$. Hence, we have

$$E(\hat{e}'\hat{e}) = E[tr(e'Me)] = E[tr(Mee')]$$

Since the trace is a linear operator, it is also $trE(Z) = E[tr(Z)]$, which implies

$$E(\hat{e}'\hat{e}) = tr[E(Mee')] = tr[M\sigma^2 I_T] = \sigma^2 tr(M) = \sigma^2 tr[\ I_T - W(W'W)^{-1}W'\ ] =$$
$$= \sigma^2\left[tr(I_T) - tr\left(W(W'W)^{-1}W'\right)\right] =$$
$$= \sigma^2[\ tr(I_T) - tr(\underset{2\times 2}{W'W}\ \underset{2\times 2}{(W'W)^{-1}})\ ] = \sigma^2[tr(I_T) - tr(I_2)] = \sigma^2[T-2]$$

Therefore,

$$E\left(\hat{e}'\hat{e}/_{T-2}\right) = \frac{E(\sigma^2(T-2))}{T-2} = \sigma^2,$$

In summary, the formula for an unbiased estimator of σ^2 is:

$$\hat{\sigma}^2 = \frac{\hat{e}'\hat{e}}{T-2} = \frac{\sum_{t=1}^{T} \hat{e}_t^2}{T-2} \tag{2.7.5}$$

The estimator $\hat{\sigma}^2$ in (2.7.5) is also useful to provide an estimator of the variance covariance matrix of the OLS estimator $\hat{\beta}$:

$$(V\hat{a}r(\hat{\beta})) = \hat{\sigma}^2 (W'W)^{-1}. \tag{2.7.6}$$

Next, we want to derive the distribution of $\hat{\sigma}^2$, and for this we need to review a few statistical formulae. Let us consider a random variable $q \sim N(\mu_q, \sigma_q^2)$ and its standardized version, $z = \frac{q - \mu_q}{\sigma_q} \sim N(0,1)$. Then, if using *p* independent standard Normal variables z, z_1, z_2, …, z_p, we construct $V = z_1^2 + z_2^2 + ... + z_p^2$, the distribution of V is $\chi^2(p)$. The parameter *p*, known as *degrees of freedom*, is very important to characterize the distribution of *V*, since *E(V)=p* and *Var(V)=2p*. The larger *p* is, the closer the $\chi^2(p)$ distribution gets to a Normal one, but for small *p* the χ^2 is highly asymmetric (in addition to being defined only on the positive axis).
Let us now use this information to derive the distribution of $\hat{\sigma}^2$. Under the strong OLS assumptions, so that the errors are normal, homoscedastic and independent, it is $\sum_{t=1}^{T} \left(\frac{e_t}{\sigma} \right)^2 \sim \chi^2(T)$. If we substitute the unobservable errors e_t with the observable OLS residuals $\hat{e}_t = y_t - \hat{\beta}_1 - \hat{\beta}_2 x_t$, we obtain:

$$V = \frac{\sum_{t=1}^{T} \hat{e}_t^2}{\sigma^2} = \frac{(T-2)\hat{\sigma}^2}{\sigma^2}.$$

In order to find the distribution of V, we should note that while the errors e_t are independent by assumption, the residuals $\hat{e}_t$ are not (since $\hat{e}_t = y_t - \hat{\beta}_1 - \hat{\beta}_2 x_t$

and each $\hat{e}_t$ depends on the same estimators $\hat{\beta}_1$ and $\hat{\beta}_2$). However, it can be shown that *T-2* residuals $\hat{e}_t$ are independent and that

$$V = \frac{(T-2)\hat{\sigma}^2}{\sigma^2} \sim \chi^2(T-2) \tag{2.7.7}$$

This finding implies that

$$E(V) = E\left(\frac{(T-2)\hat{\sigma}^2}{\sigma^2}\right) = \frac{(T-2)}{\sigma^2} E\left(\hat{\sigma}^2\right) = T-2,$$

which provides an alternative proof of the unbiasedness of $\hat{\sigma}^2$. Moreover,

$$Var\left(\frac{(T-2)\hat{\sigma}^2}{\sigma^2}\right) = 2(T-2)$$

so that

$$Var\left(\hat{\sigma}^2\right) = \frac{2(T-2)}{(T-2)^2}\sigma^4 = \frac{2\sigma^4}{(T-2)}.$$

From $Var\left(\hat{\sigma}^2\right) \xrightarrow[T\to\infty]{} 0$ and $E(\hat{\sigma}^2) = \sigma^2$, it follows that $\hat{\sigma}^2$ is also a consistent estimator for σ^2.

Finally, we want to show that $\hat{\beta}$ and $\hat{\sigma}^2$ are uncorrelated (and independent if e ~N). Since $\hat{\sigma}^2 = \hat{e}'\hat{e}/(T-k)$, k = 2, it suffices to show that $\hat{e}$ and $\hat{\beta}$ are uncorrelated, namely,

$$E(\underset{T\times 1}{\hat{e}} \underset{1\times K}{\left(\hat{\beta}-\beta\right)'}) = 0 \tag{2.7.8}$$

From

$$\hat{\beta} = (W'W)^{-1}W'y = \beta + (W'W)^{-1}W'e,$$

and

$$\hat{e} = Me = \left(I - W\left(W'W\right)^{-1}W\right) e,$$

we have

$$E\left(\hat{e}\left(\hat{\beta}-\beta\right)'\right)=E[\ \left(I-W\left(W'W\right)^{-1}W'\right)ee'W\left(W'W\right)^{-1}\ ]=\left(I-W\left(W'W\right)^{-1}W'\right)\underbrace{E(ee')}_{T\times T \to \sigma^2 I_T}W\left(W'W\right)^{-1}=$$

$$=\sigma^2[\ W\left(W'W\right)^{-1}-W\left(W'W\right)^{-1}W'W\left(W'W\right)^{-1}\ ]=0\,.$$

2.8 Maximum Likelihood Estimators for the linear model

We start with the case of N independent and identically distributed (iid) Normal variables, with unit variance and unknown mean:

$$Z_i \overset{i.i.d.}{\sim} N(\mu,1) \qquad i=1,..,N \qquad f(Z_i)=\frac{1}{\sqrt{2\pi}}e^{-\frac{1}{2}(Z_i-\mu)^2}$$

$$f(Z_1,Z_2,...,Z_N)=f(Z_1)f(Z_2)...f(Z_N)=\left(\frac{1}{\sqrt{2\pi}}\right)^N e^{-\frac{1}{2}\sum_{i=1}^{N}(Z_i-\mu)^2} \tag{2.8.1}$$

The joint density function, $f(Z_1,Z_2,\ldots,Z_N)$ in (2.8.1), considered as a function of the unknown parameter μ and for given realizations $z_1,\ldots,z_N$, is labeled *Likelihood* and is typically indicated by $L(\mu)$. Usually, we apply the logarithmic (log) transformation,

$$\log L(\mu)=N\log\left(\frac{1}{\sqrt{2\pi}}\right)-\frac{1}{2}\sum_{i=1}^{N}(z_i-\mu)^2 \tag{2.8.2}$$

A graph of the log – likelihood is reported in Figure 2.7.

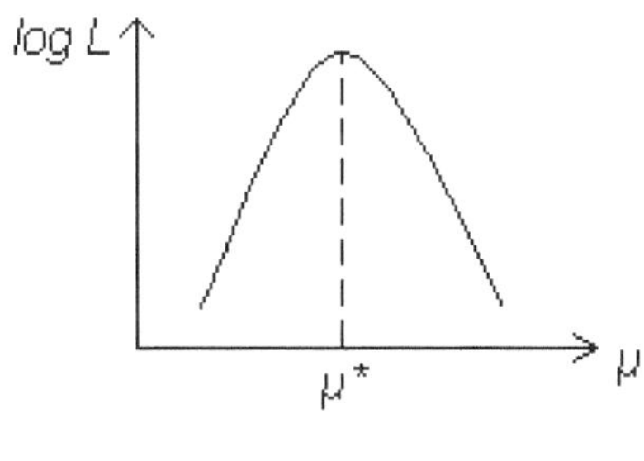

Figure 2.7: Logarithm of the likelihood function for iid. Normal variables with unknown mean μ and unit variance

The Maximum Likelihood Estimator (MLE) of the parameter μ is the maximizer of the (log-) likelihood, μ^* in Figure 2.7.
To derive analytically the MLE, we take the first derivative of the log-likelihood with respect to μ, which yields:

$$\frac{\partial \log L}{\partial \mu} = \sum_{i=1}^{N} (z_i - \mu) \tag{2.8.3}$$

By equating the first order condition in (2.8.3) to 0, we obtain that the MLE of μ coincides with the sample mean of the observations:

$$\mu^* = \frac{\sum_{i=1}^{n} z_i}{N} \tag{2.8.4}$$

Since the second derivative of the log-likelihood is always negative, μ^* is indeed a maximizer.
Let us now extend this procedure to obtain MLEs of the parameters of the linear regression model.
To start with, we have seen that we must specify a distribution (Normal in the previous example) and a data generating process (the Z were iid). Under the strong OLS hypothesis (1)-(6) for the linear model discussed in Section 2.2, we have:

$$y \sim N(W\beta, \sigma^2 I_T)$$

Specifically, it is

$$f(Y_i) = \frac{1}{\sqrt{2\pi\sigma^2}} \exp\left\{- \frac{(Y_i - \alpha - \beta X_i)^2}{2\sigma^2}\right\} \tag{2.8.5}$$

and

$$L(\alpha, \beta, \sigma^2) = f(Y_1, ..., Y_T) \overset{indip}{=} f(Y_1)...f(Y_T) = \frac{1}{(2\pi\sigma^2)^{T/2}} \exp\left[-\sum_{i=1}^{T} \frac{(Y_i - \alpha - \beta X_i)^2}{2\sigma^2}\right] \tag{2.8.6}$$

$$\log L(\alpha, \beta, \sigma^2) = -\frac{T}{2}\log 2\pi - \frac{T}{2}\log\sigma^2 - \frac{1}{2\sigma^2}\sum_{i=1}^{T}(Y_i - \alpha - \beta X_i)^2 \tag{2.8.7}$$

The first order conditions are:

$$\frac{\partial \log L}{\partial \alpha} = \frac{1}{\sigma^2}\sum_{i=1}^{T}(Y_i - \alpha - \beta X_i) = 0 \qquad (2.8.8)$$

$$\frac{\partial \log L}{\partial \beta} = \frac{1}{\sigma^2}\sum_{i=1}^{T}X_i(Y_i - \alpha - \beta X_i) = 0 \qquad (2.8.9)$$

$$\frac{\partial \log L}{\partial \sigma^2} = -\frac{T}{2\sigma^2} + \frac{1}{2\sigma^4}\sum_{i=1}^{T}(Y_i - \alpha - \beta X_i)^2 = 0 \qquad (2.8.10)$$

By solving this system of equations, we obtain:

$$\hat{\alpha}_{ML} = \overline{Y} - \overline{X}\hat{\beta}_{ML} \ ; \quad \hat{\beta}_{ML} = \frac{\sum_{i=1}^{T}(X_i - \overline{X})(Y_i - \overline{Y})}{\sum_{i=1}^{T}(X_i - \overline{X})^2}, \quad \hat{\sigma}^2_{ML} = \frac{\sum_{i=1}^{T}\hat{\varepsilon}_i^2}{T}. \qquad (2.8.11)$$

Therefore, the MLEs of α and β are equal to the OLS estimators, while the MLE for σ^2 has a different denominator, *T* instead of *T – 2* in the OLS case. The latter feature implies that $\hat{\sigma}^2_{ML}$ is slightly biased (but it remains consistent).
In general, when MLE estimators exist, they are:

- consistent. The estimators converge in probability to the true values of the parameters under correct model specification;
- efficient. The (asymptotic) variance-covariance matrix of the MLEs is given by $-E\left[\partial^2 \log L/\partial\mu\partial\mu'\right]^{-1}$), $\mu = \alpha, \beta, \sigma^2$. In our case, for $\theta = \alpha, \beta$, it is $-E\left[\partial^2 \log L/\partial\theta\ \partial\theta'\right]^{-1} = \sigma^2(W'W)^{-1}$;
- asymptotically normally distributed (when multiplied by $\sqrt{T}$)

2.9 The coefficient of determination, R^2

We now derive a measure of the goodness of the explanatory model. From $y = W\hat{\beta} + \hat{e}$ we have:

$$y'y = \hat{\beta}'W'W\hat{\beta} + \hat{e}'\hat{e} + 2\hat{\beta}'W'\hat{e} \tag{2.9.1}$$

Since $W\hat{\beta} = \hat{y}$ and $\hat{y}'\hat{e} = 0$, (2.9.1) simplifies to

$$y'y = \hat{y}'\hat{y} + \hat{e}'\hat{e} \tag{2.9.2}$$

Subtracting $T\bar{y}^2$ from both sides of (2.9.2) yields:

$$y'y - T\bar{y}^2 = \sum_{t=1}^{T}(y_t - \bar{y})^2 = \hat{y}'\hat{y} - T\bar{y}^2 + \hat{e}'\hat{e} \tag{2.9.3}$$

We now show that $\bar{y} = \bar{\hat{y}}$, i.e., the sample mean of y is the same as that of $\hat{y}$. A key result for the proof is the presence of an intercept in the regression model, which is

$$y = \beta_1 + \beta_2 X + e$$

or

$$\underset{T\times 1}{y} = \underset{1\times 1}{\beta_1}\ \underset{T\times 1}{x_1} + \beta_2 x_2 + e$$

where x_1 is a vector of 1s, or more compactly

$$y = W\beta + e.$$

We have derived the OLS estimator by minimizing the loss function

$$S = (y - \beta_1 x_1 - \beta_2 x_2)'(y - \beta_1 x_1 - \beta_2 x_2). \tag{2.9.4}$$

Taking the first derivative of (2.9.4) with respect to β_1 and equating it to 0, we have:

$$-2x'_1 y + 2x'_1 W\hat{\beta} = 0$$

and hence

$$x'_1 W\hat{\beta} = x'_1 y \tag{2.9.5}$$

Now,

$$\bar{\hat{y}} = \frac{\sum_{T=1}^{T} \hat{y}_t}{T} = \frac{x'_1 \hat{y}}{T} = \frac{x'_1 W}{T}\hat{\beta} = \frac{x'_1 y}{T} = \bar{y}, \tag{2.9.6}$$

which shows that indeed $\bar{\hat{y}} = \bar{y}$.

Inserting this result in (2.9.3), we obtain:

$$y' y - T\bar{y}^2 = \hat{y}'\hat{y} - T\bar{\hat{y}}^2 + \hat{e}'\hat{e} \tag{2.9.7}$$

or,

$$\sum_{t=1}^{T}(y_t - \bar{y})^2 = \sum_{t=1}^{T}\left(\hat{y}_t - \bar{\hat{y}}\right)^2 + \sum_{t=1}^{T}(\hat{e}_t - 0)^2 \tag{2.9.8}$$

where:

- $\sum_{t=1}^{T}(y_t - \bar{y})^2$ is the Total Sum of Squares (TSS), measuring the variability of y ; (2.9.9)

- $\sum_{t=1}^{T}\left(\hat{y}_t - \bar{\hat{y}}\right)^2$ is the Explained Sum of Squares (ESS), measuring the variability of the model fit; (2.9.10)

- $\sum_{t=1}^{T}(\hat{e}_t - 0)^2$ is the Residual Sum of Squares (RSS), measuring the variability of the residuals. (2.9.11)

A common measure of the goodness of the explanatory model is the *coefficient of determination:*

$$R^2 = \frac{ESS}{TSS} = 1 - \frac{RSS}{TSS} \tag{2.9.12}$$

with $0 \le R^2 \le 1$. R^2 represents the percentage of variability of y explained by the model.

It is worth making two comments for a proper use of the coefficient of determination:

(a) R^2 is sensible only when using OLS estimators for the model parameters, otherwise $\tilde{\beta}'W'\hat{e} \neq 0$. Moreover, $0 \leq R^2 \leq 1$ only when the model has an intercept, otherwise $\bar{\hat{y}} \neq \bar{y}$ and it can be $R^2 < 0$;

(b) it can be shown that by adding explanatory variables to the model, even if irrelevant, R^2 always increases. To avoid this issue, it is common to report the *adjusted coefficient of determination*:

(c)

$$\bar{R}^2 = 1 - \frac{RSS/T-k}{TSS/T-1} \tag{2.9.13}$$

where k is the number of regressors. It is also:

$$\bar{R}^2 = 1 - \left(\frac{T-1}{T-k}\right)\left(1-R^2\right) \tag{2.9.14}$$

As an alternative measure of the explanatory power of a model, we could compute the correlation between the dependent variable y and the model fit $\hat{y}$. For a very accurate model, this correlation should be close to 1. However, it turns out that this measure provides exactly the same information as the coefficient of determination. In fact we have:

$$corr^2(y,\hat{y}) = \frac{\mathrm{cov}^2(y,\hat{y})}{\mathrm{var}(y)\cdot\mathrm{var}(\hat{y})} = \frac{\mathrm{cov}^2(\hat{y}+\hat{e},\hat{y})}{\mathrm{var}(y)\,\mathrm{var}(\hat{y})} = \frac{\mathrm{var}^2(\hat{y})}{\mathrm{var}(y)\cdot\mathrm{var}(\hat{y})} = \frac{\mathrm{var}(\hat{y})}{\mathrm{var}(y)} =$$

$$= \frac{\sum_{t=1}^{T}(\hat{y}_t-\bar{y})^2}{\sum_{t=1}^{T}(y_t-\bar{y})^2} = \frac{ESS}{TSS} = R^2$$

2.10 Linear transformations of variables and their effects

It is common in economics to work with transformed (e.g., re-scaled) variables, because different measurement scales are possible. For example, the inflation rate can be measured as 2.15 or as 0.0215 and similarly the interest rate can be 3.14 or 0.0314. If we want to assess the effects of the interest rate on inflation, does the measurement scale of the two variables matter? To answer this

question, we need to study the consequences of linear transformations of the variables in the context of the linear regression model.
Let us start with our usual model,

$$y = W\beta + e \tag{2.10.1}$$

and suppose that all values of the explanatory variables are divided by a constant, *a*,

$$y = \frac{W}{a}\gamma + u = \omega\gamma + u \tag{2.10.2}$$

The OLS estimator of the parameters γ in the transformed model is:

$$\hat{\gamma} = (\omega'\omega)^{-1}\omega' y = a(W'W)^{-1}W' y = a\hat{\beta} \tag{2.10.3}$$

Then, the estimator coincides with what we obtained in the original model multiplied by the *a* constant. Its variance is given by:

$$\text{var}(\hat{\gamma}) = \sigma^2(\omega'\omega)^{-1} = a^2\sigma^2(W'W)^{-1} = a^2\,\text{var}(\hat{\beta}) \tag{2.10.4}$$

It is worth noting that:

$$\hat{\gamma}_i \Big/ \sqrt{\text{var}(\hat{\gamma}_i)} = a\hat{\beta}_i \Big/ a\sqrt{\text{var}(\hat{\beta}_i)} = \hat{\beta}_i \Big/ \sqrt{\text{var}(\hat{\beta}_i)} \quad \text{for each } i. \tag{2.10.5}$$

We will come back to this ratio in the next chapter, where we will see that it can be used to test the non-significance of the ith regressor as an explanatory variable for *y*. Thus, if a variable has no explanatory power for *y*, transforming its scale does not change the result.
Finally, it is:

$$\hat{u} = y - \omega\hat{\gamma} = y - \frac{W}{a}a\hat{\beta} = y - W\hat{\beta} = \hat{e}, \tag{2.10.6}$$

so that the residuals are not affected by the linear variable transformation. It is $\hat{\sigma}_u^2 = \hat{\sigma}_e^2$ and the coefficient of determination, R^2, is also unchanged.
Next we study the case where the dependent variable is divided by *a*,

$$\frac{y}{a} = W\gamma + u \tag{2.10.7}$$

The OLS estimators become

$$\hat{\gamma} = (W'W)^{-1} W' y / a = \frac{1}{a} \hat{\beta}, \tag{2.10.8}$$

so in this case they are identical to those of the original model divided by *a*. Their variance covariance matrix is given by:

$$\operatorname{var}(\hat{\gamma}) = \frac{1}{a^2} \operatorname{var}(\hat{\beta}). \tag{2.10.9}$$

The ratio

$$\hat{\gamma}_i \Big/ \sqrt{\operatorname{var}(\hat{\gamma}_i)} = \hat{\beta}_i \Big/ \sqrt{\operatorname{var}(\hat{\beta}_i)} \tag{2.10.10}$$

remains the same for each i. Instead, it is

$$\hat{u} = \frac{y}{a} - \frac{x}{a} \hat{\beta} = \frac{\hat{e}}{a}, \; \hat{\sigma}_u^2 = \hat{\sigma}_e^2 / a^2, \tag{2.10.11}$$

but the coefficient of determination, R^2, does not change as the variance of *y* decreases by the same amount.

As an exercise, you can evaluate the consequences of transforming both the dependent and the independent variables, which yields the model,

$$\frac{y}{a} = \frac{W}{a} \gamma + u \; . \tag{2.10.12}$$

2.11 The multiple regression model

So far we have explained the variable of interest *y* using a single regressor, $y_t = \beta_0 + \beta_1 x_t + e_t$. In fact, economic theory and empirical observation suggest that generally more explanatory variables are needed, whereby

$$y_t = \beta_1 + \beta_2 x_{2t} + \ldots + \beta_k x_{kt} + e_t \, , \qquad (2.11.1)$$

which is the *multiple regression model.*
We can write the multiple regression model in vector form as

$$\underset{}{y_t} = \underset{1\times k}{x_t} \; \underset{k\times 1}{\beta} + e_t , \qquad (2.11.2)$$

with $x_t = [1, \quad x_{2t}, \quad \ldots \quad x_{kt}]$, or,

$$\underset{T\times 1}{y} = \underset{T\times k}{X} \; \underset{k\times 1}{\beta} + \underset{T\times 1}{e}. \qquad (2.11.3)$$

The model in (2.11.3) is formally equal to the one we used in Section 2.5 (equation 2.5.4), with $W = X$ and $k = 2$.
The model assumptions remain the same as those of Section 2.2. In particular, we continue to require

(1) linearity and stability of parameters (2.11.4)
(2) deterministic regressors (2.11.5)
(3) homoscedasticity (2.11.6)
(4) no correlation (2.11.7)

while hypothesis (5) is replaced by

(5) the regressors are not linearly dependent (the $(X'X)$ matrix has rank *k*) (2.11.8)

We will see in the next Section that hypothesis (5) is used for the identification of the model parameters, i.e., to determine the effects of each of the regressors on the dependent variable.

As to the interpretation of the coefficients, β_j still represents the change in the expected value of the dependent variable for a marginal variation of the jth regressor, assuming that all other explanatory variables remain constant. In formulae, using the notation of a partial derivative, it is:

$$\beta_j : \left. \partial E(y_t) \middle/ \partial x_j \right|_{x_1,\ldots,x_{j-1},x_{j+1},\ldots x_k \;\; fisse} \qquad (2.11.9)$$

The OLS estimators of the parameters β_j of the multiple regression model are still determined as the minimizers of the sum of squared errors,

$$S = \sum_{t=1}^{T} e_t^2 = \underset{1\times T}{e'} \underset{T\times 1}{e} = [e_1 \quad e_2 \quad \dots \quad e_T]\begin{bmatrix} e_1 \\ e_2 \\ \vdots \\ e_T \end{bmatrix} \tag{2.11.10}$$

By following the same steps as in Section 2.5, we get:

$$\underset{k\times 1}{\hat{\beta}} = \underset{k\times k}{(X'X)^{-1}} \underset{k\times 1}{X'y}, \tag{2.11.11}$$

where

$$X'X = \begin{bmatrix} \sum_{t=1}^{T} x_{1t}^2 & \sum_{t=1}^{T} x_{1t}x_{2t} & \dots & \sum_{t=1}^{T} x_{1t}x_{kt} \\ \sum_{t=1}^{T} x_{1t}x_{2t} & \sum_{t=1}^{T} x_{2t}^2 & \dots & \sum_{t=1}^{T} x_{2t}x_{kt} \\ \vdots & & \ddots & \\ \sum_{t=1}^{T} x_{1t}x_{kt} & \sum_{t=1}^{T} x_{2t}x_{kt} & \dots & \sum_{t=1}^{T} x_{kt}^2 \end{bmatrix} \text{ and } X'y = \begin{bmatrix} \sum_{t=1}^{T} x_{1t}y_t \\ \sum_{t=1}^{T} x_{2t}y_t \\ \vdots \\ \sum_{t=1}^{T} x_{kt}y_t \end{bmatrix}$$

The variance covariance matrix of the estimators is:

$$Var(\hat{\beta}) = \sigma^2 (X'X)^{-1} \tag{2.11.12}$$

Note that if the regressors are highly correlated with each other, the determinant $|X'X|$ is small, so that all elements of $\text{var}(\hat{\beta})$ are expected to be large. Intuitively, in this case it is difficult to distinguish the contribution of each explanatory variable. Note also that for the matrix $(X'X)$ to have rank k, the number of observations in the sample, T, must be at least equal to k.

If we introduce the additional assumption:

(6) Normality: $y \sim N(X\beta, \sigma^2)$, i.e., $e \sim N(0, \sigma^2)$, (2.11.13)

then,

$$\hat{\beta} \sim N\left(\beta, \sigma^2 (X'X)^{-1}\right). \tag{2.11.14}$$

Otherwise, $\hat{\beta}$ only has an asymptotically Normal distribution.
An unbiased and consistent estimator of the variance of the error term is then given by:

$$\hat{\sigma}^2 = \frac{\sum_{t=1}^{T} \hat{e}_t^2}{T-k}, \tag{2.11.15}$$

As in the case of the model with one explanatory variable, the estimators $\hat{\beta}$ are uncorrelated with σ^2, or independent if the hypothesis of Normality of the errors holds
With regard to the properties of the OLS estimators in this context, since there are no substantial differences with respect to the case with one explanatory variable, the Gauss-Markov theorem still holds and $\hat{\beta}$ is B.L.U.E. and consistent.
We now consider an alternative interpretation of the model coefficients. Let us suppose for simplicity that there are only two explanatory variables, whereby the model becomes:

$$\underset{T\times 1}{y} = \underset{T\times 1}{x_1}\ \underset{1\times 1}{\beta_1} + \underset{T\times 1}{x_2}\ \underset{1\times 1}{\beta_2} + \underset{T\times 1}{e} \tag{2.11.16}$$

Let us define the matrix

$$\left[\ I - x_2 (x'_2 x_2)^{-1} x'_2\ \right]$$

and premultiply by it both sides of (2.11.16), which yields:

$$\begin{aligned} &\left[\ I - x_2(x'_2 x_2)^{-1} x'_2\ \right] y = \\ &= \left[\ I - x_2(x'_2 x_2)^{-1} x'_2\ \right] x_1 \beta_1 + \underbrace{\left[\ I - x_2(x'_2 x_2)^{-1} x'_2\ \right] x_2}_{=0} \beta_2 + \underbrace{\left[I - x_2(x'_2 x_2)^{-1} x'_2 e \right]}_{=0} \end{aligned} \tag{2.11.17}$$

Therefore,

$$\left[\ y - x_2 (x'_2 x_2)^{-1} x'_2\, y\ \right] = \left[\ x_1 - x_2 ((x'_2 x_2)^{-1} x'_2\, x_1)\ \right] \beta_1 + u \tag{2.11.18}$$

Let us consider the dependent and independent variables in the model in (2.11.18). If we regress y on x_2, we get

$$y = x_2\gamma + e_1, \qquad \hat{\gamma} = (x'_2\, x_2)^{-1} x'_2\, y, \tag{2.11.19}$$

and

$$\hat{e}_1 = y - x_2\hat{\gamma} = (y - x_2(x'_2\, x_2)^{-1} x'_2\, y), \tag{2.11.20}$$

which corresponds to the dependent variable in (2.11.18).
If we instead regress x_1 on x_2, we have

$$x_1 = x_2\eta + e_2, \qquad \hat{\eta} = (x'_2\, x_2)^{-1} x'_2\, x_1, \tag{2.11.21}$$

and

$$\hat{e}_2 = x_1 - x_2\hat{\eta} = x_1 - x_2((x'_2\, x_2)^{-1} x'_2\, x_1), \tag{2.11.22}$$

which corresponds to the independent variable in (2.11.18).
In summary, the dependent variable in the model (2.11.18) is the residual of a regression of y on x_2, while the explanatory variable is the residual of a regression of x_1 to x_2. It follows that (2.11.18) is equivalent to

$$\hat{e}_1 = \hat{e}_2\beta_1 + u. \tag{2.11.23}$$

For this reason, β_1 measures the ability of the part of x_1 unexplained by x_2 to explain the part of y unexplained by x_2. It represents the contribution of x_1 in explaining y, net of the effects of the presence of x_2 in the model. In formulae,

$$\beta_1 = \partial E(\hat{e}_1) / \partial \hat{e}_2 .$$

More generally, i.e. in the model with k independent variables, β_1 measures the explanatory capacity of the part of x_1 unexplained by $x_2, x_3, \ldots, x_k$ for the part of y not explained by $x_2, x_3, \ldots, x_k$. And the same applies to a generic β_j.
Finally, to assess the overall explanatory power of the model we can use the coefficient of determination, defined in Section 2.9,

$$R^2 = \frac{ESS}{TSS} = 1 - \frac{RSS}{TSS}, \tag{2.11.24}$$

Or its adjusted version,

$$\overline{R} = 1 - \frac{RSS/T-k}{TSS/T-1} \quad (2.11.25)$$

2.12 Multicollinearity

Let us consider the model with two explanatory variables

$$y = \beta_1 x_1 + \beta_2 x_2 + \varepsilon \quad (2.12.1)$$

and assume that x_1 and x_2 are perfectly collinear (correlated), specifically, $x_1 = x_2$. In this case, known as *multicollinearity*, we can no longer separately identify β_1 and β_2. We can only identify their sum, $\beta_1 + \beta_2$, since the model becomes

$$y = (\beta_1 + \beta_2)x_2 + \varepsilon \quad (2.12.2)$$

For the general case with *k* regressors, we saw in the previous Section that $\hat{\beta}_{OLS}$ is determined by the first-order conditions:

$$(X'X)\hat{\beta} = X'y \quad (2.12.3)$$

If the explanatory variables are multicollinear, the matrix $X'X$ has reduced rank (<*k*) and the system (2.12.3) admits an infinite number of solutions. For this reason, we introduced hypothesis (5) in Section 2.11 that excludes the possibility of multicollinearity.

In practice, we have situations of high but not perfect collinearity of the regressors. For example, if we try to explain stock returns with interest rates at different maturities or check the effects of the prices of a range of substitute goods on the demanded quantity of a product, the regressors are probably highly correlated among themselves. In this case, the main effect is on the variance of the parameter estimators of the regression function. In fact, if $|X'X| \approx 0$, in general $(X'X)^{-1}$ has very large elements, as well as $\sigma^2 (X'X)^{-1}$.

Possible methods to discover high collinearity problems are:

(a) calculate the correlations between all possible pairs of regressors (but it does not work if more than two variables are involved in the linear relationship, e.g. $x_1 + x_2 = x_3$);
(b) calculate the *t* tests and *F-test* for the non-significance of the regressors, described in the next chapter. If they give contrasting results, there may be problems of collinearity;
(c) calculate the determinant of the matrix $(X'X)$.

In case of high collinearity among the explanatory variables, possible remedies include:

(a) Model Re-parameterization, for example $y = \gamma(x_1 - x_2) + \delta\, x_2 + \varepsilon$, to take advantage of the fact that sometimes simple transformations of the regressors may reduce correlation. In the case of the previous examples, you could use the interest rate spreads or price differentials as regressors;
(b) replacement of the regressors with their first *principal components*. The principal components are linear combinations of the variables with two properties: they are linearly independent of each other and they are ordered to explain a decreasing percentage of the overall variability of the original regressors (whereby the first component explains the greater part of variability). From an economic point of view, the problem with this approach is that principal components are generally hard to interpret. For example, while the interest rate for a certain maturity or the price of substitute goods are clearly defined entities, linear combinations of rates or of prices are not so meaningful;
(c) increase the sample size. For example, short- and long-term interest rates typically behave differently before a slowdown in the economy or the price of substitute goods may differ on special occasions such as advertising campaigns. If the sample under investigation is large enough to include several episodes of this type, the correlation among the regressors will diminish.

2.13 An empirical analysis with simulated data

In this Section we will see how to put into practice many of the theoretical concepts developed so far using artificial data for greater control and interpretability of the results. In subsequent Sections we will analyze actual data.

The data are in the EViews workfile "example_simul_chap2.wf1". To open the workfile, after starting EViews by clicking on its icon, we click on "File" in the top left corner of the startup screen of EViews, choose "Open" and click on "EViews Workfile ..". At this point, we select the path to the file. We recommend that you use a copy of the file saved on your computer's hard drive.

Now, we have an open window called "example_simul _ Chap2", which shows our data: each series contains 130 quarterly observations, from the first quarter of the year 1970 to the second quarter of 2002 (the actual start and end dates do not really matter as these are artificial data). Clicking twice, for example, on *y* shows the data either as a vector with 130 components containing the values for each quarter, or as a graph with time on the horizontal axis and the values of the variable on the vertical axis. To switch from one version to another, we click on the "View" menu in the top left of the new window, and click "Spreadsheet" for a list of data or "Graph" and, for example, the "line & symbol" option, leaving the default "Basic type" in the chart type box. We obtain the graph in Figure 2.8.

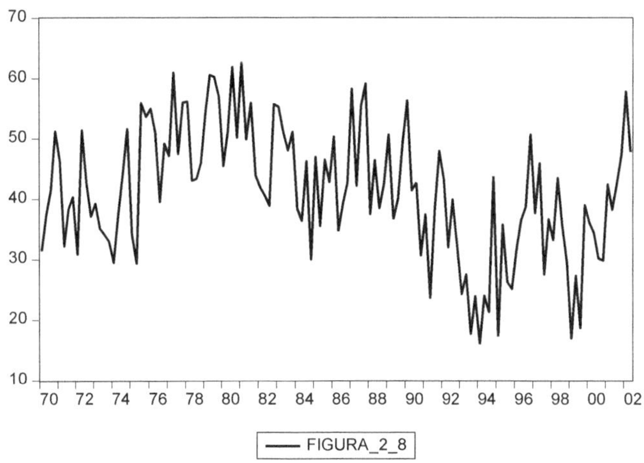

Figure 2.8: Graph of the Y variable

Next, let us assume that we want to test a theory that predicts a linear relationship between *Y* and two explanatory variables, X_1 and X_2:

$$Y = X\beta + e = c + x_1\beta_1 + x_2\beta_2 + e, \qquad (2.13.1)$$

where all the hypotheses on the model detailed in section 2.11 are satisfied.

There are several ways to perform a linear regression with EViews, all equivalent and yielding the same results. Let us see two of them.
First, we place the cursor in the white fieldbelow the toolbar "File", "Edit", etc. and type the following command:

```
ls y c x1 x2
```

where "ls" indicates the estimation method: (ordinary) least square, OLS. The syntax of the program requires writing, after `ls`, first the dependent variable (`y`), then the intercept (`c`, if present in the model), and finally the list of regressors (`x1` and `x2`). After typing the command, we press ENTER.
Second, we click on "Quick" in the toolbar of EViews and choose "Estimate Equation ...". We get "Equation Estimation" window and in the white field, as indicated by the pointer, we write

```
y c x1 x2
```

being OLS estimation already selected in the "Method" box of the "estimation settings" menu. Finally, we press ENTER.
In both cases, we obtain Table 2.1, known as the "output" of the regression. It contains several relevant pieces of information on the regression estimated by the computer:

Dependent variable: Y

Method: Least Squares

Sample: 1970q1 2002q2

Included observations: 130

Summary of the choices

mple size, T

Variable	Coefficient	Std. Error		
C	0.891067	0.228771		
X1	0.800996	0.032864		
X2	1.604939	0.011789		
R-squared	0.997498		Mean dependent var	41.27
Adjusted R-squared	0.997459		S.D. dependent var	10.70
S.E. of regression	0.539457			
Sum squared resid	36.95877			
Log likelihood	-102.7095			

Estimated fficients, $\hat{\beta}_{OLS}$ and standard errors

(The non reported values will be explained later on)

Table 2.1: Regression of the Y variable on X1 e X2

The realized value of the OLS estimator $\hat{\beta}_{OLS} = (X'X)^{-1}X'y$ is $(0.891067 \quad 0.800996 \quad 1.604939)'$. This is equal to the vector of partial derivatives of the expected value of y given X: holding constant the value of x_2, for example, the expected value of y increases by 0.800996 if the value of x_1 increases by one unit. If x_1 and x_2 are both equal to 0, then we expect y to be 0.891067.
We can save the results by clicking on the icon "Name" in the output window and choosing a name for the equation. Once we confirm with the "OK" button, the output Table is saved in the workfile, marked by the equation icon.

In the Workfile window, the series “Resid” now contains the residuals (estimated errors) $\hat{e} = y - \hat{y} = y - \left(X\hat{\beta}_{OLS}\right)$. As we know, the standard error of the regression can be estimated by $\hat{\sigma} = \sqrt{\dfrac{\hat{e}'\hat{e}}{T-k}}$, where k is the number of regressors, 3 in this example, and T is the sample size. Let us check whether the number reported in the equation output in Table 2.1 (0.539457) is equal to that resulting from the formula.

In the field for command lines (the white field below the toolbar of the main window), we type

```
scalar s
```

and press ENTER. In the bottom left corner of the screen we can read “s successfully created”. This command creates the scalar s, denoted by the symbol "#" in our workfile, without assigning a specific value to it for the time being. We then type the new command

```
s = @inner(resid,resid)
```

and press ENTER. With this command we compute the sum of squared residuals, $S = \sum_{t=1}^{T} \hat{e}_t^2 = \hat{e}'\hat{e}$. Clicking on *s*, we find the value 36.9587727085, which is indeed exactly equal to the number reported in the regression output in Table 2.1, under the heading “Sum squared resid”.

The command “@inner (.,.)” runs the *inner product* between the two specified vectors. The @ symbol is often used in the EViews programming language. For example, to compute the average of *y* we write “scalar m = @mean (y)”, and obtain m = 41.2709083964, which corresponds to the number “Mean dependent var” in the regression output in Table 2.1. There are several other functions available in EViews, please check the on-line guide (by clicking on “Help” and then choosing “Users Guide” or “Command & Programming Reference”)

To complete the calculation of the Standard Error of the regression, we use the following command:

```
scalar se = @sqrt(s/127)
```

where 127 = 130 – 3 = $T - k$. This yields 0.539457095917, which corresponds to the (rounded) value in the regression output in Table 2.1.

To obtain the Standard Deviation of the dependent variable *y* (S.D. dependent var in the output) we can use:

```
scalar sd = @sqrt(@inner(y-m,y-m)/129)
```

Can you explain the meaning of this command?

The standard error of the estimator of the coefficient j, $j = 1, 2, 3$, is obtained by multiplying the Standard Error of the regression by the square root of the jth element of the main diagonal of the matrix $(X'X)^{-1}$. The estimated variance covariance matrix $(\hat{\sigma}^2(X'X)^{-1})$ can be viewed from the output of the previously estimated equation ("eq01" in the workfile), by clicking on "View" and choosing "Covariance Matrix".

Let us try now to generate standardized variables, i.e. expressed as deviations from their mean and divided by their standard deviation. For *y* we already have the necessary quantities, and we simply type the command

```
series yst = (y-m)/sd
```

and press ENTER. The vector containing the standardized values of *y* is saved in the workfile with the name yst. For the explanatory variables we use instead the commands

```
series x1st = (x1-@mean(x1))/(@sqrt(@inner(x1--@mean(x1),x1-@mean(x1))/129))
series x2st = (x2-@mean(x2))/(@sqrt(@inner(x2--@mean(x2),x2-@mean(x2))/129))
```

At this point, in the Workfile window there are three new series: yst, x1st and x2st.

In order to obtain a figure with a chart of the three resulting series, we select all of them simultaneously (by clicking on one of them and then, while holding down the "Ctrl" key, clicking on the other two series). As usual, there are a number of alternative procedures to get the same result. For example, we can click on "View" (either in the main window or from the Workfile window), select "Open selected", then "One Window" and finally "Open Group". As a result, we obtain either the Spreadsheet (the three vectors together) or directly the graph of the three variables. The resulting chart is shown in Figure 2.9.

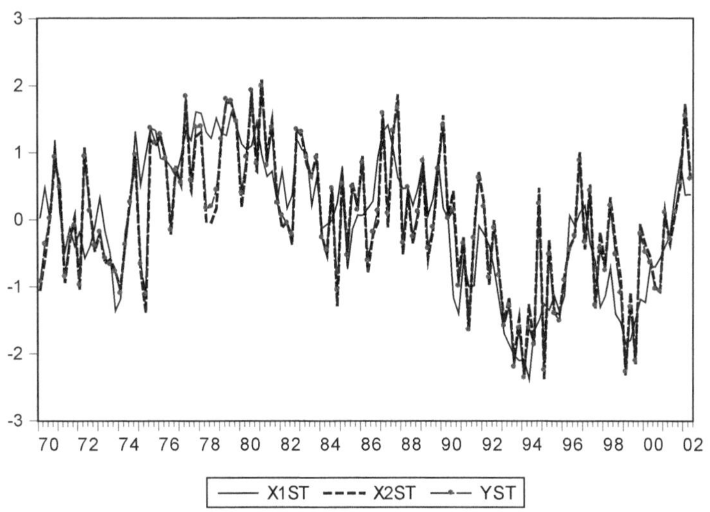

Figure 2.9: Chart of the standardized dependent and independent variables

This chart is useful to understand whether the relationship between y and the regressors is static, or rather if y depends on lagged values of x_1 and x_2. In the former case, the extreme values (peaks) of the three variables should coincide, and in the latter the peaks in x should precede those in y. Figure 2.9 seems to indicate that there are no delays. This is particularly evident for the variable x_2, with x_1 showing a similar though smoother behavior. Note, for example, that the pattern of the three variables between 1990 and 1997 (28 observations) presents a progressive reduction followed by an increase in the values of y, matched by those of the regressors. The correlation is high and positive, in line with the sign of the estimated coefficients of x_1 and x_2, both positive. To save the Group of the three series, we choose "Name" and give the object a name. Once saved, we will see its icon in the Workfile with the specific symbol.
To export the chart, we simply click on it, then select "Edit", "Copy", and, in the "Metafile Properties", we choose "Enhanced Metafile". Now the chart is in the clipboard and we can paste it into other program, such as MS Word or MS Excel.
The residual graph (we choose the mode "line & symbol" to get a better view) provides useful information to verify if the model assumptions are fulfilled, as we will discuss in more detail in Chapters 4 and 5. The values reported in Figure 2.10 seem to confirm the hypothesis of no correlation of the errors. The realization in period t does not affect the one in $t + 1$, while dependence would lead to a systematic pattern, such as clusters of positive or negative values.

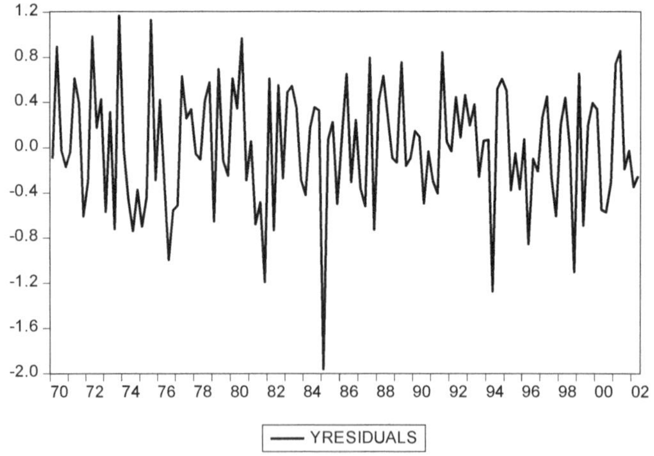

Figure 2.10: Chart of the regression residuals

The distance from the 0 value (the average of the residuals) is quite stable across time, confirming the hypothesis of homoscedasticity.

In Section 2.10 we studied the effects of linear transformations of the variables. Specifically, we analyzed what happens when the dependent variable and/or the regressors are divided by a constant number; we now reproduce the results with EViews for the case where we divide by 10.
Tables 2.2, 2.3 and 2.4 show what happens to our regression. In the first case, when we only transform the dependent variable, the estimated parameters are ten times larger than the original ones; in the second case, when we only transform the regressors, the estimated parameters are one-tenth of the original ones. If instead we transform all variables (the third case), the estimated parameters are identical to the original ones. The standard errors change accordingly, so that the *t* statistics remain unchanged.
EViews also allows us to quickly check the alternative interpretation of the coefficients of the multiple regression model proposed in Section 2.11, where the coefficients coincide with those of transformed regressions. First, we regress x_1 on a constant and x_2, resulting in the residuals $\hat{e}_1$ (orthogonal to x_2). These residuals represent x_1 after partialling out or netting out the effects of x_2. Next, a regression of y on x_2 generates the residuals $\hat{e}_2$, namely, the part of y that cannot be explained by x_2. Finally, we regress $\hat{e}_2$ on a constant and $\hat{e}_1$, obtaining that the OLS estimator for $\hat{e}_1$ is equal to that of x_1 in the full regression. The regression of $\hat{e}_2$ on $\hat{e}_1$ is called *partial regression.*

Dependent variable: Y
Method: Least Squares

Variable	Coefficient	Std. Error	*t*-Statistic	Prob.
CS	8.910674	2.287712	3.895015	0.0002
X1S	8.009958	0.328644	24.37277	0.0000
X2S	16.04939	0.117895	136.1334	0.0000

R-squared	0.997498	Mean dependent var	41.27091
Adjusted R-squared	0.997459	S.D. dependent var	10.70127
S.E. of regression	0.539457	Akaike info criterion	1.626300
Sum squared resid	36.95877	Schwarz criterion	1.692473
Log likelihood	-102.7095	Durbin-Watson stat	2.222137

Table 2.2: Regression output when the regressors are divided by 10

Dependent variable: YS
Method: Least Squares

Variable	Coefficient	Std. Error	*t*-Statistic	Prob.
C	0.089107	0.022877	3.895015	0.0002
X1	0.080100	0.003286	24.37277	0.0000
X2	0.160494	0.001179	136.1334	0.0000

Table 2.3: Regression output when the dependent variable is divided by 10

Dependent variable: YS
Method: Least Squares

Variable	Coefficient	Std. Error	*t*-Statistic	Prob.
CS	0.891067	0.228771	3.895015	0.0002
X1S	0.800996	0.032864	24.37277	0.0000
X2S	1.604939	0.011789	136.1334	0.0000

Table 2.4: Regression output when all variables are divided by 10

Each time we run a regression with EViews, the residuals are shown in the series "resid", a key object in EViews workfiles. Then, for subsequent use, we can save the complete regression residuals in a new series, for example with the command

```
Series resid0 = resid
```

Next, we run the regression

```
ls x1 c x2
```

and save the residuals with

```
series resid1 = resid
```

Then, we obtain $\hat{e}_2$ with the commands

```
ls y c x2
series resid2 = resid
```

Now we can run the partial regression

```
ls resid2 c resid1
```

and the estimated coefficient of resid1 is 0.800996, as expected.
The command

```
scalar ortog = @inner(resid1,x2)
```

shows that $\hat{e}_1$ and x_2 are orthogonal, the obtained value is close to 0.

Hence, we have that $\hat{\beta}_{i,OLS}$, i = 1, …, k, is the best estimator (in the sense of the Gauss-Markov theorem) of the contribution of that part of the ith regressor unexplained by the other regressors to explaining the part of y unexplained by the other regressors.

Note that if we perform a regression of y on a constant and x_1 or x_2 only, we obtain values of the coefficients that are different from those shown in Table 2.1, as the omitted regressor affects the outcome. The regression coefficients coincide with those of the complete regression if and only if the regressors (or blocks of them) are orthogonal. We will return to this point in the next chapter.

In the regression output in Table 2.1, we also find the values of the coefficient of determination, R^2, and of the adjusted R^2, both indicators of the goodness of the regression since they measure how much of the variation in y is explained by the model.

We can compute manually the coefficient of determination, using the command

```
scalar r2=1-@inner(resid0,resid0)/@inner(y-@mean(y),y-@mean(y))
```

Since when adding any regressor R^2 always increases, it is useful to also evaluate the adjusted R^2, denoted as mentioned by $\overline{R}^2$, with (Section 2.10):

$$\overline{R}^2 = 1 - \frac{T-1}{T-K}(1-R^2)$$

$\overline{R}^2$ can decrease by adding a regressor and can also be negative. It only grows when the contribution of the added regressor more than offsets the loss of one degree of freedom.

In our regression, both R^2 and $\overline{R}^2$ are close to 1, so that the model fits the data very well.

The last item in the regression output shown in Table 2.1 is the "Log likelihood": the maximum of the log-likelihood function, namely, the function evaluated at the MLE values. Calculating π as the value of the arctangent function multiplied by four, we can also obtain the log likelihood with the commands

```
scalar pi=@atan(1)*4
scalar l = (-65)*(1+@log(pi*2)+@log((@inner(resid0,resid0)/130)))
```

The log-likelihood value can be used to perform statistical tests, such as "Likelihood ratio tests", see for example Greene (4th edition), section 4.9.3.

Finally, further details on the implementation of the linear regression model in EViews can be obtained from the Help menu, under "Help topics", by choosing "Contents", and then "Basic single equation analysis" under the "Users guide" menu.

2.14 An empirical analysis of aggregate consumption

The Keynesian consumption function postulates a static relationship between real aggregate consumption (C) and real disposable income (Y_d):

$$C_t = c_o + c_1 Y_{d,t} \tag{2.14.1}$$

The corresponding linear regression model, with intercept and a regressor, is:

$$C_t = c_o + c_1 Y_{d,t} + \varepsilon_t . \tag{2.14.2}$$

Typically, when analyzing macroeconomic variables expressed in levels and taking positive values, as in the case of consumption and income, it is useful to work with their logarithmic (log) transformation, as we shall see in more detail in Chapter 4. Therefore, the dependent and independent variables in (2.14.2) are the log of consumption (LRC) and disposable incomes (LRYD), both expressed in real terms. The estimation sample consists of quarterly observations for Italy, for the period 1990:1-2012:2, for a total of 90 values. The series are available in the workfile "example_cons_chap2.wf1".
The results of the estimation of the model are shown in Table 2.5.

Dependent variable: LRC
Method: Least Squares
Sample (adjusted): 1990q1 2012q1
Included observations: 89 after adjustments

Variable	Coefficient	Std. Error	*t*-Statistic	Prob.
C	-1.124210	0.236425	-4.755043	0.0000
LRYD	1.112094	0.029583	37.59246	0.0000

R-squared	0.942007	Mean dependent var	7.763188
Adjusted R-squared	0.941341	S.D. dependent var	0.086122
S.E. of regression	0.020858	Akaike info criterion	-4.879901
Sum squared resid	0.037851	Schwarz criterion	-4.823977
Log likelihood	219.1556	Hannan-Quinn criterion	-4.857360
F-statistic	1413.193	Durbin-Watson stat	0.313491
Prob(*F*-statistic)	0.000000		

Table 2.5: Regression of consumption on disposable income

According to the coefficient of determination, R^2, and its adjusted version, $\overline{R}^2$, the model has a good explanatory power. However, as we shall see in Chapter 6, this could be due to the persistent trend of the two variables, see Figure 2.11. Note that the estimated negative value for the constant, and the LRYD coefficient greater than 1, are not inconsistent with economic theory, which predicts a positive autonomous consumption and a marginal propensity to consume less than 1. In fact, we are using the log transformed variables in the regression and, as we will see later on, this impairs the value of the parameters. In particular, the estimated coefficient for LRYD is consistent with an average marginal propensity to consume around 0.50.

The evolution of the model residuals, shown at the bottom of Figure 2.11, suggests, however, that the OLS assumptions are violated. In particular, there is a remarkable persistence in the residuals, and they might have higher variance in the second part of the sample. Moreover, there are some outlying values in the period starting from 2008, characterized by the global financial crisis, which we have not taken into account so far.

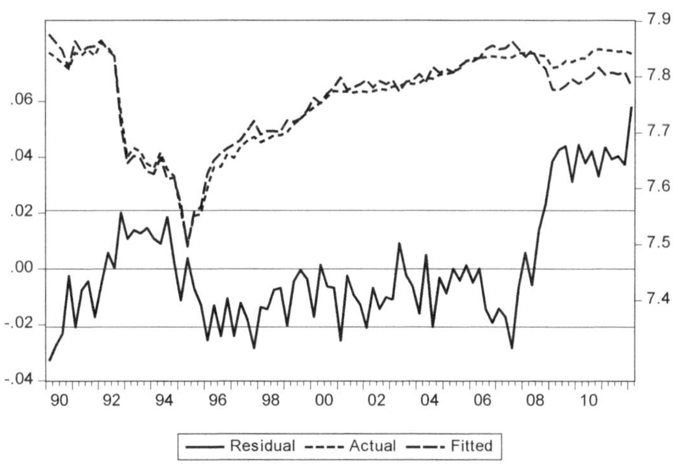

Figure 2.11: Actual and fitted values and residuals for the model of consumption

Let us now consider whether it is possible to obtain a better characterization of the consumption behavior using different or additional explanatory variables. For example, the permanent income theory of consumption suggests that current income should be replaced by permanent income as an explanatory variable, where the latter is defined as the expected value of discounted future earnings. The life cycle theory of consumption instead predicts that the level of aggregate consumption is explained by wealth and disposable income.

By regressing (the log of) real consumption on an intercept, (the log of) real disposable income and (the log of) real wealth, we get the results shown in Table 2.6.

Dependent variable: LRC
Method: Least Squares

Variable	Coefficient	Std. Error	*t*-Statistic	Prob.
C	-1.020849	0.238960	-4.272041	0.0000
LRYD	1.066816	0.037456	28.48182	0.0000
LRW	0.025530	0.013273	1.923434	0.0577

R-squared	0.944399	Mean dependent var	7.763188
Adjusted R-squared	0.943106	S.D. dependent var	0.086122
S.E. of regression	0.020542	Akaike info criterion	-4.899548
Sum squared resid	0.036290	Schwarz criterion	-4.815662
Log likelihood	221.0299	Hannan-Quinn criterion	-4.865736
F-statistic	730.3720	Durbin-Watson stat	0.297996
Prob(*F*-statistic)	0.000000		

Table 2.6: Regression of consumption on disposable income and wealth

The most interesting effect is on the coefficient of LRYD that shrinks slightly when wealth is included as a regressor, although the latter is only marginally statistically significant at the 5% level. The decrease in the coefficient of LRYD is related to the fact that, in the sample under analysis, real wealth is positively correlated with real disposable income (the value is approximately 0.63) and the estimated (marginal) effect of wealth on (the expected value of) consumption turns out to be positive and statistically significant. Therefore, the omission of wealth from the set of the regressors causes an upward distortion of the income coefficient.

We now consider three additional regressions. In the first two models we get the parts of consumption and income that cannot be explained by wealth. In the third model we use as variables the residuals we have obtained in the first two steps. In formulae:

$$c = \alpha + \beta_{wc} w + \hat{e}_c$$

$$y_d = \alpha + \beta_{wd} w + \hat{e}_{yd}$$

$$\hat{e}_c = \beta_{yd} \hat{e}_{yd} + e$$

Table 2.7 reports the results of the third regression and illustrates how the coefficient of EYD is exactly equal to that of LRYD in Table 2.6. Note that the associated standard error is slightly different.

Dependent variable: E_C
Method: Least Squares
Sample (adjusted): 1990q1 2012q1
Included observations: 89 after adjustments

Variable	Coefficient	Std. Error	*t*-Statistic	Prob.
E_YD	1.066816	0.037028	28.81110	0.0000

R-squared	0.904148	Mean dependent var	-2.52E-15
Adjusted R-squared	0.904148	S.D. dependent var	0.065592
S.E. of regression	0.020307	Akaike info criterion	-4.944492
Sum squared resid	0.036290	Schwarz criterion	-4.916530
Log likelihood	221.0299	Hannan-Quinn criterion	-4.933221
Durbin-Watson stat	0.297996		

Table 2.7: The effect of disposable income on consumption via partial regression

Finally, we consider the regression of the growth rate of consumption on that of disposable income and wealth. Note that the sample is slightly reduced, by one observation, because the growth rate refers to the period 1990: 2-2012: 1.
In this model, the coefficient of wealth loses significance even further, while that of income is considerably reduced. There is also a significant reduction in the value of R^2, linked to that in the variability of the dependent variable. Finally, from Figure 2.12, it seems that the model errors remain serially correlated and possibly heteroscedastic, and this puts into question the reliability of the obtained results.

Dependent variable: DLRC
Method: Least Squares
Sample (adjusted): 1990q2 2012q1
Included observations: 88 after adjustments

Variable	Coefficient	Std. Error	*t*-Statistic	Prob.
DLRW	0.035003	0.036128	0.968843	0.3354
DLRYD	0.824773	0.047251	17.45515	0.0000
C	0.000646	0.001055	0.611571	0.5425

R-squared	0.786997	Mean dependent var	-1.65E-05
Adjusted R-squared	0.781985	S.D. dependent var	0.021035
S.E. of regression	0.009822	Akaike info criterion	-6.374986
Sum squared resid	0.008199	Schwarz criterion	-6.290531
Log likelihood	283.4994	Hannan-Quinn criterion	-6.340961
F-statistic	157.0277	Durbin-Watson stat	2.518338
Prob(*F*-statistic)	0.000000		

Table 2.8: Regression of the growth rate of consumption on disposable income and wealth

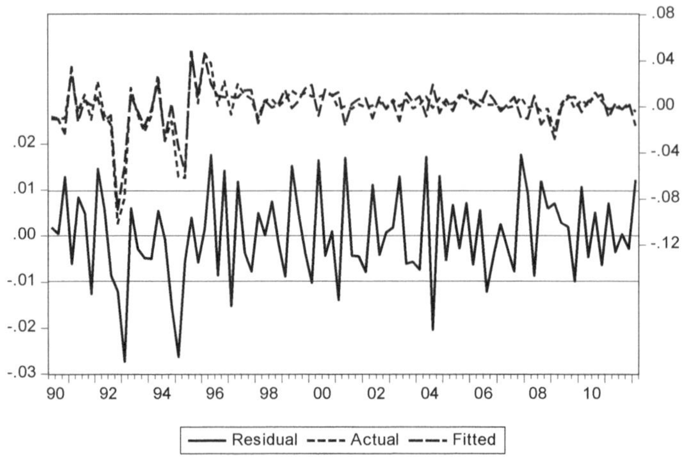

Figure 2.12: Actual and fitted values and residuals for the model of the growth rate of consumption

2.15 An empirical analysis of aggregate investment

From basic macroeconomics, we know that aggregate investments should increase when the national income increases, in anticipation of greater demand, and shrink when the interest rate increases, i.e. the cost of investments. We can then formulate a linear relation of the type:

$$\log(I) = \alpha + \beta_1 \log(Y) + \beta_2 i + \varepsilon ,$$

where it should be $\beta_1 > 0$ and $\beta_2 < 0$.

For OLS estimation of this model we use the workfile "example_inv_chap2.wf1", where LRI indicates the log of real aggregate private investment, LRYD the log of real disposable income, RR the real interest rate and all data refer to the period 1991:1-2011:4. The results are presented in Table 2.9.

Dependent variable: LRI
Method: Least Squares
Date: 08/22/12 Time: 19:33
Sample (adjusted): 1991q1 2011q4
Included observations: 84 after adjustments

Variable	Coefficient	Std. Error	*t*-Statistic	Prob.
LRYD	0.709066	0.082330	8.612446	0.0000
RR	-0.030356	0.002870	-10.57701	0.0000
C	5.559115	0.663154	8.382846	0.0000

R-squared	0.821321	Mean dependent var	11.10866
Adjusted R-squared	0.816910	S.D. dependent var	0.113882
S.E. of regression	0.048729	Akaike info criterion	-3.170030
Sum squared resid	0.192335	Schwarz criterion	-3.083215
Log likelihood	136.1413	Hannan-Quinn criterion	-3.135131
F-statistic	186.1640	Durbin-Watson stat	0.330525
Prob(*F*-statistic)	0.000000		

Table 2.9: Regression of aggregate investment on disposable income and interest rate

The regression has good explanatory power and the sign of the estimated coefficients is as expected. The residual graph in Figure 2.13 shows some persistence and change in the variability of the series, suggesting possible violations of the OLS assumptions. The chart of the actual and fitted values, also shown in Figure 2.13, suggests that the model captures quite well the evolution of investment, in line with the high value of R^2.

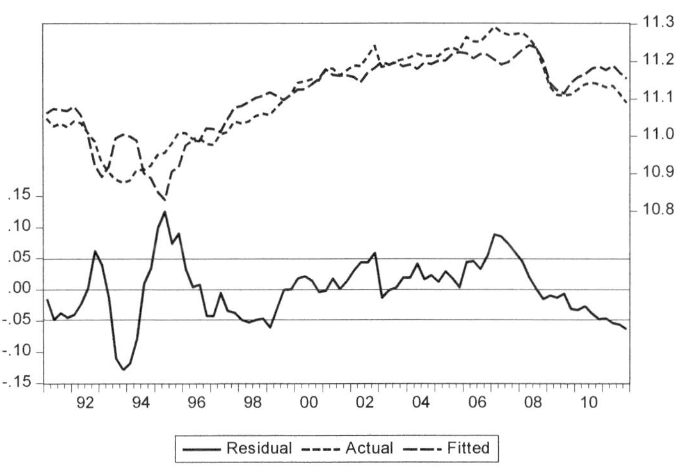

Figure 2.13: Graph of actual and fitted values and residuals for the investment model

2.16 An empirical analysis of labor productivity

Using the data in the workfile "example_prod_chap2.wf1", we now assess the relation between labor productivity, measured as product per worker, and investment. In particular, we regress the log of productivity (*LP*) on a constant and on the log of the investment to GDP ratio (*LQ*). The full dataset includes quarterly observations from 1980 until 2012, but for the purposes of our analysis we focus on the subsample 1995:1-2012:1.
A chart of the series highlights a substantially constant behavior of productivity and of the investment to GDP ratio, with a positive correlation between the two variables of approximately 0.67. Indeed, in the regression of *LP* on *LQ*, the regressor has a significant estimated coefficient, with a rather low estimated standard error, see Table 2.10.

Dependent variable: LP

Variable	Coefficient	Std. Error	*t*-Statistic	Prob.
LQ	0.238530	0.032302	7.384413	0.0000
C	3.681419	0.051986	70.81623	0.0000

R-squared	0.448694	Mean dependent var	4.065133
Adjusted R-squared	0.440465	S.D. dependent var	0.017065
S.E. of regression	0.012765	Akaike info criterion	-5.855632
Sum squared resid	0.010918	Schwarz criterion	-5.790875
Log likelihood	204.0193	Hannan-Quinn criterion	-5.829941
F-statistic	54.52955	Durbin-Watson stat	0.310907
Prob(*F*-statistic)	0.000000		

Table 2.10: Regression of labor productivity on the investment to GDP ratio

The explanatory power of the model is as good, as confirmed graphically in Figure 2.13. However, the regression residuals again suggest possible violation of the assumptions of uncorrelated and homoscedastic errors. Hence, also in this case the simple static model may be inappropriate for a good characterization of the evolution of the variable of interest.

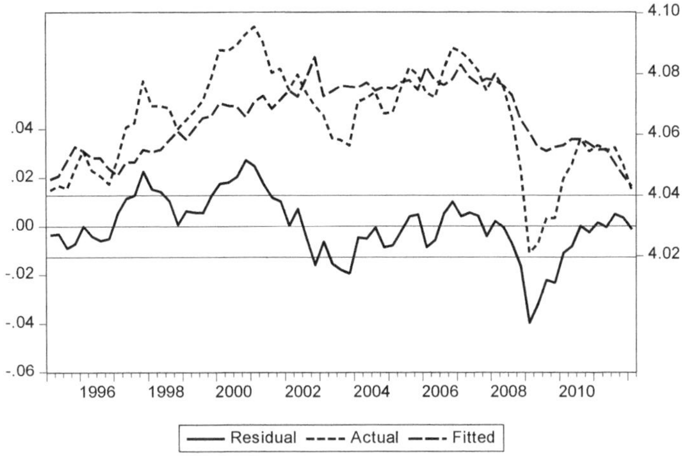

Figure 2.14: Actual and fitted values and residuals for the productivity model

The main concepts of this chapter

In this chapter we considered the linear regression model,

$$y_t = \beta_1 + \beta_2 x_t + e_t \quad ,$$

whose underlying assumptions are

(1) linearity and stability of parameters;
(2) deterministic regressors;
(3) homoscedasticity;
(4) no correlation;
(5) the regressors are not linearly dependent (the $(X'X)$ matrix has rank k).

We have then derived the OLS estimators as the minimizers of the sum of squared residuals:

$$\hat{\beta}_1 = \bar{y} - \hat{\beta}_2 \bar{x}$$

$$\hat{\beta}_2 = \frac{T\sum x_t y_t - \sum x_t \sum y_t}{T\sum x_t^2 - \left(\sum x_t\right)^2}$$

and shown that they are unbiased and consistent.

Comparing them with other linear and unbiased estimators, we demonstrated that the OLS estimators are more efficient (B.L.U.E.), with

$$V(\hat{\beta}_1) = \frac{\sigma^2 \sum x_t^2}{T\sum (x_t - \bar{x})^2} \qquad V(\hat{\beta}_2) = \frac{\sigma^2}{\sum (x_t - \bar{x})^2}.$$

Introducing the further hypothesis that $e_t \sim N$, and therefore also $y_t \sim N$, we have:

$$\hat{\beta}_1 \sim N(\beta_1, V(\hat{\beta}_1)) \quad e \qquad \hat{\beta}_2 \sim N(\beta_2, V(\hat{\beta}_2)).$$

This result holds even without the assumption $e_t \sim N$, but then only when the sample size T is very large.
We have also proposed to estimate the variance of the error term by

$$\hat{\sigma}^2 = \frac{\sum \hat{e}_t^2}{T-2},$$

and shown that this is an unbiased and consistent estimator, uncorrelated with the OLS estimator of the regression function parameters. In addition, if $e \sim \mathrm{N}$,

$$(T-2)\frac{\hat{\sigma}^2}{\sigma^2} \sim \chi^2(T-2)$$

In the case of more than one explanatory variable, the model becomes

$$\underset{1\times1}{y_t} = \beta_1 \underset{1\times1}{x_{1t}} + \beta_2 x_t + \ldots + \beta_k x_{kt} + e_t \qquad o \qquad \underset{1\times1}{y_t} = \underset{1\times k}{x_t} \underset{k\times1}{\beta} + \underset{1\times1}{e_t}$$

or

$$\underset{T\times1}{y} = \underset{T\times k}{x} \underset{k\times T}{\beta} + \underset{T\times1}{e_t}$$

Assuming that the hypotheses (1) – (5) are met, consistent and unbiased estimators of the parameters are:

$$\underset{k\times1}{\hat{\beta}} = \underbrace{\underset{k\times T\cdot T\times k}{(X'X)^{-1}}}_{k\times k} \underbrace{\underset{k\times T\cdot T\times1}{X'y}}_{k\times1} \quad , \quad \hat{\sigma}^2 = \frac{\sum \hat{e}_t^{\,2}}{T-k}.$$

In addition, the estimators of the coefficients and of the variance are uncorrelated. Adding the hypothesis of Normality (or asymptotically)

$$\underset{k\times 1}{\hat{\beta}} \sim N(\underset{k\times 1}{\beta}, \underbrace{\sigma^2 (X'X)^{-1}}_{k\times k}), \qquad (T-k)\frac{\hat{\sigma}^2}{\sigma^2} \sim \chi^2(T-k)$$

Finally, a measure of the goodness of the explanatory model is represented by the coefficient of determination:

$$R^2 = \frac{ESS}{TSS} = 1 - \frac{RSS}{TSS},$$

or by its adjusted version that takes into account the degrees of freedom:

$$\bar{R}^2 = 1 - \frac{RSS/T-k}{TSS/T-1},$$

where ESS, RSS and TSS indicate, respectively, the sum of the squares of the fitted values, residuals and dependent variable.

Exercises

1)

Given the model

$$Y = X\beta + e, \qquad e \sim \text{iid } (0, \sigma^2 I)$$

let us consider a generic linear estimator

$$\underset{Kx1}{\tilde{\beta}} = \left[\underset{KxK}{(X'X)^{-1}} \underset{KxT}{X'} + \underset{KxT}{C} \right] \underset{Tx1}{Y}$$

a) Obtain the condition on C for which $E(\tilde{\beta}) = \beta$.

b) Assuming that condition (a) is satisfied, obtain the condition on C such that $\tilde{\beta}' X' \tilde{e} = 0$, where $\tilde{e} = Y - X\tilde{\beta}$.

2)

Consider the model:

$$Y = X\beta + \varepsilon = X_1\beta_1 + X_2\beta_2 + \varepsilon \qquad \varepsilon \text{ iid } N(0, \sigma^2 I)$$

Demonstrate: (a) unbiasedness and (b) consistency of the OLS estimator:

$$\hat{\beta} = (X'X)^{-1} X'Y$$

3)

Demonstrate the Gauss-Markov theorem under weak OLS assumptions. Show that the estimator $C\hat{\beta}_{OLS}$ is the best (in the sense of Gauss-Markov) to estimate $C\beta$, where C is a non-stochastic matrix.

4)

a) Assuming weak OLS assumptions, calculate the covariance matrix between $\hat{\beta}_{OLS}$ and $\hat{Y}$.

b) Demonstrate analytically that deleting a variable from a linear model always worsens the R^2.

5)

Consider the model:

$$y_t = \alpha + \beta x_t + e_t, \qquad e_t \sim \text{iid } (0, \sigma^2)$$

Calculate the OLS estimators of α and β and their variance covariance matrix, assuming that t = 1.2 with ($x_1 = 0,_1\ y = 1$) and ($x_2 = -1, y_2 = 0$).

6)

Consider the model:

$$Y = X\beta = X_1\beta_1 + X_2\beta_2 + \varepsilon, \quad \varepsilon \text{ iid } N(0, \sigma^2 I),$$

where X_1 and X_2 are orthogonal. Suppose you estimate two separate models:

$$Y = X_1\beta_1 + u \quad \text{and} \quad Y = X_2\beta_2 + u$$

a) Are the resulting OLS estimators of β_1 e β_2 less efficient than those for the model $Y = X\beta$?

b) Would your answer change if X_1 and X_2 were not orthogonal?

7)

Under the strong OLS assumptions, prove independence of the vector of OLS estimated parameters and residuals.

8)

Use the data in the EViews workfile "example_simul2_chap2" to estimate the model:

$$Y = \beta_0 + X_1\beta_1 + X_2\beta_2 + X_3\beta_3 + \varepsilon \, .$$

a) Calculate the OLS estimator of the standard deviation of the residuals σ and the correlation of the dependent variable with the fitted values.

b) The EViews output table gives you already an indication of the values requested in step a). Explain how you could calculate them from the values reported in the regression output table.

9)

When does R^2 take the values of 0 or 1?

3. Inference in the Linear Regression Model

3.1 Interval estimators

In the previous chapter we have defined (point) estimators for the parameters of the linear regression model. We now want to derive *interval estimators* (or *confidence intervals*) for β, i.e., define a range containing β, the true value of the parameters, with a fixed probability.
To address this issue, we need to introduce a new distribution, the Student-*t*.

Let us consider two independent random variables, $z = \dfrac{q - \mu_q}{\sigma_q} \sim N(0,1)$ and $V \sim \chi^2(p)$. It can be shown that,

$$t = \frac{z}{\sqrt{V/p}} \sim t(p) \tag{3.1.1}$$

The parameter *p*, known as *degrees of freedom (df)*, completely characterizes the *t*-distribution. In fact, it is *E(t)=0* and $Var(t) = \dfrac{p}{p-2}$. When $p \to \infty$, $t \to Normal$; otherwise, the *t*-density has more probability in the tails than the Normal density. Figure 3.1 shows the *t*-density for different values of *p*.

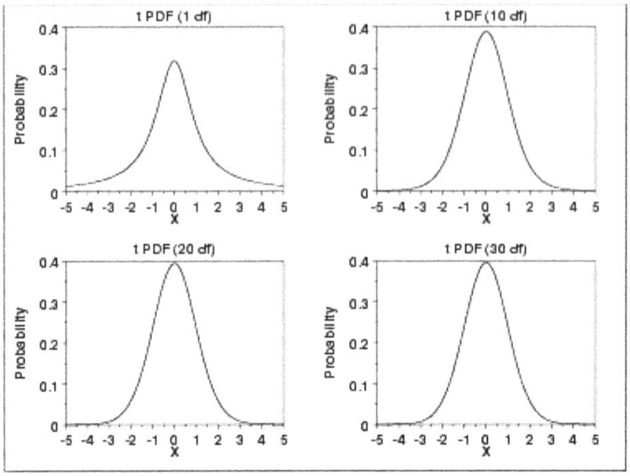

Figure 3.1: Student-t density with various degrees of freedom (df)

The *t*-distribution is quite important in the context of the linear regression model since we know that, when the errors are normally distributed, it is:

$$z = \frac{\hat{\beta}_i - \beta_i}{\sqrt{Var(\hat{\beta}_i)}} \sim N(0,1) \tag{3.1.2}$$

Moreover, we have seen in the previous chapter that z is independent of

$$V = \frac{(T-k)\hat{\sigma}^2}{\sigma^2} \sim \chi^2(T-k) \tag{3.1.3}$$

where, as usual, T indicates the sample size and k the number of regressors. From (3.1.1), (3.1.2), (3.1.3) it follows that:

$$\frac{z}{\sqrt{V/_{T-k}}} \sim t(T-k) \tag{3.1.4}$$

Let us have a closer look at the quantity in (3.1.4) for the case of $\hat{\beta}_2$ in the simple model with intercept and one explanatory variable (whose estimated coefficient is $\hat{\beta}_2$). It is:

$$\frac{z}{\sqrt{V/_{T-2}}} = \frac{\hat{\beta}_2 - \beta_2}{\sqrt{\frac{\sigma^2}{\sum (x_t - \bar{x})^2}}} \Bigg/ \sqrt{\frac{(T-2)\hat{\sigma}^2}{\sigma^2} \Big/ T-2} = \frac{\hat{\beta}_2 - \beta_2}{\sqrt{\frac{\hat{\sigma}^2}{\sum (x_t - \bar{x})^2}}} =$$
$$= \frac{\hat{\beta}_2 - \beta_2}{\sqrt{V\hat{a}r(\hat{\beta}_2)}} = \frac{\hat{\beta}_2 - \beta_2}{s.\hat{e}.(\hat{b}_2)} \sim t(T-2) \qquad (3.1.5)$$

In the general model with k regressors, we have:

$$t = (\hat{\beta}_q - \beta_q) / s.\hat{e}.(\hat{\beta}_q) \sim t(T-k), \qquad (3.1.6)$$

for $q=1,2,\ldots, k$.

We now have all the required elements to construct an interval estimator (or confidence interval) for the generic parameter β_q associated with the qth regressor. First, if we fix a probability level α, e.g. α = 0.1, 0.05, 0.01, given the result in (3.1.6), we can find the *critical value* t_c such that $(i)\ \Pr\{t \geq t_c\} = \frac{\alpha}{2}$, $(ii)\ \Pr\{t \leq -t_c\} = \frac{\alpha}{2}$ (the Student-t distribution is symmetric, so that the left critical value is equal in absolute value to the right critical value). Given α and the number of degrees of freedom T-k, we can obtain the critical value t_c from the table in Table 3.1.

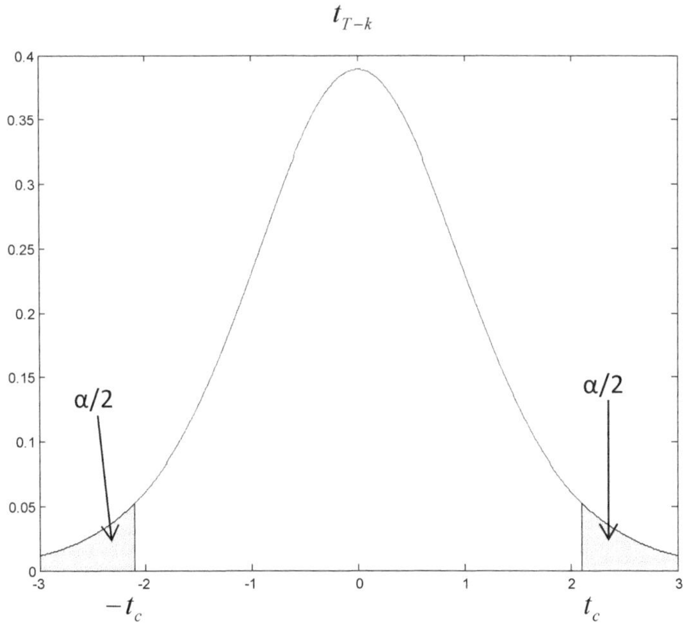

Figure 3.2: Critical values of the Student-t density

Combining the formulae in (*i*) and (*ii*) we have:

$$\Pr\{-t_c \le t \le t_c\} = 1-\alpha \qquad (3.1.7)$$

A graphical example is shown in Figure 3.2.
We can now replace *t* in (3.1.7) with its definition in terms of $\hat{\beta}_q$ in (3.1.6), which yields:

$$\Pr\left\{-t_c \le \frac{\hat{\beta}_q - \beta_q}{s.\hat{e}.(\hat{\beta}_q)} \le t_c\right\} = 1-\alpha \qquad (3.1.8)$$

wherefrom,

$$\Pr\{\hat{\beta}_q - t_c \cdot s.\hat{e}.(\hat{\beta}_q) \le \beta_q \le \hat{\beta}_q + t_c \cdot s.\hat{e}.(\hat{\beta}_q)\} = 1-\alpha \qquad (3.1.9)$$

It follows that

$$\hat{\beta}_q \pm t_c \cdot s.\hat{e}.(\hat{\beta}_q)$$

is an interval estimator or confidence interval for the parameter β_q at the (1-α) % level, while $\hat{\beta}_q$ is a point estimator.
In the simple case of the model with intercept and one explanatory variable (*k=2*), the interval estimator for β_2 becomes:

$$\frac{T\sum x_t y_t - \sum x_t \sum y_t}{T\sum x_t^2 - \left(\sum x_t\right)^2} \pm t_c \cdot \sqrt{\frac{\hat{\sigma}^2}{\sum (x_t - \bar{x})^2}} \qquad (3.1.11)$$

For example, if $\hat{\beta}_2 = 2.5,\ s.\hat{e}.(\hat{\beta}_2) = 1\ and\ T = 32$, then $p = T-2 = 30$ and the critical value for α =0.1 is t_c=1.697 (see Table 3.1). Therefore, the confidence interval at the 90% level for β_2 is $\hat{\beta}_2 \pm 1.697 \cdot s.\hat{e}(\hat{\beta}_2) = 2.5 \pm 1.697 \cong [0.8, 4.2]$.
Note that Table 3.1 shows the probability to the right of the critical value. Hence, if the chosen significance level is *a*, the appropriate column of the table is the one corresponding to *a*/2. In the former example, with *a* = 0.1, the relevant column in the table is that for *a*/2 = 0.05 and the line is that for *p* = 30.
In the previous chapter we have emphasized the difference between an estimator (a random variable) and an estimate (a number, the realization of the estimator). Similarly, we need to distinguish an interval estimator and an interval estimate. In particular, if we follow the procedure described above to

derive an interval estimator at the (1-*a*)% in a large number of different samples generated with the same parameters β, then (1-*a*)% of the intervals would contain the true value of *β*. Hence, this is a property of the statistical procedure, similar, for example, to unbiasedness for the OLS estimator $\hat{\beta}$. If we instead have one sample, as happens in practice, and one interval estimate or confidence interval, e.g. [-1.1] for β_2 with $\alpha = 0.1$, we cannot say that with 90% probability this range contains the true value of β_2. We can only say that if we could repeat many times the sampling and the interval estimation, in 90% of the cases the confidence interval would contain the true value of β_2.
In practice, however, the length of the confidence interval gives an idea of the reliability of the estimated parameter $\hat{\beta}_2$. If, in fact, $s.\hat{e}.(\hat{\beta}_2)$ is large, meaning there is a lot of uncertainty, then the confidence interval is large, and vice versa if $s.\hat{e}.(\hat{\beta}_2)$is small. We cannot, however, precisely quantify what is large or small, it depends on the specific problem we are analyzing.

3.2 Testing hypothesis on the parameters of the linear regression model

An econometric model is based on economic theory. For example, basic consumption theory suggests that $C=c_0+c_1Y$, with $c_0>0$ and $0<c_1<1$, where C indicates consumption and Y disposable income. Similarly, monetary economics suggests that real money demand (M/P) depends on the interest rate (i) and income (Y), with $M/P=d_0\, i+d_1Y$, where $d_0<0$ and $d_1>0$. In order to test whether a theory is correct, or at least compatible with the data, we can verify whether its implications for the econometric model parameters are valid. For example, it should be $c_1\neq 0$ or $d_1\neq 0$, or, more precisely, $0<c_1<1$ *or* $d_1>0$. We must therefore consider how to statistically test this kind of hypotheses on the model parameters.
To construct a hypothesis test we need 4 components:

1. the *null hypothesis*, that is, what we want to test (and assume to be valid);
2. the *alternative hypothesis*, i.e., what (we assume) is true if the null hypothesis is false;
3. a *test statistic*, i.e., a random variable which allows us to discriminate between (1) and (2);
4. a *rejection region*, i.e., a set of values such that if the test statistic in 3 takes one of these values we can reject (1).

We now illustrate these elements in the context of the regression model of aggregate consumption on disposable income:

$$C_t = \beta_1 + \beta_2 Y_t + \varepsilon_t \qquad (3.2.1)$$

1. The null hypothesis specifies a value for a parameter, for example, $H_0 : \beta_2 = 0.9$, or more generally $H_0 : \beta_2 = c$. This is what we believe is the true value of the parameter β_2 and we want to test whether or not we can reject our guess.
2. The alternative hypothesis tells us what we expect to be true about β_2 if our null hypothesis is incorrect. For example, it can be $H_1 : \beta_2 \neq c$ *or* $H_1 : \beta_2 < c$.

It is helpful to make two comments before proceeding. First, though testing H_0 against H_1 it turns out that H_0 is rejected, we do NOT necessarily accept H_1. This is because the testing procedure is meant to reject or not H_0, not to accept or reject H_1. If H_1 is of particular interest, such as $H_1 : \beta_2 < c$, then we should repeat the procedure by putting $\beta_2 < c$ as the null hypothesis.

Degrees of freedom	Probability to the right of the critical value					
	0.10	0.05	0.025	0.01	0.005	0.001
	3.078	6.314	12.706	31.821	63.657	318.313
2.	1.886	2.920	4.303	6.965	9.925	22.327
3.	1.638	2.353	3.182	4.541	5.841	10.215
4.	1.533	2.132	2.776	3.747	4.604	7.173
5.	1.476	2.015	2.571	3.365	4.032	5.893
6.	1.440	1.943	2.447	3.143	3.707	5.208
7.	1.415	1.895	2.365	2.998	3.499	4.782
8.	1.397	1.860	2.306	2.896	3.355	4.499
9.	1.383	1.833	2.262	2.821	3.250	4.296
10.	1.372	1.812	2.228	2.764	3.169	4.143
11.	1.363	1.796	2.201	2.718	3.106	4.024
12.	1.356	1.782	2.179	2.681	3.055	3.929
13.	1.350	1.771	2.160	2.650	3.012	3.852
14.	1.345	1.761	2.145	2.624	2.977	3.787
15.	1.341	1.753	2.131	2.602	2.947	3.733
16.	1.337	1.746	2.120	2.583	2.921	3.686
17.	1.333	1.740	2.110	2.567	2.898	3.646
18.	1.330	1.734	2.101	2.552	2.878	3.610
19.	1.328	1.729	2.093	2.539	2.861	3.579
20.	1.325	1.725	2.086	2.528	2.845	3.552

21.	1.323	1.721	2.080	2.518	2.831	3.527
22.	1.321	1.717	2.074	2.508	2.819	3.505
23.	1.319	1.714	2.069	2.500	2.807	3.485
24.	1.318	1.711	2.064	2.492	2.797	3.467
25.	1.316	1.708	2.060	2.485	2.787	3.450
26.	1.315	1.706	2.056	2.479	2.779	3.435
27.	1.314	1.703	2.052	2.473	2.771	3.421
28.	1.313	1.701	2.048	2.467	2.763	3.408
29.	1.311	1.699	2.045	2.462	2.756	3.396
30.	1.310	1.697	2.042	2.457	2.750	3.385
31.	1.309	1.696	2.040	2.453	2.744	3.375
32.	1.309	1.694	2.037	2.449	2.738	3.365
33.	1.308	1.692	2.035	2.445	2.733	3.356
34.	1.307	1.691	2.032	2.441	2.728	3.348
35.	1.306	1.690	2.030	2.438	2.724	3.340
36.	1.306	1.688	2.028	2.434	2.719	3.333
37.	1.305	1.687	2.026	2.431	2.715	3.326
38.	1.304	1.686	2.024	2.429	2.712	3.319
39.	1.304	1.685	2.023	2.426	2.708	3.313
40.	1.303	1.684	2.021	2.423	2.704	3.307
50.	1.299	1.676	2.009	2.403	2.678	3.261
60.	1.296	1.671	2.000	2.390	2.660	3.232
70.	1.294	1.667	1.994	2.381	2.648	3.211
80.	1.292	1.664	1.990	2.374	2.639	3.195
90.	1.291	1.662	1.987	2.368	2.632	3.183
100.	1.290	1.660	1.984	2.364	2.626	3.174
∞	1.282	1.645	1.960	2.326	2.576	3.090

Table 3.1: Critical values (tc) of the Student-t *distribution with* p *degrees of freedom*

The second comment concerns how to formulate the null and alternative hypotheses. Typically, in the statistical literature, the null hypothesis is the most important one, such as a new drug has no side effects (H_0), against the same drug has side effects (H_1). In econometrics things are a little reversed, e.g., income (*Y*) should affect consumption (C), but often the null hypothesis is formulated as $H_0: \ \beta_2 = 0$, i.e., *Y* has no effect on *C*, versus $H_1: \ \beta_2 > 0$, i.e., *Y* has a positive effect on *C*. This is because often we are not able to formulate a precise null hypothesis, e.g. $\beta_2 = 0.77$, and therefore we are satisfied with less, namely, if we can at least reject that an explanatory variable has no effects on the dependent variable, i.e., $H_0: \ \beta_2 = 0$. We should keep in mind that hypothesis testing is built assuming that H_0 is the key hypothesis of interest and hence it is more problematic to reject H_0 by mistake than not to reject it by mistake (the procedure is conservative in this sense, it favors H_0). On this issue we will return shortly.

3. The test statistic is a random variable that must be informative on the null hypothesis of interest. For example, if $H_0 : \beta_2 = c$ against $H_1 : \beta_2 \neq c$, then we need something that summarizes the information in the sample on β_2. For this we can use $\hat{\beta}_2$, the OLS estimator of β_2. Since we know that under the assumptions of the linear regression model, $\hat{\beta}_2$ is a good estimator of β_2, we could use $\hat{\beta}_2 - c$ to estimate the distance between β_2 and *c*. If $\hat{\beta}_2 - c$ is large, probably $\beta_2 - c$ is also large, and then we can reject H_0. But we must consider that $\hat{\beta}_2$ is an estimator and has a variance, so that $\hat{\beta}_2 - c$ could be large just by randomness. To take this feature into account, we could use $\frac{\hat{\beta}_2 - c}{s.e.(\hat{\beta}_2)}$ as a statistic. If the parameter estimator is accurate, $s.e.(\hat{\beta}_2)$ is small and therefore even a small difference between $\hat{\beta}_2$ and *c* is magnified. The opposite is true if $s.e.(\hat{\beta}_2)$ is great, i.e., if there is a lot of uncertainty.
   A second important characteristic of a test statistic is that its distribution should be fully known when the null hypothesis is true. In this regard, we know from the previous section that

$$t = \frac{\hat{\beta}_2 - \beta_2}{s.\hat{e}.(\hat{\beta}_2)} \sim t(T-2)$$

Therefore, if $H_0 : \beta_2 = c$ is true, it is:

$$t = \frac{\hat{\beta}_2 - c}{s.\hat{e}.(\hat{\beta}_2)} \sim t(T-2) \qquad (3.2.2)$$

Note that if H_0 is false, *t* does not have the *t (t-2)* distribution. In fact, if $\beta_2 \neq c$, then from

$$t = \frac{\hat{\beta}_2 - \beta_2}{\sqrt{Var(\hat{\beta}_2)}} \sim N(0,1) \qquad (3.2.3)$$

,

it follows that

$$\frac{\hat{\beta}_2 - c}{\sqrt{Var(\hat{\beta}_2)}} + \frac{c - \beta_2}{\sqrt{Var(\hat{\beta}_2)}} \sim N(0,1) \tag{3.2.4}$$
,

wherefrom

$$\frac{\hat{\beta}_2 - c}{\sqrt{Var(\hat{\beta}_2)}} \sim N\left(\frac{\beta_2 - c}{\sqrt{Var(\hat{\beta}_2)}}, 1\right) \tag{3.2.5}$$

Hence, if H_0 is true, the density of the *t*-statistic is similar to that graphed in Figure 3.3.

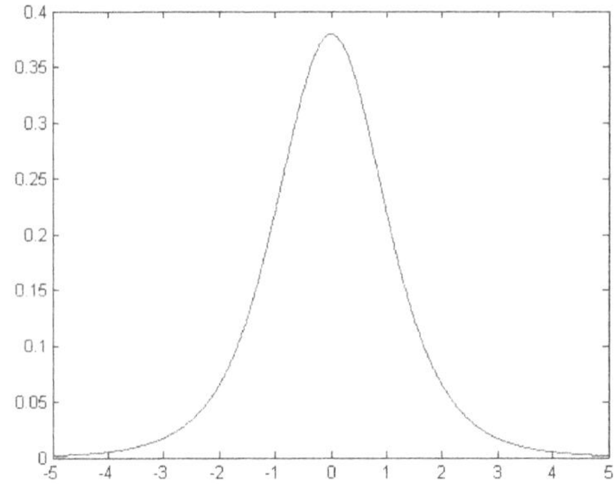

Figure 3.3: Student-t density with 5 degrees of freedom

If instead H_0 is false, e.g. $\beta_2 > c$, then the density of the *t*-statistic is similar to that graphed in Figure 3.4, which is centered not on 0 but on $\frac{\beta_2 - c}{\sqrt{Var(\hat{\beta}_2)}}$.

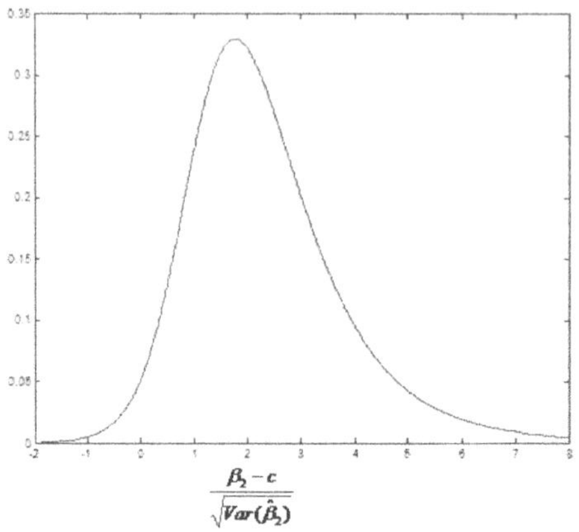

Figure 3.4: Non-central Student-t density with 5 degrees of freedom and non-centrality parameter $\frac{\beta_2 - c}{\sqrt{Var(\hat{\beta}_2)}}$

In other words, if H_0 is true, it is likely that t is close to 0. If H_0 is false it is likely that t is positive, if $\beta_2 > c$, or negative, if $\beta_2 < c$. The idea is not to reject H_0 if the calculated value for the t statistic is close to 0, and to reject it otherwise. To formalize the meaning of "close to 0", we introduce the notion of the rejection region.

4. The rejection region is defined as the set of values such that, if the test statistic is one of these values, then H_0 is rejected. In the case of the t-test, the rejection region is represented by values larger than the critical value t_c and smaller than $-t_c$. To determine the critical value t_c we must define the *significance level* of the test, α, with typically α = 0.05, or 0.01 0.10. Then, we choose t_c such that $\Pr\{t < -t_c\} = \Pr\{t > t_c\} = \frac{\alpha}{2}$. If $t < -t_c$ *or* $t > t_c$ then we reject the null hypothesis. The rejection region is then $(-\infty, -t_c) \cup (\infty, t_c)$. The critical value t_c is again determined using Table 3.1, for given values of α and T-k. Figure 3.5 reports a graphical example.

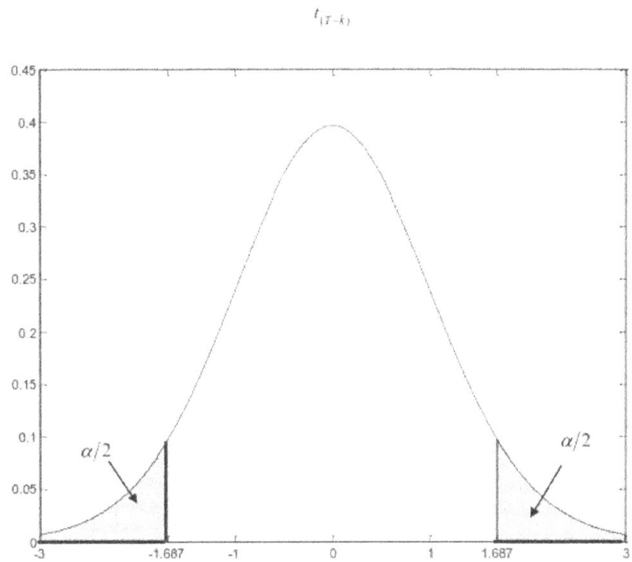

Figure 3.5: Rejection region of the t-test, for a=0.1 and T-k=37

What happens if $-t_c < t < t_c$? We do not reject *H₀*. Can we then say that we accept *H₀*, namely, that *H₀* is true? No, strictly speaking. For example, if $H_0 : \beta_2 = 0.9$ but the true value of the p
arameter is $\beta_2 = 0.88$ or $\beta_2 = 0.92$, then in practice we never reject *H₀* unless the number of available observations, *T*, is very large.

As an example of hypothesis testing, let us suppose that we want to test H_0: $\beta_2 = 1$ against H_1: $\beta_2 \neq 1$, and we have estimated $\hat{\beta}_2 = 2.5$, with $s.e.(\hat{\beta}_2) = 1, T = 39, T-k = 37$ and a = 0.1. The value of the *t*-statistic is $t = (\hat{\beta}_2 - 1)/1 = 1.5$ while $t_c = 1.687$. Then the null hypothesis $H_0 : \beta_2 = 1$ is not rejected since the *t-test* does not fall in the rejection region $(-\infty, -1.687] \cup [1.687, \infty)$.

If the null hypothesis is $H_0 : \beta_2 = 0$, then $t = (\hat{\beta}_2 - 0)/1 = 2.5$ and therefore H_0 is rejected since $t > t_C$.

3.3 Type I and type II errors

Let us consider Table 3.2.

	Test does not reject	Test rejects
H_0 true	1	3 (type I error)
H_0 false	2 (type II error)	4

Table 3.2: Error types in hypothesis testing

In cases 1 and 4 the testing procedure leads to the right decision, do not reject H_0 when it is true or reject it when it is false.
In cases 2 and 3 the test leads instead to mistakes. In case 3 we reject H_0 when it is true, i.e., we commit what is called a *type I error*. This case is represented graphically in Figure 3.6, which shows how this type of mistake happens when the test value falls in the rejection region even though the null hypothesis is true, so that the distribution of the test statistic is that under the null hypothesis.

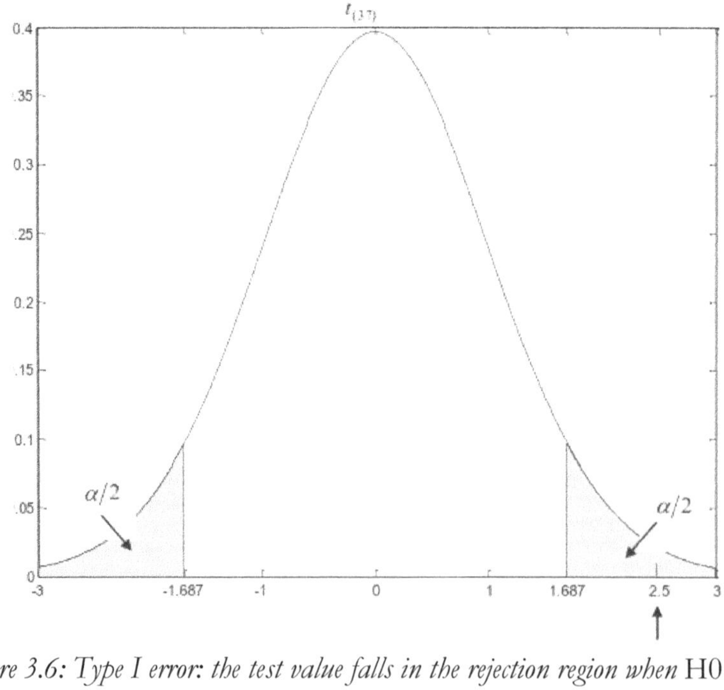

Figure 3.6: Type I error: the test value falls in the rejection region when H0 *is true*

As a consequence, the probability of making a type I error is equal to a, the significance level of the test. Remember that we set a, and the fact that it usually takes low values such as 0.10, 0.05 or 0.01 reflects the choice of having a low probability of committing a type I error.
We take the wrong decision also in case 2, where we do not reject H_0 when it is false. This is called a *type II error* and is illustrated graphically in Figure 3.7. We commit a type II error when the value of the test falls into the acceptance region although the distribution of the test statistic is that under the alternative hypothesis (not centered on 0).

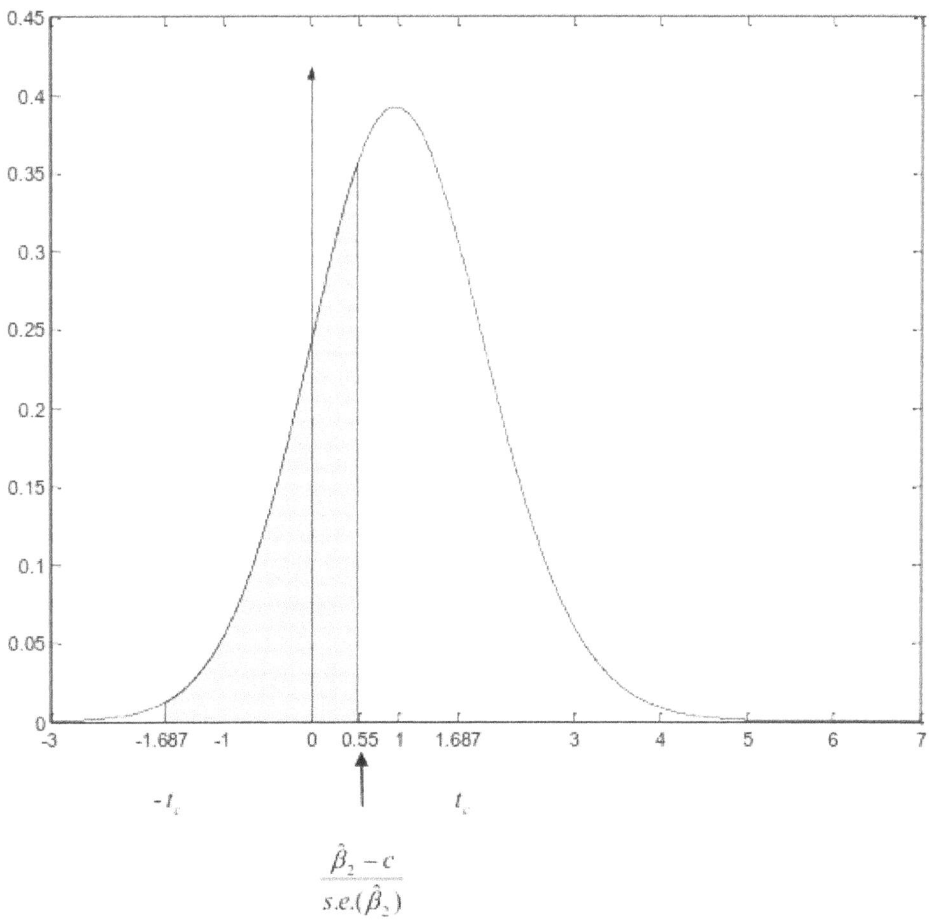

Figure 3.7: Type II error: the test value falls in the acceptance region when H_0 is false

The probability of making a type II error cannot be controlled. It depends:

(a) inversely from *a*, the significance level of the test. When *a* is reduced, the acceptance region widens, and the probability of a type II error increases. With reference to Figure 3.7, the interval $[-t_c, t_c]$ widens and therefore also the gray area, representing the probability of type II error, increases;
(b) inversely from the degree of the violation of the null hypothesis, i.e., in this example, how β_2 differs from *c*. The more different β_2 is from *c*, the more different the distribution of the test statistic under H_1 is from that under H_0, and, therefore, the probability of type II error is reduced. With reference to Figure 3.7, if β_2 is very close to *c* ("mild" violation of the null hypothesis), the distribution of the test statistic is centered at a point close to 0 and thus the gray area increases;
(c) inversely from the sample size, *T*, so if T increases the probability of type II error decreases. This is because when *T* increases the parameter estimator of the parameter becomes more accurate, the $s.\hat{e}.(\hat{\beta}_2)$ is reduced, and therefore the (absolute value of the) statistic increases. In the limiting case where *T* tends to infinity, consistency of the OLS estimator implies that $\hat{\beta}_2 \to \beta_2$ (and $s.\hat{e}.(\hat{\beta}_2) \to 0$). Hence, if $\beta_2 \neq c$, it is

$$\frac{\hat{\beta}_2 - c}{s.\hat{e}.(\hat{\beta}_2)} \to \pm\infty,$$

and therefore the null hypothesis is always properly rejected and the probability of type II error goes to 0. A test with this property is said to be *consistent*. With reference to Figure 3.7, when *T* increases the distribution of the test statistic becomes more and more concentrated, and is centered on a value that tends to move rightward of t_c (or leftward of $-t_c$). Then the gray area shrinks.

So far we proposed to use the *t*-test for testing the hypothesis $H_0: \beta_2 = c$. We could also build alternative tests, based on different statistics. However, it can be shown that the *t*-test is the one that minimizes the probability of type II error for a given probability of type I error, and in this sense it is optimal.

3.4 The concept of p-value

We have defined the rejection rule of a testing procedure based on whether the calculated value of the statistic, $\hat{t}$, falls inside or outside a certain range of values. In particular, we reject H_0 if $\hat{t} < -t_c$ *or* $\hat{t} > t_c$. To calculate t_c we need tables

with the distribution of the test statistic, that might not be readily available. We therefore define an equivalent rejection rule that is more easily verifiable.
Let us consider the value obtained for the test, $\hat{t}$, and calculate the probability

$$\Pr\{t \leq -\hat{t}\} + \Pr\{t \geq \hat{t}\}. \tag{3.4.1}$$

This probability is called *p-value* (probability value) of the test.
If *p-value*> α we do not reject H_0, if *p-value*< α we reject H_0. This rule is very useful and convenient in particular because many econometric software packages, such as EViews, report directly the *p*-value of a test or let you calculate it easily.
The *p*-value based rejection rule is shown graphically in Figure 3.8. The gray colored area is the *p*-value associated with the value $\hat{t}$ of the test statistic (divided by two). It is therefore clear that the *p*-value is greater than the significance level of the test if and only if $\hat{t}$ falls in the acceptance region.

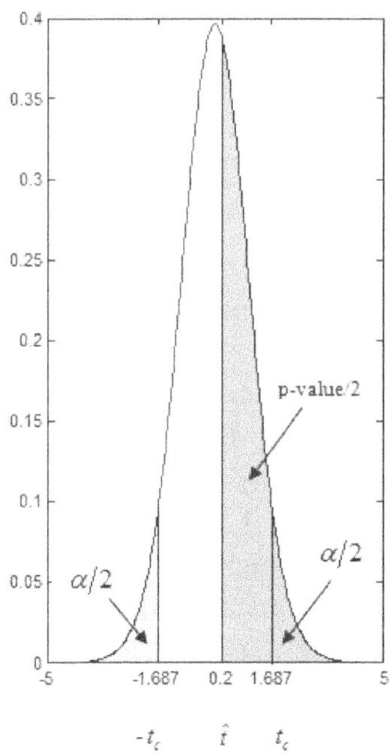

Figure 3.8.1: p-value > α
⇒ No rejection of H_0

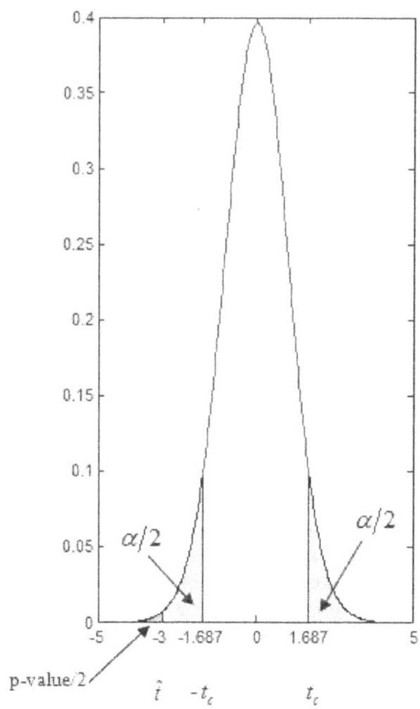

Figure 3.8.2: p-value < α
⇒ Rejection of H_0

3.5 Significance testing

A very common null hypothesis in empirical applications is $\beta_2 = 0$ against the alternative $\beta_2 \neq 0$, i.e., under H_0 the explanatory variable X is not significant in the regression model (for this reason this is called *significance testing*). In this case, the expression of the t in test (3.2.2) simplifies to

$$t = \frac{\hat{\beta}_2 - \beta_2}{s.\hat{e}.(\hat{\beta}_2)} = \frac{\hat{\beta}_2}{s.\hat{e}.(\hat{\beta}_2)}. \tag{3.5.1}$$

Note that the statistical significance of a coefficient does not necessarily mean that it is also economically relevant, and vice versa. For example, let us suppose we want to estimate a Phillips curve to evaluate the existence of a trade-off between inflation and unemployment, having a sample of 100 observations. In this case, the dependent variable is the rate of inflation and the explanatory variable is the rate of unemployment. If the estimated coefficient is $\hat{\beta} = -0.001$, with $s.\hat{e}. = 0.0001$, then the null hypothesis of no significance of the unemployment rate is rejected for most commonly used significance levels, since the calculated value of the statistic is equal to -10. From an economic point of view, however, the unemployment rate is pretty much irrelevant to explain inflation, given the estimated value of the parameter.

If, instead, it were $\hat{\beta} = -0.5$, with $s.\hat{e}. = 0.5$, then the null hypothesis of no significance of the unemployment rate would not be statistically rejected since the t-test is -1, while the estimated value of the parameter would imply a significant economic importance of unemployment.

The previous example also emphasizes the importance of the choice of the null hypothesis that we want to test. If we set $\beta_2 = -1$ instead of $\beta_2 = 0$ as the null hypothesis, implying a one-to-one trade-off between inflation and unemployment, the value of the t-test is 1 and the null hypothesis is not rejected. On the other hand, even a null hypothesis such as $\beta_2 = 0.4$, a positive relationship between inflation and unemployment as in stagflation periods, would not be rejected at a 5% significance level, being the t-test equal to -1.8 and the critical value -1.984 (from Table 3.1, using the critical values in the row for 100 degrees of freedom as an approximation).

If instead we set the significance level of the test at 10%, then the null hypothesis $\beta_2 = 0.4$ is rejected, because the critical value is equal to -1.660. Therefore, the type I error probability we are willing to accept is also an important choice from an economic point of view, because it can determine whether or not the hypothesis of interest is rejected.

The basic problem in this example is the substantial uncertainty around the estimated value of the parameter. In an experimental context, the uncertainty could be reduced by repeating the experiment a few more times in order to gather additional data, but this is generally not possible in economics.

3.6 The relationship between confidence intervals and hypothesis testing

We now assess the relationship between parameter interval estimators and hypothesis testing. In Section 3.1 we built the confidence interval for β at the (1-α)% level as:

$$\hat{\beta} - t_c s.\hat{e}.(\hat{\beta}) \leq \beta \leq \hat{\beta} + t_c s.\hat{e}.(\hat{\beta}) \quad , \tag{3.6.1}$$

with

$$\Pr\left\{ -t_c \leq \frac{\hat{\beta} - \beta}{s.\hat{e}.(\hat{\beta})} \leq t_c \right\} = 1 - \alpha \quad . \tag{3.6.2}$$

The expression in (3.6.2) can be used to derive the acceptance (more properly, non-rejection) region of a test. In fact, if we replace β with c, then in Section 3.2 to calculate the acceptance region for $H_0 : \beta = c$ we used:

$$\Pr\left\{ -t_c \leq \underbrace{\frac{\hat{\beta} - c}{s.\hat{e}.(\hat{\beta})}}_{t} \leq t_c \right\} = 1 - \alpha \quad . \tag{3.6.3}$$

As a result, if $\hat{t}$ falls in the acceptance region for $H_0 : \beta = c$ at the a% level, and therefore H_0 is not rejected, then c is part of a (1-a)% confidence interval for β. Conversely, if c is part of a (1-a)% confidence interval for β, then the null hypothesis $H_0 : \beta = c$ is not rejected at a significance level a.
For example, we saw in Section 3.1, that if $\hat{\beta}_2 = 2.5, s.e.(\hat{\beta}_2) = 1, T = 32, T - k = 30$ and α = 0.1, then a 90% confidence interval for $\hat{\beta}_2$ is [0.8, 4.2]. Thus, the null hypothesis $H_0 : \beta = 0$ would be rejected at a 10% significance level of 10%, while $H_0 : \beta = 1$ would not be rejected.

3.7 One-way tests

So far we have always considered alternative hypotheses of the type $\beta \neq c$. We now extend the testing procedure to allow for a different formulation of the alternative hypothesis, maintaining the same null hypothesis:

$$H_0 : \beta = c \text{ and } H_1 : \beta > c.$$

The only difference with respect to the testing procedure we have used so far is in the choice of the rejection region, which is now in the right tail of the distribution of the statistics (left if $H_1 : \beta < c$), as shown graphically in Figure 3.9. In fact, in this case, if H_0 is false we expect very large positive values for the *t-statistic* ($\beta > c$).

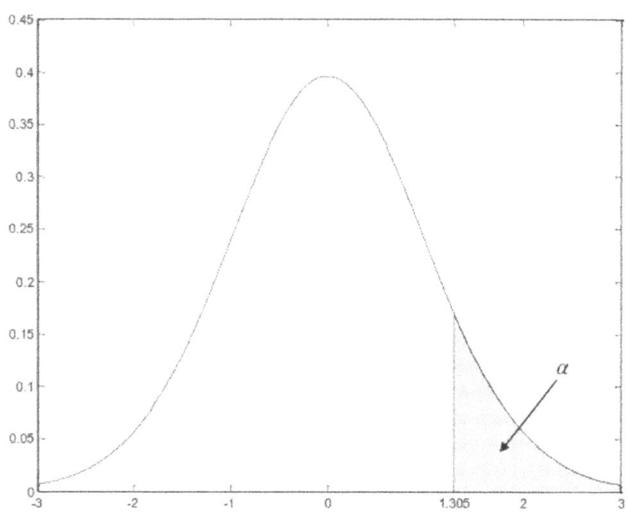

Figure 3.9: An example of rejection region for one-way tests

If, for example,

$$H_0 : \beta_2 = 1,\ H_1 : \beta_2 > 1,\ \hat{\beta}_2 = 2.5,\ s.\hat{e}.(\hat{\beta}_2) = 1,$$
$$T = 39,\ T - k = 37$$

and $\alpha = 0.1$, then

$$t = \frac{\hat{\beta}_2 - 1}{1} = 1.5, t_c = 1.310 \text{ (from Table 3.1)}$$

and therefore H_0 is rejected. Note that if instead it is $H_1 : \beta_2 \neq 1$ then H_0 is not rejected.

3.8 The F-test

So far we have learned to test hypotheses on individual parameters of the linear regression model, for example, $\beta_k = 0 \ or \ \beta_k = 1$. However, there are hypotheses of interest that involve multiple parameters, such as $\beta_1 = 0 \ and \ \beta_2 = 1$. These are often called *composite hypotheses.* For example, from an economic point of view, we may want to test if both income and the interest rate are jointly relevant to explain money demand, or if both disposable income and wealth matter to explain aggregate private consumption, or if a production function has constant returns to scale, or if the demand for a good is homogeneous, etc.

To test more complex (composite) hypotheses it is useful to introduce the *F-distribution.* Given two random variables $v_1 \sim \chi^2(p), v_2 \sim \chi^2(q)$ with v_1 and v_2 independent, we have:

$$\frac{v_1/p}{v_2/q} \sim F(p,q) \tag{3.8.1}$$

Figure 3.10 shows the *F*-density for selected values of the *p* and *q* degrees of freedom.

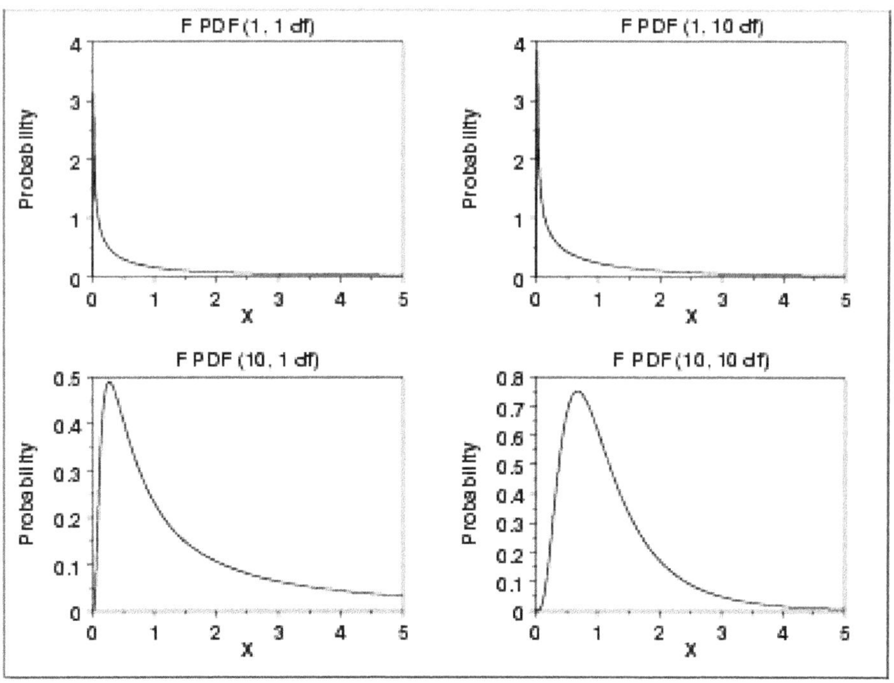

Figure 3.10: The F-density for selected values of the degrees of freedom

Let us suppose that we want to test the hypotheses:

$$H_0: \beta_1 = c_1, \beta_2 = c_2, \ldots, \beta_k = c_k \quad vs \quad H_1: \beta_1 \neq c_1, \beta_2 \neq c_2, \ldots, \beta_k \neq c_k \qquad (3.8.2)$$

or, using the vector notation,

$$H_0: \underset{k\times1}{\beta} = \underset{k\times1}{c} \quad vs \quad H_1: \underset{k\times1}{\beta} \neq \underset{k\times1}{c}, \qquad (3.8.3)$$

in the model

$$\underset{T\times1}{y} = \underset{T\times k}{X}\underset{k\times1}{\beta} + \varepsilon. \qquad (3.8.4)$$

Note that the alternative hypothesis states that *at least one* of the components of H_0 is false. In addition, the alternative hypothesis is formulated in terms of "different from", it cannot be formulated in terms of "greater than" or "less than", as we will see shortly.

Again, the idea is to use the OLS estimator of the parameters, $\hat{\beta}$, and assess if they are "very different" from the values assumed under the null hypothesis, grouped in the vector c. Let us then consider the variable

$$\underset{1\times k}{\left(\hat{\beta}-c\right)'}\left[\sigma^2\left(X'X\right)^{-1}\right]^{-1}\left(\hat{\beta}-c\right), \qquad (3.8.5)$$

namely,

$$v_1 : \frac{\left(\hat{\beta}-c\right)' X'X\left(\hat{\beta}-c\right)}{\sigma^2}. \qquad (3.8.6)$$

It can be shown that, if H_0 is true, $v_1 \sim \chi^2(k)$.

The variable v_1 cannot be used directly as a test statistic because the σ^2 parameter in the denominator is unknown. However, we know that

$$v_2 : (T-k)\frac{\hat{\sigma}^2}{\sigma^2} \sim \chi^2(T-k) \qquad (3.8.7)$$

Moreover, since $\hat{\beta}$ and $\hat{\sigma}^2$ are independent, as we have shown in Chapter 2, v_1 and v_2 are also independent.

Therefore, we can construct the *F-statistic* or *F-test* as:

$$\lambda = \frac{v_1/k}{v_2/(T-k)} = \frac{(\hat{\beta}-c)'(X'X)(\hat{\beta}-c)}{k\hat{\sigma}^2} \sim F(k,T-k). \tag{3.8.8}$$

We now have three of the four elements that we need to construct the test: the null hypothesis H_0, the alternative hypothesis H_1, the test statistic λ. The missing element is the region of rejection, or its complement, the acceptance region.

If H_0 is false, λ is expected to be large and positive. Hence, the acceptance region is $\{\lambda < F_c\}$, where F_c is the critical value such that $P_{H_0}\{\lambda \geq F_c\} = \alpha$ and α is the significance level of the test.

Note that λ assumes positive values for both $\beta > c$ and $\beta < c$, so that we cannot conduct directional tests in the case of composite hypotheses. Therefore, the alternative hypothesis should be formulated as $\beta \neq c$.

A graphical example of critical region for the F-test is presented in Figure 3.11.

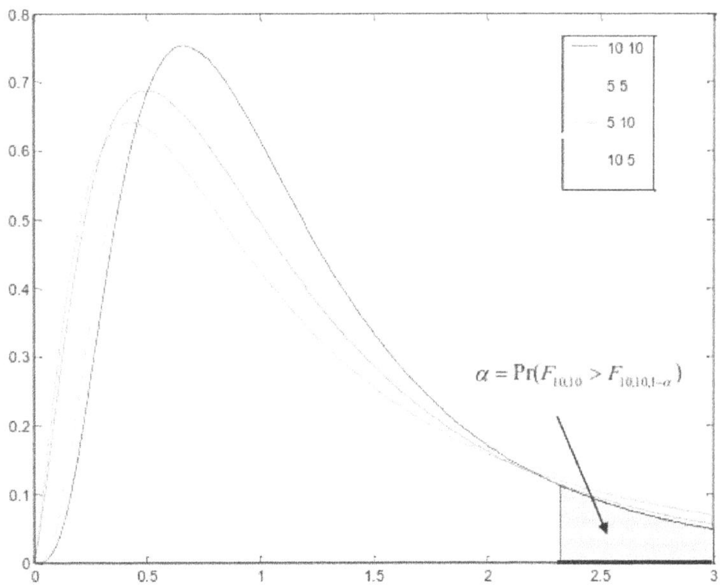

Figure 3.11: Example of rejection region for the F-test

Table 3.3 reports the proper critical values for α = 0.05 and different combinations of degrees of freedom. For example, if k = 5 and T-k = 20, then $F_{(c)}$= 2.71, or, if k = 2 and T-k = 40, then $F_{(c)}$= 3.23.

	v_1															
v_2	1	2	3	4	5	6	7	8	9	10	11	12	15	20	30	inf
1	161.4	199.5	215.7	224.6	230.2	234.0	236.8	238.9	240.5	241.9	243.0	243.9	245.9	248.0	250.1	254.3
2	18.51	19.00	19.16	19.25	19.30	19.33	19.35	19.37	19.38	19.40	19.40	19.41	19.43	19.45	19.46	19.50
3	10.13	9.55	9.28	9.12	9.01	8.94	8.89	8.85	8.81	8.79	8.76	8.74	8.70	8.66	8.62	8.53
4	7.71	6.94	6.59	6.39	6.26	6.16	6.09	6.04	6.00	5.96	5.94	5.91	5.86	5.80	5.75	5.63
5	6.61	5.79	5.41	5.19	5.05	4.95	4.88	4.82	4.77	4.74	4.70	4.68	4.62	4.56	4.50	4.36
6	5.99	5.14	4.76	4.53	4.39	4.28	4.21	4.15	4.10	4.06	4.03	4.00	3.94	3.87	3.81	3.67
7	5.59	4.74	4.35	4.12	3.97	3.87	3.79	3.73	3.68	3.64	3.60	3.57	3.51	3.44	3.38	3.23
8	5.32	4.46	4.07	3.84	3.69	3.58	3.50	3.44	3.39	3.35	3.31	3.28	3.22	3.15	3.08	2.93
9	5.12	4.26	3.86	3.63	3.48	3.37	3.29	3.23	3.18	3.14	3.10	3.07	3.01	2.94	2.86	2.71
10	4.96	4.10	3.71	3.48	3.33	3.22	3.14	3.07	3.02	2.98	2.94	2.91	2.85	2.77	2.70	2.54
11	4.84	3.98	3.59	3.36	3.20	3.09	3.01	2.95	2.90	2.85	2.82	2.79	2.72	2.65	2.57	2.40
12	4.75	3.89	3.49	3.26	3.11	3.00	2.91	2.85	2.80	2.75	2.72	2.69	2.62	2.54	2.47	2.30
13	4.67	3.81	3.41	3.18	3.03	2.92	2.83	2.77	2.71	2.67	2.63	2.60	2.53	2.46	2.38	2.21
14	4.60	3.74	3.34	3.11	2.96	2.85	2.76	2.70	2.65	2.60	2.57	2.53	2.46	2.39	2.31	2.13
15	4.54	3.68	3.29	3.06	2.90	2.79	2.71	2.64	2.59	2.54	2.51	2.48	2.40	2.33	2.25	2.07
16	4.49	3.63	3.24	3.01	2.85	2.74	2.66	2.59	2.54	2.49	2.46	2.42	2.35	2.28	2.19	2.01
17	4.45	3.59	3.20	2.96	2.81	2.70	2.61	2.55	2.49	2.45	2.41	2.38	2.31	2.23	2.15	1.96
18	4.41	3.55	3.16	2.93	2.77	2.66	2.58	2.51	2.46	2.41	2.37	2.34	2.27	2.19	2.11	1.92
19	4.38	3.52	3.13	2.90	2.74	2.63	2.54	2.48	2.42	2.38	2.34	2.31	2.23	2.16	2.07	1.88
20	4.35	3.49	3.10	2.87	2.71	2.60	2.51	2.45	2.39	2.35	2.31	2.28	2.20	2.12	2.04	1.84
21	4.32	3.47	3.07	2.84	2.68	2.57	2.49	2.42	2.37	2.32	2.28	2.25	2.18	2.10	2.01	1.81
22	4.30	3.44	3.05	2.82	2.66	2.55	2.46	2.40	2.34	2.30	2.26	2.23	2.15	2.07	1.98	1.78
23	4.28	3.42	3.03	2.80	2.64	2.53	2.44	2.37	2.32	2.27	2.24	2.20	2.13	2.05	1.96	1.76
24	4.26	3.40	3.01	2.78	2.62	2.51	2.42	2.36	2.30	2.25	2.22	2.18	2.11	2.03	1.94	1.73
25	4.24	3.39	2.99	2.76	2.60	2.49	2.40	2.34	2.28	2.24	2.20	2.16	2.09	2.01	1.92	1.71
26	4.23	3.37	2.98	2.74	2.59	2.47	2.39	2.32	2.27	2.22	2.18	2.15	2.07	1.99	1.90	1.69
27	4.21	3.35	2.96	2.73	2.57	2.46	2.37	2.31	2.25	2.20	2.17	2.13	2.06	1.97	1.88	1.67
28	4.20	3.34	2.95	2.71	2.56	2.45	2.36	2.29	2.24	2.19	2.15	2.12	2.04	1.96	1.87	1.65
29	4.18	3.33	2.93	2.70	2.55	2.43	2.35	2.28	2.22	2.18	2.14	2.10	2.03	1.94	1.85	1.64

30	4.17	3.32	2.92	2.69	2.53	2.42	2.33	2.27	2.21	2.16	2.13	2.09	2.01	1.93	1.84	1.62
40	4.08	3.23	2.84	2.61	2.45	2.34	2.25	2.18	2.12	2.08	2.04	2.00	1.92	1.84	1.74	1.51
60	4.00	3.15	2.76	2.53	2.37	2.25	2.17	2.10	2.04	1.99	1.95	1.92	1.84	1.75	1.65	1.39
120	3.92	3.07	2.68	2.45	2.29	2.18	2.09	2.02	1.96	1.91	1.87	1.83	1.75	1.66	1.55	1.25
inf	3.84	3.00	2.60	2.37	2.21	2.10	2.01	1.94	1.88	1.83	1.79	1.75	1.67	1.57	1.46	1.00

Table 3.3: Right-tail critical values for the F-distribution (95th percentiles)

We now introduce a general way to express hypotheses about the parameters of interest. For example, the hypothesis $H_0: \beta_1 = \beta_2 \quad vs \quad H_1: \beta_1 \neq \beta_2$ can be rewritten as

$$H_0: \beta_1 - \beta_2 = 0 \quad or \quad [1 \quad -1]\begin{bmatrix}\beta_1 \\ \beta_2\end{bmatrix} = 0 \quad or \quad R\beta = r$$

As another example, if $H_0: \dfrac{\beta_1}{\beta_2} = 4 \quad vs \; H_1: \dfrac{\beta_1}{\beta_2} \neq 4$, then we can re-write H_0 as

$$H_0: \beta_1 = 4\beta_2 \quad or \quad [1 \quad -4]\begin{bmatrix}\beta_1 \\ \beta_2\end{bmatrix} = 0 \quad or \quad R\beta = r.$$

Also, if: $H_0: \beta_1 = \beta_2 = \beta_3 = 3 \quad vs \quad H_1: \beta_1 \neq \beta_2 \neq \beta_3 \neq 3$, then:

$$H_0: \begin{bmatrix}1 & 0 & 0 \\ 0 & 1 & 0 \\ 0 & 0 & 1\end{bmatrix}\begin{bmatrix}\beta_1 \\ \beta_2 \\ \beta_3\end{bmatrix} = \begin{bmatrix}3 \\ 3 \\ 3\end{bmatrix} \quad or \quad \underset{3\times3}{R}\,\underset{3\times1}{\beta} = \underset{3\times1}{r}.$$

Therefore, a general hypothesis of interest is:

$$H_0: \underset{j\times k}{R}\,\underset{k\times1}{\beta} = \underset{j\times1}{r} \quad vs \quad H_1: R\beta \neq r \tag{3.8.9}$$

To propose a test statistic for the null hypothesis in (3.8.9), we note that if $\hat{\beta} \sim N\left(\beta, \sigma^2 (X'X)^{-1}\right)$ then $R\hat{\beta} \sim N\left(R\beta, \sigma^2 R(X'X)^{-1} R'\right)$, and it can be shown that

$$v^{*}{}_{1}: \left(R\hat{\beta} - R\beta\right)' \left[\sigma^{2} R(X'X)^{-1} R\right]^{-1} \left(R\hat{\beta} - R\beta\right) \sim \chi^{2}(j), \tag{3.8.10}$$

Hence, if H_0 holds,

$$\lambda = \frac{v^{*}{}_{1}/j}{v_{2}/(T-k)} = \frac{\left(R\hat{\beta} - r\right)' \left[R(X'X)^{-1} R'\right]^{-1} \left(R\hat{\beta} - r\right)}{j\hat{\sigma}^{2}} \sim F(j, T-k) \tag{3.8.11}$$

The rejection region is one-sided, as before: $\{\lambda \geq F_C\}$, where F_C is the critical value such that $P_{H_0}\{\lambda \geq F_c\} = \alpha$.
It can be shown (see, e.g., Judge *et al.* (1988) p. 257-258) that the λ-statistic can be re-written as:

$$\lambda = \frac{(RSS_R - RSS_U)/j}{RSS_U/(T-k)}, \tag{3.8.12}$$

where $RSS_U = \sum_{t=1}^{T} \hat{e}_t^2$, $RSS_R = \sum_{t=1}^{T} \tilde{e}_t^2$, $\hat{e}_t$ are the OLS residuals of the unrestricted model, while $\tilde{e}_t$ are the residuals from the model whose coefficients are estimated under the constraints in $H_0: R\beta = r$, namely, they are the residuals of the restricted model. In general, $RSS_R \geq RSS_U$ but, if the restrictions in the null hypothesis hold, the difference is expected to be small and the test statistic falls in the acceptance region.
It is also worth mentioning that, if we test a single restriction on the model parameters, so that $j = 1$, then

$$\lambda = t^2 = \left(\frac{\hat{\beta} - c}{s.\hat{e}.(\hat{\beta})}\right)^2 \tag{3.8.13}$$

Moreover, if $v \sim t(p)$ then $v^2 \sim F(1, p)$. Hence, under $H_0: \underset{1\times 1}{\beta} = c$ vs $: H_1 \ \underset{1\times 1}{\beta} \neq c$, then

$$t^2 = \lambda \sim F(1, T-k) \tag{3.8.14}$$

and the t-test and F-test have the same p-value.
The single restriction of interest can be formulated in a more general manner. For example, to test:

$$H_0 : \underset{1\times k}{R} \underset{k\times 1}{\beta} = \underset{1\times 1}{r} \text{ vs. } H_1 : R\beta \neq r$$

we can use the test statistic:

$$\frac{R\hat{\beta} - r}{s.\hat{e}.(R\hat{\beta})} \overset{H_0}{\sim} t(T-k). \quad (3.8.15)$$

In the case of these more general single hypotheses that can still be tested by means of t-statistics, we can also consider unidirectional tests, while we have seen that this is not feasible for the F-test.
An important application of the F-statistic is to test for the *model significance*, namely, the joint significance of the explanatory variables in a model. Given the model:

$$y_t = \beta_1 + \beta_2 x_{2t} + \beta_3 x_{3t} + \ldots + \beta_k x_{kt} + e_t,$$

the hypothesis of interest is:

$$H_0 : \; \beta_2 = 0, \, \beta_3 = 0, \ldots, \beta_k = 0 \quad vs\, H_1 : at\, least\, one\, element\, of\; \beta \neq 0 .$$

Why do we use the F-statistic to verify this composite hypothesis instead of a set of *k-1* t-statistics? Because $\hat{\beta}_1, \hat{\beta}_2, \ldots, \hat{\beta}_k$ are generally correlated and this feature is taken into account in the construction of the F-test, but not in the set of t-tests. As a result, it may happen that, for the same significance level, the hypothesis $H_0 : \; \beta_2 = 0, \, \beta_3 = 0, \ldots, \beta_k = 0$ is not rejected by the F-test, while some of its components are rejected by the associated t-tests.

3.9 Restricted OLS estimators

How do we obtain $\widetilde{e}_t$, the residuals in the restricted model whose parameters are estimated imposing the constraint $R\beta = r$, which appears in the alternative formulation of the F-statistic in (3.8.12)? We must find proper parameter estimators, $\widetilde{\beta}$, that minimize the sum of squared errors and satisfy the required constraints. We then get $\widetilde{e}$ as $\widetilde{e} = y - X\widetilde{\beta}$.
In formulae, we want:

$$\min_{\beta} (y - X\beta)'(y - X\beta) = \underset{1\times T}{e'} \underset{T\times 1}{e} = \sum_{t=1}^{T} e_t^2 \tag{3.9.1}$$

$$s.t. \quad \underset{j\times k}{R} \; \underset{k\times 1}{\beta} = \underset{j\times 1}{r} \quad \textit{that is } R\beta - r = 0 .$$

To solve this problem, we construct the Lagrangian function

$$\begin{aligned} L &: e'e + 2\left(\underset{1\times j}{r' - \beta'R'}\right)\underset{j\times 1}{\lambda} \\ &= (y - X\beta)'(y - X\beta) + 2\left(\underset{1\times j}{r' - \beta'R'}\right)\underset{j\times 1}{\lambda} \\ &= y'y - 2\beta'X'y - \beta'X'X\beta + 2\left(\underset{1\times j}{r' - \beta'R'}\right)\underset{j\times 1}{\lambda} \end{aligned} \tag{3.9.3}$$

The first order conditions are:

$$\frac{\partial L}{\partial \beta} = -2X'y + 2X'X\beta^* - 2R'\lambda^* = 0 \tag{3.9.4}$$

$$\frac{\partial L}{\partial \lambda} = 2(r - R\beta^*) = 0 \tag{3.9.5}$$

From (3.9.4) we get:

$$X'X\beta^* = X'y + R'\lambda^*$$
$$(X'X)^{-1} X'X\beta^* = (X'X)^{-1} X'y + (X'X)^{-1} R'\lambda^*$$

and therefore

$$\beta^* = \hat{\beta} + (X'X)^{-1} R'\lambda^* \tag{3.9.6}$$

Multiplying both sides by R, and given (3.9.5), yields:

$$R\beta^* = R\hat{\beta} + R(X'X)^{-1} R'\lambda^* = r \tag{3.9.7}$$

Solving for λ^* we obtain:

$$\lambda^* = \left[R(X'X)^{-1} R'\right]^{-1}\left[R\beta^* - R\hat{\beta}\right] = \left[R(X'X)^{-1} R'\right]^{-1}\left[r - R\hat{\beta}\right] \tag{3.9.8}$$

Finally, if we replace λ^* in (3.9.6), we obtain:

$$\tilde{\beta} = \hat{\beta} + (X'X)^{-1} R'\left[R(X'X)^{-1} R'\right]^{-1}\left[r - R\hat{\beta}\right] \tag{3.9.9}$$

$\tilde{\beta}$ is known as the *Restricted Ordinary Least Squares estimator* (restricted OLS o ROLS)

The ROLS estimator is very useful to insert *a priori* information in the model. For example, if we want to estimate the parameters of the linear regression model by imposing the constraint $\beta_2 + \beta_3 + \beta_4 + \beta_5 = 0$, then we can estimate them by standard OLS and then use the formula in (3.9.9), formulating the restriction as:

$$\begin{bmatrix} 1 & 1 & 1 & 1 \end{bmatrix} \begin{bmatrix} \beta_2 \\ \beta_3 \\ \beta_4 \\ \beta_5 \end{bmatrix} = [0], \quad R\beta = r.$$

Let us now study the properties of the ROLS estimator.

(a) **Unbiasedness**. If $R\beta = r$ then, $E(\tilde{\beta}) = \beta$.

In fact, it is:

$$E(\tilde{\beta}) = E(\hat{\beta}) + (X'X)^{-1} R'\left[R(X'X)^{-1} R'\right]^{-1}\underbrace{\left[r - RE(\beta)\right]}_{r - R\beta = 0}. \tag{3.9.10}$$

However, if $R\beta \neq r$, then $E(\tilde{\beta}) \neq \beta$. In words, if we impose invalid restrictions, then the resulting estimator is biased. This feature is important to understand, for example, what happens if we omit relevant explanatory variables from the model, which is the same as setting their coefficients equal to 0 by mistake. We will return to this issue in the next Section.

(b) **Efficiency**. $V(\hat{\beta})>V(\tilde{\beta})$ even if $\tilde{\beta}$ is biased.

Imposing restrictions, even if incorrect, reduces uncertainty and thus the variance of the estimators. For a formal proof of this result see, e.g., Judge *et al.* (1988, pp. 238-240). Note that this result is not in contrast with the Gauss-Markov theorem about the efficiency of the OLS estimator. In fact, $\hat{\beta}$ is the best linear unbiased estimator while, as we have seen, if the restrictions are invalid $\tilde{\beta}$ is biased. Moreover, even when the restrictions are valid, in the construction of $\tilde{\beta}$ we are using more information than for $\hat{\beta}$, and therefore the two estimators are not directly comparable in the context of the Gauss-Markov theorem.

3.10 Omitted variables and irrelevant variables

An important distinction in econometrics is that between the true *Data Generating Process*, the so-called *DGP*, and the econometric model, which is what we suppose to generate the data. So far we assumed that the DGP and the model coincide but in practice, in general, this is not the case.

The *DGP* and the model can differ for many reasons, and in this Section we focus on the role of the choice of explanatory variables. Table 3.3 shows that we can make two types of errors: omit from the econometric model relevant regressors or include non-significant (irrelevant) explanatory variables.

	DGP	
MODEL	$y = X_1\beta_1 + X_2\beta_2 + \varepsilon_t$	$y = X_1\beta_1 + \varepsilon_t$
$y = \underset{1\times k_1}{X_1}\gamma_1 + \underset{1\times k_2}{X_2}\gamma_2 + e_t$	**OK**	**Caso II: Irrelevant Variables**
$y = X_1\eta_1 + u_t$	**Case I: Omitted Variables**	**OK**

Table 3.3: DGP and econometric model

Let us start with Case I: *Omitted Variables*. The OLS estimator is:

$$\hat{\eta}_1 = (X_1'X_1)^{-1}X_1'y \overset{DGP}{=} (X_1'X_1)^{-1}X_1'(X_1\beta_1 + X_2\beta_2 + \varepsilon_t) \\ = \beta_1 + (X_1'X_1)^{-1}X_1'(X_2\beta_2) + (X_1'X_1)^{-1}X_1'\varepsilon_t \tag{3.10.1}$$

Hence,

$$E(\hat{\eta}_1) = \beta_1 + (X_1'X_1)^{-1}X_1'X_2\beta_2 \tag{3.10.2}$$

Therefore, $\hat{\eta}_1$ is biased for β_1, unless $X_1'X_2 = 0$.
Some examples of omitted variables:

a) Model: $C = \alpha Y$, $DGP: C = \gamma Y + \eta W$, where *C*, *Y* and *W* indicate aggregate consumption, disposable income and wealth.
b) Model: $I = \alpha i + \beta Y$, $DGP: I = \gamma i + \eta Y + \delta\pi$, where *I*, *i*, *Y* *e* π indicate aggregate investment, the interest rate, income and profits.

Off estimated parameter signs or values may therefore be due to the omission of significant variables from the econometric model. To overcome this problem, the initial set of regressors should be sufficiently large, and possibly later restricted according to the outcome of significance tests (i.e. based on the fact that the null hypothesis $\beta_i = 0$ is rejected or not). A problem with this approach is that if, e.g., $\hat{\beta}_i = 0.1$ b*ut* $s.\hat{e}.(\hat{\beta}_i) = 0.1$, then the hypothesis $H_0: \beta_i = 0$ is not rejected but X_i could still be relevant. There are also statistical problems related to the sequential use of significance tests (check first if $\beta_k = 0$, then if $\beta_{k-1} = 0$, etc.).
The opposite strategy, to include all potentially relevant variables in the regression, can lead to Case II: *Irrelevant Variables*. In this case it can be shown that $E(\hat{\gamma}_1) = \beta_1$ *and* $E(\hat{\gamma}_2) = 0$ but the variance of $\hat{\gamma}_1$ is greater than if estimating the correct model $y = X_1\gamma_1 + \varepsilon_t$ unless $X_1'X_2 = 0$. Therefore, we get more inaccurate estimates, reflected for example in wider confidence intervals for $\hat{\gamma}_1$.
In practice, it is important to follow the hints of economic theory and institutional knowledge for the specification of the econometric model, keeping in mind that "strange" or highly uncertain estimated parameter values may be due to the omission of relevant variables or the inclusion of non-significant regressors.

3.11 An empirical analysis with artificial data

Using the workfile "example_simul_chap3", we return to the output of the regression of *y* on a constant, *x1* and *x2* that we studied in the previous chapter and that is shown in Table 3.4.
We start by deriving the estimated confidence intervals for the parameters. Since the sample size is T=130 and there are 3 parameters in the regression model, the degrees of freedom are 127. This brings us to use, in Table 3.1 (reporting the critical values of the Student-t), the values in the row labeled "∞ ", which is equivalent to using the critical values of a standard Normal distribution. We choose $\alpha = 10\%$, so as to construct 90% confidence intervals. From Section 1, we know that we need the 5% and 95% percentiles, as shown in Figure 3.12. Table 3.1 indicates that the proper critical value is 1.645.

We have the OLS estimates of β_1 and its standard error. From Section 1, the confidence interval for the parameter β_1 is

$$I.C._{90\%} = \left[\ \hat{\beta}_1 - s.\hat{e}.(\hat{\beta}_1)\cdot t_c, \hat{\beta}_1 + s.\hat{e}.(\hat{\beta}_1)\cdot t_c\ \right] \tag{3.11.1}$$

which for us becomes

$$I.C._{90\%} = [\,0.8 - 1.645\cdot 0.033; 0.8 + 1.645\cdot 0.033\,] = [\,0.75, 0.85] \tag{3.11.2}$$

You can try to construct an 80% confidence interval for β_2. You should get [1.585, 1.615].
In EViews, confidence intervals are available by clicking on the "View" button of the regression object, by choosing "Regression Coefficient Tests" and then "Confidence Ellipse". In the window that appears we should indicate at least two coefficients, using the notation described in the "View", "Representations" in the regression object. Hence, "c(1)" denotes the constant, "c(2)" the coefficient β_1 and "c(3)" the coefficient of β_2. For example, let us choose "c(2)" and "c(3)", and select 0.90 in the "Confidence levels" box. By clicking "OK", there appears a confidence ellipse, at the chosen confidence level, for the parameter vector (β_1, β_2). To see the extremes of the individual confidence intervals, we simply click twice on the dotted line. Then, we can read under "Position", "Data value" the requested values for c (2). To calculate the confidence interval for c(3) at 80%, we just repeat the operation with 0.80 in the box "Confidence levels". The red dot inside the ellipse represents the realization of the estimators (i.e., the estimates of β_1, β_2).

Dependent variable: Y
Method: Least Squares
Sample: 1970q1 2002q2
Included observations: 130

Variable	Coefficient	Std. Error	*t*-Statistic	Prob.
C	0.891067	0.228771	3.895015	0.0002
X1	0.800996	0.032864	24.37277	0.0000
X2	1.604939	0.011789	136.1334	0.0000

R-squared	0.997498	Mean dependent var	41.27091
Adjusted R-squared	0.997459	S.D. dependent var	10.70127
S.E. of regression	0.539457	Akaike info criterion	1.626300
Sum squared resid	36.95877	Schwarz criterion	1.692473
Log likelihood	-102.7095	*F*-statistic	25317.94
Durbin-Watson stat	2.222137	Prob(*F*-statistic)	0.000000

Table 3.4: Output of the regression of y on c, x1 and x2

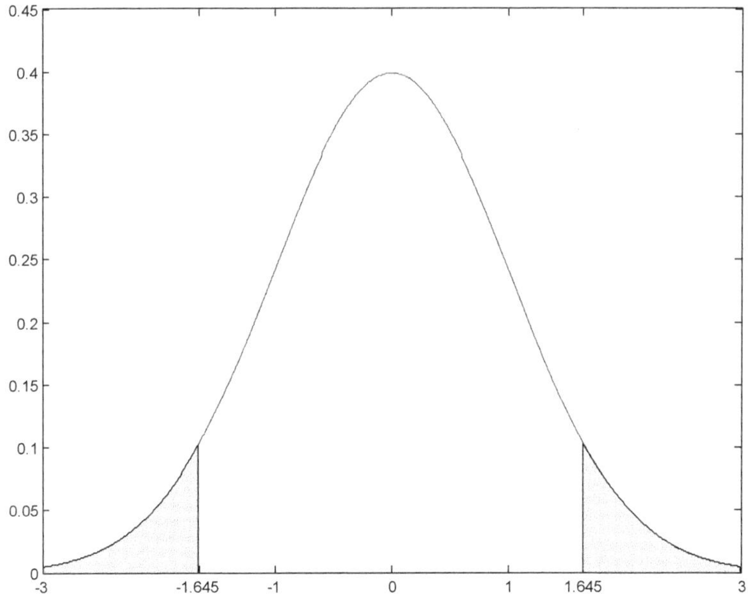

Figure 3.12: Critical values to construct confidence intervals in the example based on simulated data

We now proceed to the implementation of hypothesis testing. Let us suppose we want to test the null hypothesis $H_0 : \beta_1 = 0$, namely, the non-significance of $x1$ to explain y, against the alternative hypothesis $H_0 : \beta_1 = 0$, with a significance level of 10%. In the output Table, column "*t*-statistic", we can read the value of the *t*-statistic for the $x1$ coefficient, calculated as

$$t = \frac{\hat{\beta}_1 - 0}{s.e.(\hat{\beta}_1)} = \frac{0.800996}{0.032864} = 24.37277 \tag{3.11.3}$$

The critical value for comparison is, as we have seen, 1.645, which is lower than the *t*-statistic. Thus, the null hypothesis (which says that the regressor $x1$ does not explain *y*) is rejected.
We showed in Section 3.4 the equivalence of decision rules for testing based either on the rejection region or on *p*-values. In the output Table, in the column "prob.", we can read the *p*-values associated with the test for the null hypothesis of no significance of each regressor. EViews calculates the *p*-value as the probability that a *t*-distributed random variable takes a value larger than the computed test statistic (in absolute value). In our example, the computed *p*-value is equal to 0.00< 10%= α so that, as expected, we reject H_0. Note that with a *p*-value so low the null hypothesis can be rejected at a much lower significance level: 1% for example.
Denoting by t_c the *(100-c)*th percentile of the distribution, in our case c = 0.95, Figure 3.13 provides a graphical representation.

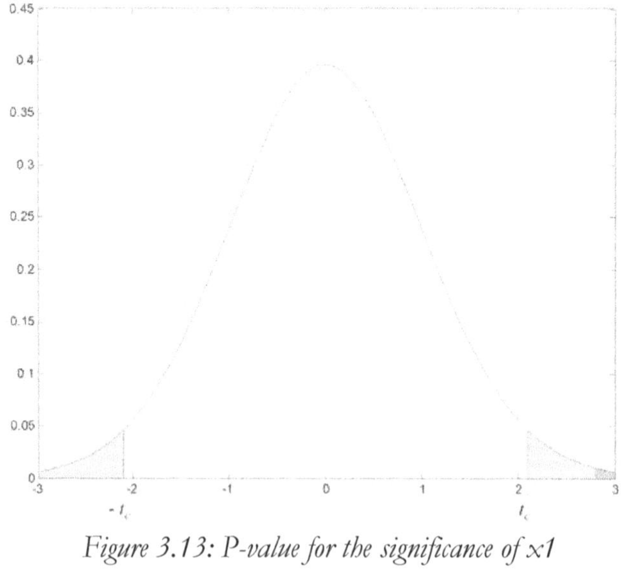

Figure 3.13: P-value for the significance of x1

Another way to test hypotheses in EViews is to click on "View" from the regression window, choose "Regression Coefficient Tests" and then "Wald – Coefficient restrictions". In the window that appears we can write the command "c(2) = 0", where we said that "c(2)" in the EViews language represents β_2. This procedure runs a type of F-test, and the resulting value of the statistic is the square of the t-statistic (as seen in (3.8.14))
The Wald-type test computed by EViews is a more general procedure compared to the F-test. It becomes an F-test in the case of linear hypotheses about the parameters of the linear regression model.
Let us try now to test the hypothesis $H_0 : \beta_2 = 1.6$ against $H_1 : \beta_1 \neq 1.6$ at the $\alpha = 10\%$ significance level. The test statistic is:

$$t = \frac{\hat{\beta}_2 - 1.60}{s.e.(\hat{\beta}_2)} = \frac{1.6049 - 1.60}{0.0118} = 0.4153 \quad (3.11.4)$$

so that the null hypothesis is not rejected. The situation is represented graphically in Figure 3.14.

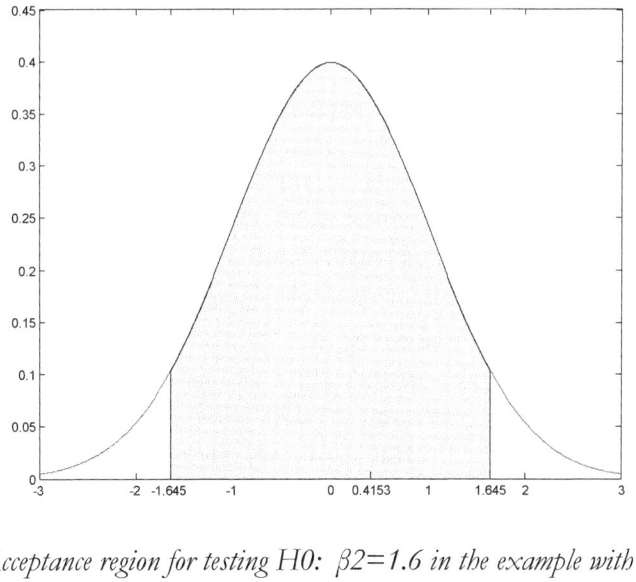

Figure 3.14: Acceptance region for testing H0: β2=1.6 in the example with simulated data

Repeating the Wald test procedure in EViews, and indicating in the appropriate field of the Wald test "c(3) = 1.6", we obtain Table 3.5, where the value of the F-statistic is the square of the t-statistic. Note also the degrees of freedom (df), equal to 1 and 127, where the former indicates the number of restrictions and the latter the number of observations minus that of estimated parameters. The column "Probability" reports the *p*-value, 0.6760, which being larger than conventional significance levels (1%, 5% or 10%) leads to non-rejection of the null hypothesis.

Wald Test:

Equation: LS1

Test Statistic	Value	df	Probability
F-statistic	0.175489	(1, 127)	0.6760
Chi-square	0.175489	1	0.6753

Null Hypothesis Summary:

Normalized Restriction (= 0)	Value	Std. Err.
-1.6 + C(3)	0.004939	0.011789

Restrictions are linear in coefficients.

Table 3.5: Output of the F-test for *H0: β2=1.6 in the example with simulated data*

Confidence intervals and two-tailed tests can be connected, as we saw in Section 3.6. Testing $H_0: \beta_k = b_k$ against $H_1: \beta_k \neq b_k$, k = 1, …, K, H_0 is rejected at the $\alpha = 10\%$ significance level if and only if b_k does not belong to the $(1 - \alpha)\%$ confidence interval for β_k .
For example, using the regression output in Table 3.4, we get that a 90% confidence interval for the intercept is $[0.891067 \pm 0.228771 \cdot 1.645]$, namely, about [0.51; 1.27]. Since the 0 value is outside this interval, we can reject, at the 10% significance level, the null hypothesis H_0: $c = 0$ vs H_1: $c \neq 0$.
We now implement a one-sided test, as in Section 3.7. Let us suppose that we want to verify $H_0: \beta_2 = 1.6$ versus $H_1: \beta_1 > 1.6$ with $\alpha = 10\%$. The critical value of the Student-t with $T - k$ degrees of freedom is t_c = 1,282 (remember that T – k = 127 whereby we use the line ∞ in Table 3.1 or, equivalently, the tables of the standard Normal distribution).

The calculations are the same as seen above and the realization of the t-statistic is 0.415. Hence, we do not reject the null hypothesis. In this case we cannot use the F-test seen above, given that the alternative hypothesis is one-sided.
Let us return to the EViews output in Table 3.4. At the bottom right of the Table, we find values labeled as "F-statistic" and "Prob (F-statistic)", equal respectively to 25317.94 and 0.000000. The latter is the p-value associated with the former, and the former is the F-statistic for the null hypothesis

$$H_0: \begin{pmatrix} \beta_1 \\ \beta_2 \end{pmatrix} = \begin{pmatrix} 0 \\ 0 \end{pmatrix}, \qquad (3.11.5)$$

against the alternative hypothesis

$$\mathrm{H}_1: \begin{pmatrix} \beta_1 \\ \beta_2 \end{pmatrix} \neq \begin{pmatrix} 0 \\ 0 \end{pmatrix}, \qquad (3.11.6)$$

namely, at least one of the regression coefficients is different from 0.
The null hypothesis can be interpreted as "the regression is not significant" or "the chosen regressors do not explain the dependent variable". The low p-value leads to reject the null hypothesis, whereby the regressors are *jointly statistically significant*
Next, we use EViews to test the following composite hypothesis:

$$H_0: \begin{pmatrix} \beta_1 \\ \beta_2 \end{pmatrix} = \begin{pmatrix} 0.8 \\ 1.6 \end{pmatrix}, \qquad (3.11.7)$$

against

$$H_1: \begin{pmatrix} \beta_1 \\ \beta_2 \end{pmatrix} \neq \begin{pmatrix} 0.8 \\ 1.6 \end{pmatrix}. \qquad (3.11.8)$$

We use the F-test (Wald) procedure described above. Table 3.6 reports the most interesting results, already discussed above. In particular, the realization of the test is *0.206017*, to be compared with the appropriate critical value from an F-distribution with (2, 127) degrees of freedom, where $q=2$ is the number of restrictions and $T-K=127$.
The table with critical values of the $F(q,T\text{-}k)$ distribution (Table 3.2) is built so that the columns vary according to the degrees of freedom of the numerator (for us, $q=2$) and the rows change depending on the degrees of freedom of the denominator (for us, $T-K=127$). The table reports generally percentiles up to $q = 30$ and $T-K = 120$ and then the percentiles for $q = +\infty$, $T-K = +\infty$, as for $q>30$ and T-K>120 the distribution is less sensitive to changes in degrees of

freedom. In our case, we use the column q =2 in Table 3.2 and the last line, $T - K = +\infty$. It follows that the critical value is $F_c \approx 3.00$ and the region of rejection is $[3.00, +\infty)$. Therefore, we do not reject the null hypothesis in (3.11.7). Alternatively, we can use the fact that the *p*-value of the test is *0.8141* to conclude that the null hypothesis is not rejected (see Figure 3.15).

Wald Test:
Equation: LS1

Test Statistic	Value	df	Probability
F-statistic	0.206017	(2, 127)	0.8141
Chi-square	0.412034	2	0.8138

Null Hypothesis Summary:

Normalized Restriction (= 0)	Value	Std. Err.
-0.8 + C(2)	0.000996	0.032864
-1.6 + C(3)	0.004939	0.011789

Table 3.6: Outcome of the F-test for a composite *hypothesis*

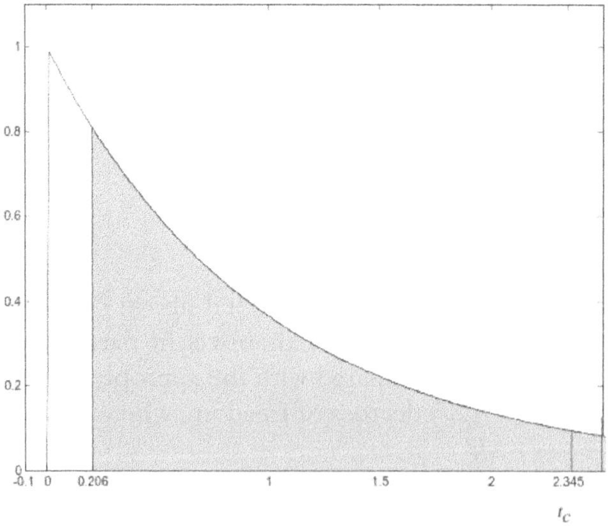

Figure 3.15: Rejection region, realization of the F-test *and its p-value*

Next, we consider empirically the effects of variable omission, that we studied theoretically in Section 3.10. If we run a regression of *y* on *X1* only, the OLS estimator is biased and inconsistent. Actually, the coefficient of *X1* reported in Table 3.7 is very different from that in Table 3.4. The test statistics are also unreliable, as they are based on bad parameter estimators.

Dependent variable: Y
Method: Least Squares
Sample: 1970q1 2002q2
Included observations: 130

Variable	Coefficient	Std. Error	*t*-Statistic	Prob.
C	1.168764	2.762020	0.423156	0.6729
X1	4.049239	0.272860	14.84000	0.0000

R-squared	0.632422	Mean dependent var	41.27091
Adjusted R-squared	0.629551	S.D. dependent var	10.70127
S.E. of regression	6.513278	Akaike info criterion	6.600828
Sum squared resid	5430.118	Schwarz criterion	6.644944
Log likelihood	-427.0538	*F*-statistic	220.2257
Durbin-Watson stat	2.181696	Prob(*F*-statistic)	0.000000

Table 3.7: Effects of variable omission

3.12 An empirical analysis of the determinants of aggregate consumption

Let us reconsider the regression of (the log of) consumption on (the logs of) disposable income and wealth, as discussed in Section 2.14. The results, using the workfile "example_cons_chap3.wf1", are summarized in Table 3.8.

To start, we want to run a one-sided test for the hypothesis that the elasticity of consumption to income is 0 against the alternative hypothesis that it is less than 0, at a 5% confidence level. In this case, we can accept (not reject) the null hypothesis by noting that the coefficient of the regressor is statistically significantly positive (the rejection region of the test is one sided, of the type $(-\infty,-t_c]$, while the realization of the statistic is positive).

Dependent variable: LRC
Method: Least Squares
Sample (adjusted): 1990q1 2012q1
Included observations: 89 after adjustments

Variable	Coefficient	Std. Error	*t*-Statistic	Prob.
C	-1.020849	0.238960	-4.272041	0.0000
LRYD	1.066816	0.037456	28.48182	0.0000
LRW	0.025530	0.013273	1.923434	0.0577

R-squared	0.944399	Mean dependent var	7.763188
Adjusted R-squared	0.943106	S.D. dependent var	0.086122
S.E. of regression	0.020542	Akaike info criterion	-4.899548
Sum squared resid	0.036290	Schwarz criterion	-4.815662
Log likelihood	221.0299	Hannan-Quinn criterion	-4.865736
F-statistic	730.3720	Durbin-Watson stat	0.297996
Prob(*F*-statistic)	0.000000		

Table 3.8: Regression of consumption on income and wealth

The values given in Table 3.8 allow us to test other economically relevant hypotheses. We can for example test the null hypothesis that the elasticity of consumption relative to income is equal to 1 against the alternative that it is different from 1, at a 10% significance level. The t-statistic for this hypothesis is equal to:

$$t = \frac{\hat{\beta}_{LRYD} - 1}{s.e.(\hat{\beta}_{LRYD})} = \frac{1.066816 - 1}{0.037456} = 1.784 \qquad (3.11.9)$$

The rejection region with 86 degrees of freedom is given by the intervals $(-\infty, -1.664]$ and $[\ 1.664, \infty)$, so we cannot reject the null hypothesis of unit elasticity at the 5% significance level.
Another test is for the significance of the constant. In this case the constant is significantly negative, as we can see from the t -statistic and the associated p-value close to 0. We can also reject non significance of real wealth, but only at 10%.
The two hypotheses of irrelevant wealth and elasticity of consumption relative to disposable income equal to 1 can be jointly tested by means of an F-test. Table 3.9 confirms what previously found: the two restrictions jointly considered are clearly rejected.

Wald Test:
Equation: EQ_COMPLETE

Test Statistic	Value	df	Probability
F-statistic	9.251364	(2, 86)	0.0002
Chi-square	18.50273	2	0.0001

Null Hypothesis: C(2)=1, C(3)=0
Null Hypothesis Summary:

Normalized Restriction (= 0)	Value	Std. Err.
-1 + C(2)	0.066816	0.037456
C(3)	0.025530	0.013273

Table 3.9: F-test for H_0: $\beta_{yd}=1$ e $\beta_w=0$.

An alternative way to calculate this statistic is to use the sum of the squared residuals in the unrestricted model and in the restricted model. To get the latter, we must estimate the restricted model $C_t = c_o + Y_{d,t}$ (see Table 3.10).

Dependent variable: LRC
Method: Least Squares
Sample (adjusted): 1990q1 2012q1
Included observations: 89 after adjustments
LRC=C(1)+LRYD

	Coefficient	Std. Error	*t*-Statistic	Prob.
C(1)	-0.228401	0.002373	-96.25546	0.0000

R-squared	0.932437	Mean dependent var	7.763188
Adjusted R-squared	0.932437	S.D. dependent var	0.086122
S.E. of regression	0.022386	Akaike info criterion	-4.749626
Sum squared resid	0.044098	Schwarz criterion	-4.721664
Log likelihood	212.3584	Hannan-Quinn criterion	-4.738356
Durbin-Watson stat	0.218356		

Table 3.10: Results for the restricted model $C_t = c_o + Y_{d,t}$

As it can be seen from equation (3.11.10), we get the same value for the F-test as shown in Table 3.9 (small differences are due to rounding).

$$F-statistic=\frac{(RSS_R-RSS_U)/j}{RSS_U/(T-k)}= \frac{(0.044098-0.036290)/2}{0.036290/(89-3)}=9.2512 \tag{3.11.10}$$

Finally, we can test the hypothesis that both income and wealth are not statistically significant. This hypothesis is also strongly rejected, with a value of the F-statistic equal to that reported in Table 3.8.

Wald Test:
Equation: EQ_COMPLETE

Test Statistic	Value	Df	Probability
F-statistic	730.3720	(2, 86)	0.0000
Chi-square	1460.744	2	0.0000

Table 3.11: F-test for H_0: $\beta_{yd}=\beta_w=0$

3.13 An empirical analysis of the determinants of aggregate investment

We reconsider the aggregate investment function where (the log of real) investment is explained by (the log of real) aggregate disposable income (as a proxy for current and expected profits) and the real interest rate (as a cost measure). The OLS regression output, reported in Table 3.12 and obtained from the EViews workfile "example_inv_chap3.wf1", includes the estimated parameters, their associated standard error, the t -tests for the significance of the coefficients, i.e. the ratio of the estimated coefficient to their standard errors, and the p-values associated with the t-statistics (the area to the right of $|t|$ calculated on the basis of a Student-t distribution with $T-K$ degrees of freedom, in this case 84-3).
Disposable income and the real interest rate are highly significant, i.e., the associated t-tests strongly reject their non-significance. The F-test for the joint non-significance of the two regressors, also reported in the regression output, strongly rejects the null hypothesis. Its realized value is 186.1640, with a close to 0 a p-value (calculated on the basis of an F distribution with 2 and 81 degrees of freedom).

Dependent variable: LRI
Method: Least Squares
Sample (adjusted): 1991q1 2011q4
Included observations: 84 after adjustments

Variable	Coefficient	Std. Error	*t*-Statistic	Prob.
C	5.559115	0.663154	8.382846	0.0000
LRYD	0.709066	0.082330	8.612446	0.0000
RR	-0.030356	0.002870	-10.57701	0.0000

R-squared	0.821321	Mean dependent var	11.10866
Adjusted R-squared	0.816910	S.D. dependent var	0.113882
S.E. of regression	0.048729	Akaike info criterion	-3.170030
Sum squared resid	0.192335	Schwarz criterion	-3.083215
Log likelihood	136.1413	Hannan-Quinn criterion	-3.135131
F-statistic	186.1640	Durbin-Watson stat	0.330525
Prob(*F*-statistic)	0.000000		

Table 3.12: Regression output for the aggregate investment model

If we want to test the joint hypothesis that all the three coefficients are equal to 0 against the hypothesis that at least one of them is different from 0, we click on "View", "Coefficient Tests", "Wald – Coefficient Restrictions". In the resulting window we write

```
c(1) = 0, c(2) = 0, c(3) = 0
```

The test, again, strongly rejects the null hypothesis.

We can also test for nonlinear restrictions on the parameters. For example, we can test the hypothesis "*at least* one between LRYD and RR are meaningless" against the alternative hypothesis that they are jointly significant. This can be formalized as:

```
c(2)*c(3)=0
```

It can be shown that the test statistic is asymptotically distributed as χ^2, with as many degrees of freedom as the number of restrictions (one in our example). From the output in Table 3.13, the test rejects the null hypothesis at a 1% significance level.

Wald Test:
Equation: EQINV

Test Statistic	Value	Df	Probability
t-statistic	-9.382472	81	0.0000
F-statistic	88.03077	(1, 81)	0.0000
Chi-square	88.03077	1	0.0000

Null Hypothesis: C(2)*C(3)=0
Null Hypothesis Summary:

Normalized Restriction (= 0)	Value	Std. Err.
C(2) * C(3)	-0.021525	0.002294

Delta method computed using analytic derivatives

Table 3.13: Wald test for a nonlinear restriction

EViews provides confidence intervals and confidence ellipses for the parameters, at the chosen level $1-\alpha$. As we have seen before, we click on "View", "Coefficient Diagnostics", "Confidence Ellipse", and we obtain the window in Figure 3.16.

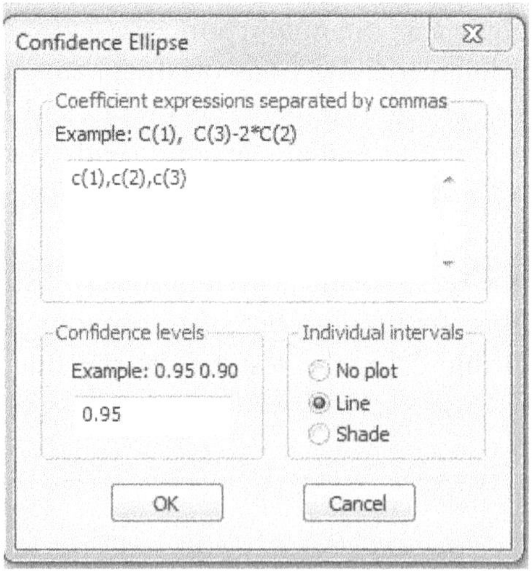

Figure 3.16: EViews window to compute confidence ellipses and intervals

In the top editor we list the coefficients of which we want to obtain the confidence ellipses (EViews computes an ellipse for each possible pair) and intervals. In the bottom left part of the window we select the confidence level (1 – α), whose default value is 95%. In the bottom right we select the representation of the individual confidence intervals. Ellipses for c(1) and c(3), and c(2) and c(3), are reported, respectively, in the left and right panels of Figure 3.17.

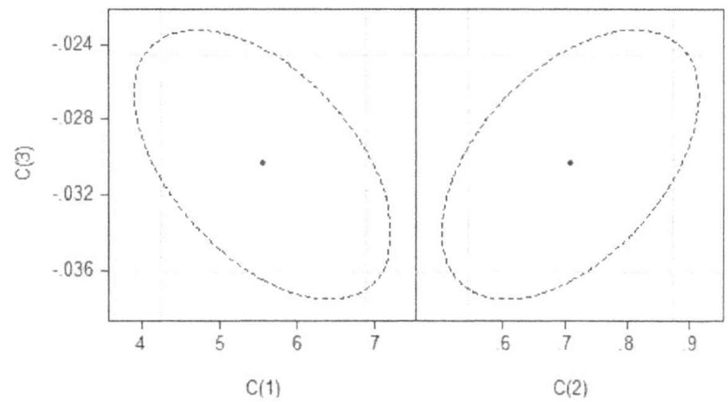

Figure 3.17: Confidence ellipses for selected parameter pairs

In the center of the ellipses are the estimated parameters, while the dashed lines correspond to the extremes of the univariate confidence intervals. Clicking on the line, for example on the lower bound for c(2), we obtain Figure 3.18 that shows in its middle right part the value of interest. Therefore, the estimated confidence interval at the 95% level for the real investment elasticity to real income is about [0.545, 0.872].

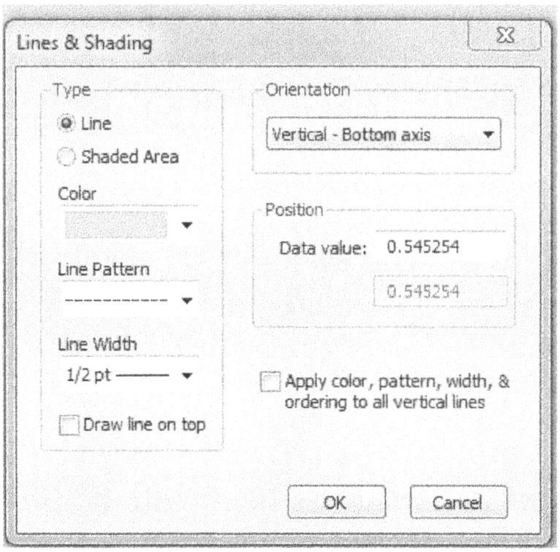

Figure 3.18: EViews window reporting the extremes of the confidence intervals

3.14 An empirical analysis of the determinants of labor productivity

In Chapter 2, we saw that a model where labor productivity is explained by aggregate investment is not so good. From Figure 2.14, it seems that productivity presents two different patterns over the sample under analysis: first it grows moderately from the beginning of the sample until 2006, then it declines and increases again over the period 2007-2012. Using the EViews workfile “example_prod_chap3.wf1£, we conduct a separate analysis for the two parts of the sample. In this way, we are able to isolate the most problematic part of the sample, the one containing the global financial crisis.
In the first subsample, the one ending in 2006, we regress productivity on a constant, a linear trend (called “@TREND” in EViews, with values equal to 1.2, ...,T), and investment. In the second subsample, that includes the years of crisis, we substitute the linear trend with a quadratic trend (“@TREND^2” in EViews). The latter permits a more general pattern, for example first descending and then ascending.
The deterministic trends are proxies for unmodelled variables, such as technological progress or changes in the institutional context.

Dependent Variable: LP
Method: Least Squares
Sample (adjusted): 1991Q12006Q4
Included observations: 64 after adjustments

Variable	Coefficient	Std. Error	t-Statistic	Prob.
C	4.383544	0.095392	45.95290	0.0000
LQ	-0.336031	0.064220	-5.232544	0.0000
@TREND	0.002649	0.000190	13.95906	0.0000

R-squared	0.773182	Mean dependent var	4.044511
Adjusted R-squared	0.765745	S.D. dependent var	0.047002
S.E. of regression	0.022749	Akaike info criterion	-4.682848
Sum squared resid	0.031569	Schwarz criterion	-4.581650
Log likelihood	152.8511	Hannan-Quinn criter.	-4.642981
F-statistic	103.9689	Durbin-Watson stat	0.149342
Prob(F-statistic)	0.000000		

Dependent Variable: LP
Method: Least Squares
Sample (adjusted): 2007Q12012Q1
Included observations: 21 after adjustments

Variable	Coefficient	Std. Error	t-Statistic	Prob.
C	2.441586	0.196146	12.44782	0.000000
LQ	0.838854	0.095706	8.764866	0.000000
@TREND^2	1.96E-05	3.26E-06	6.014818	0.000000

R-squared	0.863853	Mean dependent var	4.055875
Adjusted R-squared	0.848726	S.D. dependent var	0.019528
S.E. of regression	0.007595	Akaike info criterion	-6.791068
Sum squared resid	0.001038	Schwarz criterion	-6.641851
Log likelihood	74.30622	Hannan-Quinn criter.	-6.758684
F-statistic	57.10504	Durbin-Watson stat	1.383607
Prob(F-statistic)	0.000000		

Table 3.14: Linear and quadratic trends in a model for labor productivity

The results of the two regressions are presented in Table 3.14. In both samples the deterministic trends are strongly significant. Moreover, on the basis of the F-test, the two explanatory variables are strongly jointly significant. Regarding the economic interpretation of the results, investment is significant in both subsamples, but with a negative coefficient in the pre-crisis period, a positive one after 2006.
The omission of the deterministic trends can distort the role of other regressors. In fact, the coefficient of LQ in the two subsamples is rather different from that in Table 2.10. The coefficient of determination also increases by more than 50% with respect to Table 2.10, indicating an

improvement in the explanatory power of the model. However, the result must be interpreted with caution since the number of observation is lower than in Table 2.10. In addition, the persistent pattern of the residuals, shown in Figure 3.19, suggests that the model is still not correctly specified.

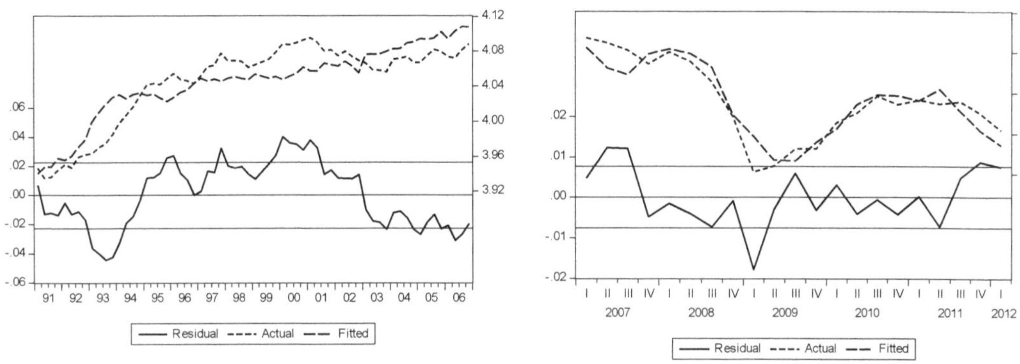

Figure 3.19: Actual and fitted values and residuals for the labor productivity model in the two subsamples 1991-2006 and 2007-2012

3.15 An empirical analysis of the CAPM

Since individuals are normally risk averse, they require higher yields on risky assets, i.e., a risk premium. Until the 1950s, this premium was measured as a function of the variance of the returns. Later, it was observed that asset diversification within a portfolio can reduce the variance of its return, and therefore the risk premium. While idiosyncratic risks (related to individual assets) may be substantially reduced, other risks cannot be avoided even with a well-diversified portfolio (e.g. shocks that affect the entire financial market).
There emerged a new model for assessing the risk premium that took into account the concepts of risk diversification and unavoidable risk. It was supposed to measure the risk premium associated with the ith asset according to the unavoidable risk that it adds to the market portfolio.
This is the Capital Asset Pricing Model (CAPM), whose solution makes it possible to assess the risk premium with a simple regression as:

$$\left(R_{it} - R_{ft}\right) = \beta_i \left(R_{mt} - R_{ft}\right) + e_t \qquad t = 1,\dots,T \qquad (3.15.1)$$

R_{it} indicates the time-*t* yield of a generic asset *i* (for example the ENI stock) R_{ft} is the interest rate on a safe asset (proxied, e.g., by the 3 months EURIBOR) and R_{mt} is the yield of a financial market index (e.g., the MIB 30).
In regression (3.15.1), the parameter β_i measures how the risk premium for asset *i* moves relative to the market risk premium, i.e., how much risk asset *i* adds to the market index risk.
The EViews workfile "example_chapm_chap3.wf1" contains daily data from 31 December 2004 to 29 May 2009 on the returns of a major Italian stock market index (MIB30), a risk-free rate (3-month EURIBOR, indicated by RF) and the yields of five securities in the MIB30: Assicurazioni Generali, ENEL, ENI, Fiat and Telecom Italia. Table 3.15 reports the regression output for the model in (3.15.1), for each of the five stocks (see eqX1 in the workfile for stock X).

Assicurazioni Generali

Variable	Coefficient	Std. Error	*t*-Statistic	Prob.
MIB30R-RF	0.964832	0.007649	126.1370	0.0000

Enel

Variable	Coefficient	Std. Error	*t*-Statistic	Prob.
MIB30R-RF	0.976401	0.008374	116.5939	0.000000

Eni

Variable	Coefficient	Std. Error	*t*-Statistic	Prob.
MIB30R-RF	1.004027	0.007801	128.6991	0.000000

Fiat

Variable	Coefficient	Std. Error	*t*-Statistic	Prob.
MIB30R-RF	1.029503	0.016119	63.86990	0.000000

Telecom Italia

Variable	Coefficient	Std. Error	*t*-Statistic	Prob.
MIB30R-RF	0.998894	0.011119	89.83827	0.000000

Table 3.15: Estimated CAPM for five stocks in the MIB30

Table 3.15 suggests that all the βs are strongly significant and close to 1, which is not surprising since all the stocks are part of the Italian MIB30 stock index. Based on these results, we can rank the stocks according to their risk level with respect to the market index. We have three titles that reduce the index risk ($\beta<1$): Telecom Italia, ENEL and Assicurazioni Generali. Instead, Fiat and ENI have $\beta>1$, and are therefore riskier than the index. It should be noted that in (3.15.1) there is no intercept, a debated restriction. In 1972, Black, Jensen and Scholes wrote an article proving that the null hypothesis of constant equal to 0 was rejected empirically in most cases. They also showed theoretically that there should be no constant only if all the agents can freely borrow any amount at the risk-free rate. Table 3.16 reports the same regression results as in Table 3.15 but adding an intercept to the model (in the EViews workfile such equations are indicated by eqX2 for stock X).

Assicurazioni Generali

Variable	Coefficient	Std. Error	*t*-Statistic	Prob.
C	-0.482369	0.059445	-8.114483	0.0000
MIB30R-RF	0.857624	0.015164	56.55732	0.0000

Enel

Variable	Coefficient	Std. Error	*t*-Statistic	Prob.
C	-0.371723	0.066017	-5.630704	0.000000
MIB30R-RF	0.893784	0.016840	53.07457	0.000000

Eni

Variable	Coefficient	Std. Error	*t*-Statistic	Prob.
C	0.187611	0.062097	3.021254	0.002600
MIB30R-RF	1.045724	0.015840	66.01730	0.000000

Fiat

Variable	Coefficient	Std. Error	*t*-Statistic	Prob.
C	0.656119	0.127347	5.152229	0.000000
MIB30R-RF	1.175329	0.032485	36.18117	0.000000

Telecom Italia

Variable	Coefficient	Std. Error	*t*-Statistic	Prob.
C	-0.264898	0.088510	-2.992877	0.002800
MIB30R-RF	0.940019	0.022578	41.63487	0.000000

Table 3.16: Estimated CAPM with intercept for the 5 stocks in the MIB30

Table 3.16 highlights that all the βs remain strongly significant, are rather different from those in Table 3.15, but Fiat and ENI remain riskier than the index, while Telecom Italia, ENEL, Fiat, and Assicurazioni Generali are less risky.
An additional hypothesis of interest is whether β=1, against the alternative hypothesis that $\beta \neq 1$. The outcome of the *t*-tests for the five stocks are reported in Table 3.17.

	No intercept			Intercept		
Stock	Estim β	Std. Error	*t*-test for β =1	Estim β	Std. Error	*t*-test for β =1
ASSGEN	0.9648325	0.0076491	-4.597614202	0.8576236	0.0151638	-9.389229195
ENEL	0.9764009	0.0083744	-2.818008868	0.8937837	0.0168402	-6.307322266
ENI	1.0040267	0.0078014	0.516148627	1.045724	0.0158402	2.886587814
FIAT	1.0295035	0.0161188	1.830380618	1.175329	0.0324845	5.397305841
Telecom	0.9988938	0.0111188	-0.099490039	0.9400189	0.0225777	-2.656653735

Table 3.17: t-tests for H_0*:* $\beta = 1$ *versus* H_1*:* $\beta \neq 1$.

Since the 5% acceptance region is (-1.96, 1.96), then Table 3.17 implies that the null hypothesis is not rejected for ENI, Fiat and Telecom with no constant in the model, while it is rejected for all the stocks when there is a constant in the model.
Hence, the proper specification of the deterministic component of the model matters. Excluding the intercept by mistake leads to wrong statistical results, with relevant economic implications: in our example, too often we accept that a given stock is as risky as the market index (β =1).

To conclude, we note that so far in the empirical examples we have not verified whether the assumptions underlying the linear regression model are valid. Therefore, the estimates and test results we obtained may be questionable. We will address this issue in the next chapter.

The main concepts of this chapter

In this chapter we have introduced the interval estimators for the linear regression model parameters. We have shown that, under strong OLS assumptions, $(\hat{\beta}_q - \beta_q)/s.\hat{e}.(\hat{\beta}_q) \sim t(T-k)$ for q = 1.2, ..., k, where k is the number of regressors. Knowing that the number of degrees of freedom is equal to $T - k$, and for a given confidence level $a \in (0.1)$, using the tables of the

Student-t distribution we can find a critical value t_c such that $\Pr\{t_{T-k} \geq t_c\} = \frac{\alpha}{2}$ e $\Pr\{t_{T-k} \leq -t_c\} = \frac{\alpha}{2}$. It follows that $\Pr\{\hat{\beta}_k - t_c \cdot s.\hat{e}.(\hat{\beta}_k) \leq \beta_k \leq \hat{\beta}_k + t_c \cdot s.\hat{e}.(\hat{\beta}_k)\} = 1 - \alpha$ and therefore the (1 - α)% interval estimator, or confidence interval, is

$$\hat{\beta}_k \pm t_c \cdot s.\hat{e}.(\hat{\beta}_k)$$

We have then studied hypothesis testing for the parameters of the linear regression model. This requires formulating the null hypothesis, the alternative hypothesis, the test statistic and the decision rule.

The test statistic is a random variable related to the hypothesis of interest, whose distribution under the null hypothesis is fully known. The rejection region is the set of realizations of the test statistic that lead to rejection of the null hypothesis (decision rule). The complement of the rejection region is the acceptance region.

Rejection of the null hypothesis when true implies a type I error, while non-rejection of the null hypothesis when false leads to a type II error. The probability of a type I error is denoted by a and is called the significance level of the test. The level of the test is selected by the researcher. It is typically low in order to have a low probability of rejecting the null hypothesis when true. The probability of type II error cannot be controlled, but it goes to 0 asymptotically for consistent tests.

The *p*-value of a test is the probability that the test statistic (considered as a random variable) is larger than the absolute value of the realization of the statistic. Therefore, the *p*-value is greater than the significance level of the test if and only if the realization of the statistic falls in the acceptance region. For this reason, the decision rule can also be based on whether the *p*-value is larger than the significance level of the test (non-rejection of the null hypothesis).

The most common hypotheses in practice concern single coefficients of the linear regression model, e.g., $H_0: \beta_q = c$, $H_1: \beta_q \neq c$. To test hypotheses of this type we can use the *t*-statistic or *t*-ratio, already used to construct confidence intervals:

$$(\hat{\beta}_q - \beta_q) / s.\hat{e}.(\hat{\beta}_q) \underset{H_0}{\sim} t(T-k)$$
.

If H_0 is true, it is likely that the t-test is close to 0. If H_0 is false, it is likely that t is positive, if $\beta_q > c$, or negative, if $\beta_q < c$. The decision rule is therefore to reject H_0 if t is far from 0. The rejection region is $(-\infty, -t_c) \cup (\infty, t_c)$.
The relationship between the confidence interval for a coefficient and the t test for its significance is given by

$$\Pr\left\{ -t_c \leq \underbrace{\frac{\hat{\beta} - c}{s.\hat{e}.(\hat{\beta})}}_{t} \leq t_c \right\} = 1 - \alpha$$

Therefore, if $\hat{t}$ falls in the α% acceptance region for $H_0 : \beta = c$, and hence H_0 is not rejected, then c is part of a (1-a)% confidence interval for β. Conversely, if c is part of a (1-α)% confidence interval for β, then the null hypothesis $H_0 : \beta = c$ is not rejected at a significance level α.
To test joint hypotheses on multiple coefficients, formulated as H_0: $R\beta - r = 0$ versus H_1: $R\beta - r \neq 0$, we have suggested to use the F-statistic:

$$\lambda = \frac{\left(R\hat{\beta} - r\right)' \left[R(X'X)^{-1} R'\right]^{-1} \left(R\hat{\beta} - r\right)}{j\hat{\sigma}^2} \underset{H_0}{\sim} F(j, T-k),$$

where R is the $j \times k$ matrix of linear restrictions on the parameters. If H_0 is false, λ is expected to be large and positive. Thus, the acceptance region is $\{\lambda < F_c\}$, where F_c is the critical value such that $P_{H_0}\{\lambda \geq F_c\} = \alpha$ and α is the significance level of the test.
To explicitly consider linear restrictions on the parameters of the linear regression model, we have derived the restricted OLS (ROLS) estimator:

$$\tilde{\beta} = \hat{\beta} + (X'X)^{-1} R' \left[R(X'X)^{-1} R'\right]^{-1} \left[r - R\hat{\beta}\right].$$

The ROLS estimator is unbiased if the constraint is valid. It is more efficient than the OLS estimator even if the constraint is invalid, since it uses more information (even though possibly false) when estimating the parameters of interest.
Using the ROLS estimator, an alternative formulation of the F-test for $R\beta - r = 0$ is:

$$\lambda = \frac{(RSS_R - RSS_U)/j}{RSS_U/(T-k)} \sim F(j, T-k),$$

where RSS_R is the residual sum of squares in the model estimated by ROLS and RSS_U that resulting from standard OLS estimation.
Finally, looking at the relationship between the data generating process (DGP) and the econometric model, we have considered the cases of omitted and irrelevant variables. In the former, the OLS estimator is biased and inconsistent. In the latter is remains unbiased and consistent, but it is not efficient.

Exercises

1)
Let us consider the model: $Y = X\beta = X_1\beta_1 + X_2\beta_2 + \varepsilon$, ε iid N (0, σ^2I)
with X_1 and X_2 orthogonal. Let us also suppose that $\hat{\beta}_1 = 1, \hat{\beta}_2 = 2$, $se(\hat{\beta}_1) = 1$, $se(\hat{\beta}_2) = 1$.
a) If the 10% critical value for testing H_0: $\beta_1 = 0$ against H_1:$_1\beta \neq 0$ is 1.647, do you reject H_0?
b) Derive a 90% confidence interval for β_2. Based on this interval, do you reject H_0: $\beta_2 = 0$ vs H_{12}: $\beta \neq 0$?

2)
Under strong OLS assumptions, derive the (a = 0.05) rejection region for the hypothesis H_0: $\beta_1 = 4\beta_2$ against H_1: $\beta_1 \geq 4\beta_2$ in the model $y = \beta_0 + \beta_1 X_1 + \beta_2 X_2 + \varepsilon$.

3)
Regressing the Y variable on the *orthogonal* regressors X and Z we get:

Regressor	Coefficient	Std. Error
X	0.80	0.80
Z	0.70	0.10

where the number of observations (T) is 32, the residual sum of squares (RSS) is 60, while the sum of squared residuals when the coefficients of X and Z are 0 (TSS) is 160.
Construct an F-test for the joint significance of X and Z.

4)
Demonstrate that the following formulae to calculate the F-statistics are equivalent:

$$F_1 = \frac{(RSS_R - RSS_U)/j}{RSS_U/(T-k)}$$

$$F_2 = \frac{(R\hat{\beta} - r)\left(R(X'X)^{-1}R'\right)^{-1}(R\hat{\beta} - r)/j}{\hat{\varepsilon}'\hat{\varepsilon}/(T-k)}$$

5)
a) Assuming that you have the correct critical value, construct a 95% confidence interval for the error variance σ^2 under strong OLS assumptions. Which particular OLS assumption is used?

b) Now let us assume that we want to test $H_0 : \sigma^2 = 1$ vs. $H_1 : \sigma^2 \neq 1$ at the 5% significance level. Can we use the result of the previous point to carry out the test?

6)
What is the formula for the (1-α)% confidence interval for the generic parameter β_j in a linear regression model under OLS assumptions with unknown error variance? Can this confidence interval be transformed into an acceptance region to test a hypothesis about β_j?

7)
Let us consider the model $y_t = \alpha + \beta x_t + e_t$, $e_t \sim$ iid $(0, \sigma^2)$
Construct the F-test for the null hypothesis H_0: $a = 0$, $\beta = 0$
Assuming $t = 1, 2$ with ($x_1 = 0$,$_1 y = 1$) and ($x_2 = -1$, $y_2 = 0$), compute the F-test for the null hypothesis and its p-value.

8)
Given the model: $y_t = b_0 + b_1 * x_{1,t} + b_2 * x_{2,t} + b_3 * x_{3,t} + b_4 * x_{4,t} + b_5 * x_{5,t} + e_t$
that can be re-written in vector notation as $Y = X\beta + \varepsilon$,
write in the form $R\beta = q$ the following linear restrictions:
$b_3 = 0$
$b_2 = b_3$
$b_2 + b_3 - b_4 = 1$
$b_0 = 0$, $b_1 = 0$, $b_5 = 7$
$b_1 = 0$, $b_2 = b_3$, $b_4 - b_5 = 1$

9)
a) Derive the constrained OLS estimator with the generic constraint $R\beta = q$
b) Under which assumptions is the expected value of $\hat{\varepsilon}_{RLS}$ equal to 0?
c) Under which assumptions is the sum of the elements of $\hat{\varepsilon}_{RLS}$ equal to 0?

10)
Show that the ROLS estimator is more efficient than the OLS estimator.

11)
Use the EViews workfile "example_simul2_chap3" to estimate the model:
$Y = \gamma_0 + X_1\gamma_1 + X_2\gamma_2 + X_3\gamma_3 + \varepsilon$.
Impose the following restrictions on the parameters:

$\gamma_1 = 0.9$

$\gamma_2 = -0.1$

Comment on the results by making use of the appropriate tests for the restrictions.

4. The Generalized Linear Regression Model

4.1 Heteroscedasticity and serial correlation of the errors

Given the linear regression model

$$\underset{Tx1}{Y} = \underset{Txk}{X}\,\underset{kx1}{\beta} + \underset{Tx1}{\varepsilon} \tag{4.1.1}$$

or

$$Y_i = \beta_1 + X_{2i}\beta_2 + X_{3i}\beta_3 + \ldots + X_{ki}\beta_k + \varepsilon_i\,, \tag{4.1.2}$$

where i=1,…,T, in Chapter 2 we made a number of additional assumptions on the variables and the errors. Let us write them as follows.

1) The explanatory variables X are a) deterministic and b) not perfectly linearly dependent.
2) The errors ε_i are homoscedastic, namely, $V(\varepsilon_i)=\sigma^2$,
3) The errors ε_i are uncorrelated over time
4) The errors ε_i are Normal
   (and given 3 they are also independent)

   (2–4: $\varepsilon \sim N(0, \sigma^2 I_T)$)
5) The model is linear (in the parameters)
6) The parameters are stable over time (or across the index i)

We have seen that, under these (strong OLS) assumptions, the OLS parameter estimators,

$$\hat{\beta}_{OLS} = (X'X)^{-1}X'Y\,,$$

are unbiased, consistent, efficient and normally distributed.

In this chapter we examine what happens if assumptions 2-5 are not satisfied. The case of non-constant parameters (assumption 6) is covered in the next chapter, that of stochastic regressors (assumption 1a) in Chapter 6, while in Chapter 2 we saw that if the explanatory variables are perfectly collinear (assumption 1b) then the model parameters are not identified.
We begin by assuming that assumption 2 is violated, that is, the errors are heteroscedastic, so $V(\varepsilon_i^2) = V(y_i^2) = \sigma_i^2$, meaning that the variance of the error term, which coincides with that of the dependent variable as seen in Chapter 2, potentially changes for each observation. For example, if the dependent variable is the volume of sales of a firm and the explanatory variable is its size, then it is likely that the variability of sales increases with the firm's size. Similarly, the variability of consumption or investment may increase with disposable income or the level of profits.
Let us assume that assumption 3 is also not satisfied, so the errors are correlated over time or between units, corr $(\varepsilon_i, \varepsilon_j) \neq 0$, e.g. $\varepsilon_t = \rho\varepsilon_{t-1} + u_t$ in the case of first-order serial correlation. For example, if ε_t is an exogenous shock to labor productivity due to the introduction of new technologies or changes in legislation, then typically the shock will be positively correlated in time. Or, if we omit from the set of explanatory variables a relevant indicator that is correlated over time, part of its effects on the dependent variable will enter the error term, creating serial correlation.
To introduce formally heteroscedasticity and serial correlation in the linear model, we assume that the error variance-covariance matrix (henceforth, variance for simplicity) is not $\sigma^2 I$ but Ω, where Ω is a $T \times T$ matrix. For the moment, we assume that Ω is known. For example, if the errors are uncorrelated but their variance changes at each point in time, it is:

$$\Omega = \begin{bmatrix} \sigma_1^2 & 0 & \cdots & 0 \\ 0 & \sigma_2^2 & \ddots & \vdots \\ \vdots & \ddots & \ddots & 0 \\ 0 & \cdots & 0 & \sigma_T^2 \end{bmatrix}. \qquad (4.1.3)$$

If instead the errors are serially correlated with $\varepsilon_t = \rho\varepsilon_{t-1} + u_t$, and $u_t \sim iid(0, \sigma_u^2)$, then,

$$\Omega=\frac{\sigma_u^2}{1-\rho^2}\begin{bmatrix} 1 & \rho & \rho^2 & \dots & \rho^{T-1} \\ \rho & 1 & \rho & \ddots & \vdots \\ \rho^2 & \rho & \ddots & \ddots & \rho^2 \\ \vdots & \ddots & \ddots & 1 & \rho \\ \rho^{T-1} & \dots & \rho^2 & \rho & 1 \end{bmatrix} \tag{4.1.4}$$

If we replace assumptions 2 and 3 with

$$V(\varepsilon)=\Omega, \tag{4.1.5}$$

the resulting model is called *generalized linear regression model.*
Next, we evaluate the consequences of hypothesis (4.1.5), heteroscedasticity and correlated errors, on the properties of the OLS estimators of the parameters.

1) $\hat{\beta}_{OLS}$ remains unbiased. In fact, it is:

$$\hat{\beta}=(X'X)^{-1}X'Y=\beta+(X'X)^{-1}X'e, \tag{4.1.6}$$

and therefore,

$$E(\hat{\beta})=\beta \tag{4.1.7}$$

2) The formula we derived for the variance (matrix) of $\hat{\beta}_{OLS}$ is no longer valid. In fact, it is:

$$Var(\hat{\beta})=Var\left[(X'X)^{-1}X'e\right]=(X'X)^{-1}X'\Omega X(X'X)^{-1}\neq\sigma^2(X'X)^{-1} \tag{4.1.8}$$

3) The OLS estimator is no longer B.L.U.E.. There is an alternative linear unbiased estimator with lower variance, known as the *Generalized Least Squares estimator (GLS)*. We will derive it in the next Section.

4) $\hat{\beta}_{OLS}$ remains consistent. Actually, from the formula of its variance in (4.1.8), $Var(\hat{\beta})$ converges to 0 when the sample size, *T*, diverges. Since $\hat{\beta}_{OLS}$ is unbiased, $\hat{\beta}_{OLS}$ converges to β. It can be shown that this is no longer necessarily true in dynamic models (where lags of the dependent variable appear as explanatory variables, see Chapter 7) with correlated errors.

5) In general, $\hat{\sigma}^2_{OLS}=\dfrac{\hat{\varepsilon}'\hat{\varepsilon}}{T-k}$ is biased since it is derived under the assumption $V(\varepsilon)=\sigma^2 I$ while it is $V(\varepsilon)=\Omega$.

6) The results in 2) and 5) imply that the formulae we derived in Chapter 3 to compute confidence intervals and test statistics for simple and composite hypotheses are no longer valid.
To overcome these serious drawbacks, there are basically three types of remedies.
First, we can improve the specification of the model when the rejection of the assumptions about errors is due to incorrect specification, e.g. omitted explanatory variables or parameter instability. This possibility should be considered in detail for each empirical analysis.
Second, we can transform the variables so that, in the model for the transformed variables, the homoscedasticity and no correlation assumptions are satisfied. We will see how this is possible in the next Section.
Finally, we can use $\hat{\beta}_{OLS}$ combined with an alternative estimator for its variance. Since we have seen that in general $\hat{\beta}_{OLS}$ remains unbiased and consistent, using an appropriate estimator for its variance it is still possible to calculate confidence intervals and conduct hypothesis testing as in Chapter 3. Specifically, the formula in (4.1.8) suggests using $(X'X)^{-1}X'\hat{\Omega}X(X'X)^{-1}$ instead of $\hat{\sigma}^2(X'X)^{-1}$. This is an example of a *HAC* variance estimator, where HAC stands for Heteroscedasticity and Autocorrelation Consistent.
Since there is a more efficient estimator than OLS, GSL, the combination of $\hat{\beta}_{OLS}$ and a HAC estimator for its variance is not optimal. In addition, if the violation of the assumptions on the errors is due to incorrect model specification, the OLS estimator is generally biased and inconsistent, so that the use of a HAC estimator for its variance does not really solve the problem.

4.2 The Generalized Least Squares estimators (GLS and FGLS)

The error variance matrix Ω is positive definite, so there exists a matrix H such that

$$H\Omega H' = I, \qquad (4.2.1)$$

and H is invertible. Therefore,

$$\Omega = H^{-1}(H')^{-1} = (H'H)^{-1}. \qquad (4.2.2)$$

For example, in the case of heteroscedastic errors with variance matrix as in (4.1.3), it is:

$$H = \begin{bmatrix} 1/\sigma_1 & & 0 \\ & \ddots & \\ 0 & & 1/\sigma_T \end{bmatrix}. \tag{4.2.3}$$

If instead the errors are serially correlated with a variance matrix as in (4.1.4), it can be shown that

$$H = \begin{bmatrix} \sqrt{1-\rho^2} & 0 & \dots & 0 \\ -\rho & 1 & 0 & \vdots \\ 0 & -\rho & 1 & 0 \\ 0 & 0 & -\rho & 1 \end{bmatrix}. \tag{4.2.4}$$

Let us now re-consider the linear model in (4.1.1),

$$\underset{Tx1}{Y} = \underset{Txk}{X}\,\underset{kx1}{\beta} + \underset{Tx1}{\varepsilon} \tag{4.2.5}$$

with $V(\varepsilon)= \Omega$, and pre-multiply both sides of (4.2.5) by the H matrix. We get:

$$HY = HX\beta + H\varepsilon \tag{4.2.6}$$

or

$$\widetilde{Y} = \widetilde{X}\beta + \widetilde{\varepsilon} \tag{4.2.7}$$

with

$$E(\widetilde{\varepsilon}\widetilde{\varepsilon}') = H\Omega H' = I \tag{4.2.8}$$

The variable transformation makes the errors in the model in (4.2.7) uncorrelated and homoscedastic, while the parameters remain equal to β. We can then apply the OLS estimation method in this modified model to obtain the *Generalized Least Squares estimators* (GLS):

$$\hat{\beta}_{GLS} = (\tilde{X}'\tilde{X})^{-1}\tilde{X}'\tilde{Y} = (X'\Omega^{-1}X)^{-1}X'\Omega^{-1}Y. \tag{4.2.9}$$

Let us now evaluate the properties of the GLS estimators.

1) $\hat{\beta}_{GLS}$ is unbiased. In fact, it is:

$$E(\hat{\beta}_{GLS}) = E((X'\Omega^{-1}X)^{-1}X'\Omega^{-1}Y) = \beta + (X'\Omega^{-1}X)^{-1}X'\Omega^{-1}E(\varepsilon) = \beta. \tag{4.2.10}$$

2) The variance of $\hat{\beta}_{GLS}$ is:

$$\text{var}(\hat{\beta}_{GLS}) = E(\hat{\beta}_{GLS} - \beta)(\hat{\beta}_{GLS} - \beta)' = (X'\Omega^{-1}X)^{-1} \tag{4.2.11}$$

Moreover, $\hat{\beta}_{GLS}$ is efficient. There is no other linear unbiased estimator with a lower variance. This result can be easily shown by applying the Gauss-Markov theorem to the model transformed as in (4.2.7).

3) $\hat{\beta}_{GLS}$ is consistent. In fact, it is unbiased and its variance converges to 0 when the number of observations diverges.

4) If the errors are Normal, (assumption 4), then $\hat{\beta}_{GLS}$ is also Normal:

$$\hat{\beta}_{GLS} \sim N(\beta, (X'\Omega^{-1}X)^{-1}) \tag{4.2.12}$$

Otherwise, Normality holds asymptotically. If the distribution of $\hat{\beta}_{GLS}$ is Normal, we can use the formulae in Chapter 3 to construct confidence intervals and test statistics for hypothesis testing using $\hat{\beta}_{GLS}$ instead of $\hat{\beta}_{OLS}$ as parameter estimator.

5) If $\Omega = \sigma^2 I$, then $\hat{\beta}_{GLS} = \hat{\beta}_{OLS}$.

All the results we have obtained so far require the variance matrix Ω to be known. In practice, though, Ω is not known and must be estimated. This is particularly important for the calculation of the GLS estimator that otherwise would depend on the unknown parameters of Ω and therefore would not be a proper estimator. When the Ω matrix is replaced by an estimator, $\hat{\Omega}$, the GLS estimator is called the *Feasible Generalized Least Squares estimator*, (FGLS or feasible GLS):

$$\hat{\beta}_{FGLS} = (X'\hat{\Omega}^{-1}X)^{-1}X'\hat{\Omega}^{-1}Y \tag{4.2.13}$$

Estimation of the Ω matrix is complex, since it has $T(T+1)/2$ elements, many more than the available observations T. Therefore, it is necessary to impose a priori constraints to estimate Ω.

For example, in the case of heteroscedasticity, it can be assumed that the variance of the error takes only two values σ_1^2 and σ_2^2, or, in the case of correlation, we can assume that $\varepsilon_t = \varrho\varepsilon_{t-1} + u_t$, so that the parameters to be estimated are only ϱ and σ_u^2.

In the first case, we can split the sample into two subsamples, T_1 and T_2, and estimate σ^2 separately in each subsample. The resulting $\hat{\Omega}$ matrix is:

$$\hat{\Omega} = \begin{bmatrix} \hat{\sigma}_1^2 & 0 & 0 & \ldots\ldots & & 0 \\ 0 & \ddots & & & & 0 \\ \cdot & & \hat{\sigma}_1^2 & & & \cdot \\ \cdot & & & \hat{\sigma}_2^2 & & \cdot\cdot \\ \cdot & & & & \ddots & \\ 0 & 0 & 0 & \ldots\ldots & & \hat{\sigma}_2^2 \end{bmatrix} \tag{4.2.14}$$

In the second case, we can estimate the model by OLS and then use the OLS residuals in the regression $\hat{\varepsilon}_t = \rho\hat{\varepsilon}_{t-1} + u_t$. The OLS estimator of $\rho, \hat{\rho}$, can be used to build

$$\hat{\Omega} = \begin{bmatrix} 1 & \hat{\rho} & \hat{\rho}^2 & \cdots & \hat{\rho}^{T-1} \\ \hat{\rho} & 1 & \hat{\rho} & \cdots & \hat{\rho}^{T-2} \\ \vdots & & & & \\ \hat{\rho}^{T-1} & & \cdots & & \hat{\rho} \end{bmatrix} \frac{\hat{\sigma}_u^2}{1-\hat{\rho}^2} \tag{4.2.15}$$

where

$$\hat{\sigma}_u^2 = \frac{\hat{u}'\hat{u}}{T-1} \tag{4.2.16}$$

To construct the estimator $\hat{\Omega}$ we use observations on Y and X, so that in general $\hat{\Omega}$ and ε are correlated, and this creates problems for the FGLS estimator. In particular, the FGLS estimator is biased in finite samples as

$$E\left[(X'\hat{\Omega}^{-1}X)^{-1}X'\hat{\Omega}\varepsilon\right] \neq 0 \Rightarrow E(\hat{\beta}_{FGLS}) \neq \beta \tag{4.2.17}$$

In addition, $\hat{\beta}_{FGLS}$ is no longer a linear estimator, because Y is also used to estimate $\hat{\Omega}$, and it has not necessarily the lowest variance. However, in general, $\hat{\beta}_{FGLS}$ remains consistent (as long as $\hat{\Omega}$ is consistent), $\hat{\beta}_{FGLS} \to \beta$ and the results seen for $\hat{\beta}_{GLS}$ remain asymptotically valid also for $\hat{\beta}_{FGLS}$.

4.3 Tests for homoscedasticity

Given the complications that arise when the assumptions underlying the simple regression model are violated, it is convenient to have proper tools to verify their validity. Many hypotheses are formulated in terms of errors, which are not observable. However, the model residuals are observable and can provide useful information, as we will see in this Section and in the next one.
An informal method for assessing the presence of heteroscedasticity is to consider the series $\hat{\varepsilon}_i^2$, i=1,…,T, where $\hat{\varepsilon}$ are the OLS residuals $\hat{\varepsilon} = y - X'\hat{\beta}_{OLS}$. If the variability of $\hat{\varepsilon}_i^2$ tends to change over time (or between units in the case of cross-sectional data), we could have problems of heteroscedasticity.
Other methods to verify the hypothesis of homoscedasticity are more formal and based on hypothesis testing procedures. We will consider, in turn, the Goldfeld-Quandt test, the Breusch-Pagan test and the White test.
The *Goldfeld-Quandt test* for homoscedasticity has the null hypothesis H_0: $\sigma_i^2=\sigma^2$ and the alternative hypothesis H_1: $\sigma_i^2=cX_i^2$, $c>0$, that is, the variance increases proportionally to the value of the explanatory variable X.
The required steps for the construction of the test are as follows.

- We re-order the sample observations (y_i, x_i), i = 1, ..., T, ranking them on the basis of the value of x_i, , so that in the first part of the sample we have relatively low values of x_i, and in the second part higher values. It follows that, under the alternative hypothesis, the error variance should increase with i.
- We delete the d observations in the middle of the re-ordered sample (e.g. $d = 0.2\ T$) and divide the remaining values into two subsamples of equal size $\frac{T-d}{2}$. Hence, under the alternative hypothesis, the error variance should be lower in the first subsample.
- For each subsample we calculate the residual sum of squares (RSS $=\sum \hat{\varepsilon}_i^2$), knowing that there are $p = \frac{T-d}{2} - k$ degrees of freedom.

- We compute the ratio of the sums of squared residuals in the second and first subsample,

$$GQ = \frac{RSS_2}{RSS_1}. \tag{4.3.1}$$

- Under the null hypothesis, and assuming that other assumptions on the linear model are met, the test statistic is distributed as $F(p, p)$.

The reasoning underlying the Goldfeld and Quandt test is that if the variance of the errors is constant, then the numerator and the denominator of the GQ statistic in (4.3.1) should be similar, and so the test should fall in the acceptance region. Instead, under the alternative hypothesis, RSS_2 should be larger than RSS_1, and the test should have large values and therefore fall in the rejection region. Usually, the test is one-sided, whereby if under H_1 it is assumed that the variance decreases over time rather than increases, the statistic to consider becomes RSS_1/RSS_2.
The *Breusch-Pagan test* for homoscedasticity has as null hypothesis H_0: $\sigma_i^2=\sigma^2$ and as alternative hypothesis the existence of a generic relationship between the explanatory variable and the variance, that is, H_1: $\sigma_i^2=f(\gamma+\delta Z)$, with $Z=X$, or, e.g., $Z=X^2$. In the case of multiple explanatory variables, Z can include a subset of them (or their products). Z can also contain different variables from those used in the regression model.
The test is conducted as follows.

- Using the OLS estimator of the parameters, we calculate the OLS residuals $\hat{\varepsilon}$ and $\hat{\sigma}^2$, with the usual formulae.
- Next, we run the regression

$$\frac{\hat{\varepsilon}_i^2}{\hat{\sigma}^2} = \gamma + \delta Z_i + v_i. \tag{4.3.2}$$

- We compute the sum of squared fitted values:

$$\text{BP} = \frac{\sum_{i=1}^{T} \left(\hat{\gamma} + \hat{\delta} Z_i\right)^2}{2} \tag{4.3.3}$$

- Under the null hypothesis, and assuming that other assumptions on the linear model are met, BP$\sim \chi^2(q)$, where q is the dimension of Z.

The intuition is that, if the error variance is constant, then the Z variables should not be significant in the regression (4.3.2). Hence, the statistic BP should take a low value, falling in the acceptance region.

The *White test* for homoscedasticity has the null hypothesis H_0: $\sigma_i^2=\sigma^2$ and the same alternative hypothesis as the Breusch-Pagan test, i.e., H_1: $\sigma_i^2=f(\gamma+\delta Z)$. However, it does not require the assumption of Normality of the errors to derive the distribution of the test statistic.
To construct the test, we proceed as follows.

- We consider the regression
$$\hat{\varepsilon}_i^2 = \gamma + \delta Z_i + v_i \tag{4.3.4}$$
- We compute the statistic
$$W = T\,R^2, \tag{4.3.5}$$
where R^2 is the coefficient of determination in the model in (4.3.4).
- Under the null hypothesis, and assuming that other hypotheses on the linear model are met (except for Normality of the errors), $W \sim \chi^2(q)$, asymptotically, where q is the dimension of Z.

The intuition is similar to that of the Breusch-Pagan test: under H_0, R^2 should be very low.

4.4 Tests for no correlation

One of the most famous tests in econometrics is the *Durbin-Watson* (DW) test, whose null hypothesis is H_0: ε_t uncorrelated and while under the alternative hypothesis is first-order serial correlation, H_1: $\varepsilon_t = \varrho\varepsilon_{t-1} + u_t$, $\varrho\neq 0$.
The test statistic is based on the OLS residuals:

$$DW = \frac{\sum_{t=2}^{T}(\hat{\varepsilon}_t - \hat{\varepsilon}_{t-1})^2}{\sum_{t=1}^{T}\hat{\varepsilon}_t^2} \approx 2 - 2\hat{\rho} \tag{4.4.1}$$
,

so that $0 \le DW \le 4$, with $DW \approx 2$ under H_0.
Despite its reputation, this test has many problems. First, DW does not have a standard distribution under the null hypothesis and requires the use of appropriate tables to determine the critical values. Second, it cannot be used in dynamic models. Third, the alternative hypothesis is very specific, for example, we might want to test against correlation of a generic order *m*. Therefore, DW is often used as a kind of first, almost informal, model check, with values very different from two signaling possible correlation in the errors.

An alternative test for no correlation that solves the problems of DW is based on the Lagrange Multiplier testing principle (LM, see, e.g., Spanos (1986) for technical details). The *LM test*, also called the *Breusch-Godfrey test,* has the null hypothesis H_0 : ε_t is uncorrelated and the alternative hypothesis H_1: ε_t is correlated up to order *m*, where the order *m* is chosen by the user.
By defining

$$\hat{r}_j = \frac{\sum_{t=j+1}^{T} \hat{\varepsilon}_t \hat{\varepsilon}_{t-j}}{\sum_{t=1}^{T} \hat{\varepsilon}_t^{\,2}}, \tag{4.4.2}$$

the test statistic is:

$$LM = T\left(\sum_{j=1}^{m} \hat{r}_j^2\right). \tag{4.4.3}$$

Under the null hypothesis of no correlation, LM is distributed asymptotically as $\chi^2(m)$, assuming that all the other assumptions on the linear model are met. There are no rules of thumb for choosing *m*, though m=1 is a common choice, or *m*=*q* where *q* is the sampling frequency, e.g., *q* =4= for quarterly data, since there could be some correlation with the same quarter of the previous year.
The reasoning underlying the construction of the LM test is that if the errors are uncorrelated, then the residual sample correlations should also be low, and therefore the values of $\hat{r}_j$ should be close to 0 for j = *1, ..., m*. Conversely, if there exists a positive or negative correlation, the statistic should have very high and positive values. Therefore, the test is one directional (the rejection region is in the right tail of the $\chi^2(m)$).

4.5 The assumptions of Normality of the errors

We saw in Chapter 2 and the previous Section 2 that the assumption of Normality is not required for the OLS or GLS estimators to be B.L.U.E., when the other (weak assumptions) on the model are satisfied. Normality is, however, important in finite samples to ensure that the estimators have a Normal distribution, so that we can apply the theory of hypotheses testing developed in Chapter 3. In addition, non-Normality is often due to abnormal observations (*outliers*) or to changes of regime, which can distort noticeably the

OLS estimators unless suitably modelled, as we shall see in the next chapter. Therefore, it is useful to have a procedure to verify whether or not the assumption of Normality is fulfilled.

One of the most common statistics for the null hypothesis of Normality against the alternative of non-Normality is the *Jarque and Bera test.* The statistic is based on the characteristics of a Normal random variable. In particular, if $e \sim N(0, \sigma^2)$ then its third and fourth moments are:

$$\mu_3 = E(e^3) = 0 \tag{4.5.1}$$

and

$$\mu_4 = E(e^4) = 3\sigma^4 . \tag{4.5.2}$$

These moments can be estimated by their sample counterparts, whose asymptotic distributions are:

$$\sqrt{T}\hat{\mu}_3 = \sqrt{T}\sum \frac{\hat{e}^3}{T} \underset{a}{\sim} N\left(0, 6\sigma^6\right) \tag{4.5.3}$$

and

$$\sqrt{T}\left(\hat{\mu}_4 - 3\hat{\sigma}^4\right) \underset{a}{\sim} N\left(0, 24\sigma^8\right) \tag{4.5.4}$$

Moreover, it can be shown that $\hat{\mu}_3$ and $\left(\hat{\mu}_4 - 3\hat{\sigma}^4\right)$ are independent.

The *Jarque and Bera test* for Normality combines the third and fourth estimated moments into the statistic:

$$JB = T\left(\frac{\hat{\mu}_3^2}{6\hat{\sigma}^6} + \frac{\left(\hat{\mu}_4 - 3\hat{\sigma}^4\right)^2}{24\hat{\sigma}^8}\right) \tag{4.5.5}$$

Under the null hypothesis, and if the other hypotheses on the model are met, the asymptotic distribution of JB is $\chi^2(2)$, being the sum of the squares of two independent standard Normal variables. The test is one-directional, since non-Normality is associated with large and positive values of the statistic.

4.6 The hypothesis of linearity in the parameters

The basic assumption of the linear regression model is that the expected value of the dependent variable is a linear function (in the parameters) of the dependent variables, $E(Y)=X\beta$.

All the results we have seen remain valid if linearity is preserved after a suitable transformation of the dependent variable and/or of the regressors, such as $E(Y)=g(X)\beta$, where $g()$ is a generic function. We will study in detail some of these functions in the next Section.

What happens if instead $E(Y)=f(X;\beta)$, where f is a generic function, whereby the linearity assumption is no longer satisfied? In general, the OLS estimators $\hat{\beta}$ and $\hat{\sigma}^2$, derived under the assumption of a linear model, are biased and inconsistent. There are estimation methods for the parameters of nonlinear models, but they are too complex for an introductory course. Instead, we now discuss how to test whether or not the assumption of linearity is satisfied.

Linearity tests depend on the specific functional form $f(X;\beta)$ assumed under the alternative hypothesis. A procedure that we can use for a generic alternative hypothesis, such as H_1:$E(Y)=f(X;\beta)$, is as follows.

- Let us consider the regression model:

$$Y = X\beta + W\gamma + Z\delta + u \tag{4.6.1}$$

  where W contains the cross products of the elements of X (e.g., X_1*X_2, X_3*X_6, etc.), and Z the products of three elements of X (e.g., $X_1*X_2*X_3$, $X_3*X_4*X_6$, etc.).
- If $E(Y)$ is linear in X, then the regressors in W and Z are not significant. If, instead, it is $E(Y)=f(X;\beta)$, then we can approximate f by means of a Taylor expansion, so that the additional variables in W and Z become significant in (4.6.1).
- Then, the test for the null hypothesis of linearity is an *F-test* for H_0: $\gamma=0$, $\delta=0$, versus H_1: $\gamma\neq 0$, $\delta\neq 0$. The test is one sided and the critical value should be determined using the $F(q, T-p)$ distribution, where q is the dimension of (the number of parameters in) γ and δ, and p that in γ, δ and β.

The problem of this test is that when k, the number of regressors in X, is large, the model in (4.6.1) includes $k + k^2 + k^3$ regressors, which quickly exhausts the degrees of freedom (for example, for $k=5$, $k + k^2 + k^3 = 155$). Alternatively, we can use a two-stage procedure, known as *Ramsey test* or *RESET test*.

In the first stage, we regress the dependent variable y on X, which yields the fitted values $\hat{y}$. In the second stage, we regress y on X and a number of powers of $\hat{y}$, for example:

$$y = X\beta + \gamma_1 \hat{y}^2 + \gamma_2 \hat{y}^3 + v \quad (4.6.2)$$

Under the null hypothesis of linearity, it should be $\gamma_1 = 0, \gamma_2 = 0$, a hypothesis that we can easily test by means of an F-statistic.
Finally, it is useful to note once again that the distribution of a test for the validity of one of the assumptions of the linear model is derived assuming that the other assumptions are satisfied. For example, if the errors are not Normal, then the Ramsey test does not have an F-distribution, at least in finite samples. Often, however, in the event of incorrect specification of the model, more than one hypothesis is violated. For example, the omission of an explanatory variable can determine serial correlation, heteroscedasticity and non-Normality of the errors. In these cases, the distributions of the tests that we have seen are no longer reliable, so that the statistics must be interpreted more as diagnostic checks than as formal statistical tests.
For this reason, model specification should be considered as a sequential procedure, where a given model is progressively improved, by eliminating or correcting the most apparent flaws, until reaching a formulation that passes all or most diagnostic checks. From a statistical point of view, this sequential procedure creates considerable technical problems, such as determining the overall significance level, but empirically it is a strategy that often leads to satisfactory results in terms of the economic interpretability of the resulting model.

4.7 Nonlinear transformations of the variables

In Chapter 2 we studied the effects of linear transformations of the dependent variable and of the regressors on the OLS estimators in the linear regression model. In this Section we assess the consequences of more general transformations on the estimators, on the interpretation of the estimated parameters, and on the functional form of the model.
For example, let us suppose we want to estimate the parameters of a Cobb-Douglas production function where output depends on two inputs (typically, capital and labor), and on shocks to their productivity:

$$Y = \gamma_1 X_1^{\alpha} X_2^{\delta} \varepsilon^* \quad (4.7.1)$$

This is a highly nonlinear (in the parameters) model, so we cannot estimate α and δ by the Ordinary Least Squares method. However, if we take the logarithms of both sides of (4.7.1), we get:

$$\log Y = \log \gamma_1 + \alpha \log X_1 + \delta \log X_2 + \log \varepsilon^* \tag{4.7.2}$$

If $\log \varepsilon^*$ satisfies assumptions 2) e 3), then the parameters α and δ can be estimated by OLS in the transformed model (4.7.2).
More generally, the logarithmic transformation is useful to improve the characteristics of Normality when the original variable has a lognormal or gamma, or χ^2 distribution. Also, it usually reduces problems of heteroscedasticity when the variance increases with the expected value of the variable. Thirdly, if Y and X are expressed as *log*, then the β coefficients are interpretable as elasticity $\left(\frac{\partial \log(y)}{\partial \log(x_i)} = \frac{\partial y}{\partial x_i}\frac{x_i}{y}\right)$, and the first differences of the variables as growth rates:

$$\log X_t - \log X_{t-1} = \log \frac{X_t}{X_{t-1}} = \log\left(1 + \frac{X_t - X_{t-1}}{X_{t-1}}\right) \approx \frac{X_t - X_{t-1}}{X_{t-1}} \tag{4.7.3}$$

Finally, if the relationship between Y and X is similar to that in Figure 4.1, then the relationship between *logY* and *logX* becomes close to linear.

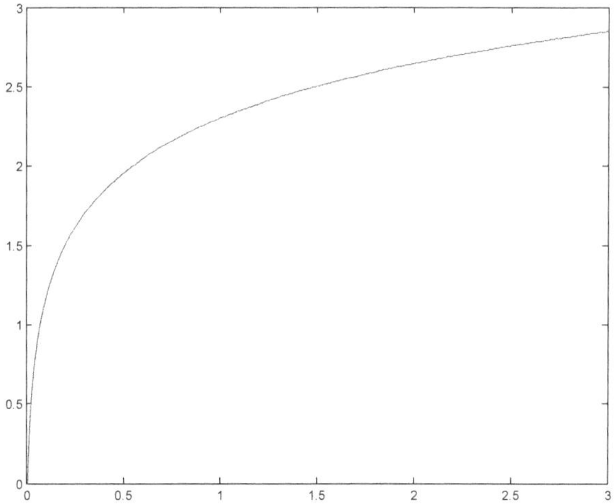

Figure 4.1: An example of nonlinear relationship between Y and X that can be linearized by the log transformation of the variables

Other variable transformations are possible, such as the square root or the reciprocal. In general, the choice of the transformation should lead to a model that satisfies all the assumptions about errors.
The panels of Figure 4.2 illustrate, for the univariate case, some nonlinear transformations that make the relationship between the explanatory and the dependent variables linear.
In the model for the transformed variables, $\tilde{y}$ and $\tilde{x}$, the coefficients still measure how $E(\tilde{y})$ changes in the presence of a marginal change in $\tilde{x}_i$, assuming that the other regressors remain constant. However, the derivation of the relationship between the baseline variables y and x from the model for $\tilde{y}$ and $\tilde{x}$, requires great care, due to the nonlinear transformation, see Table 4.1.

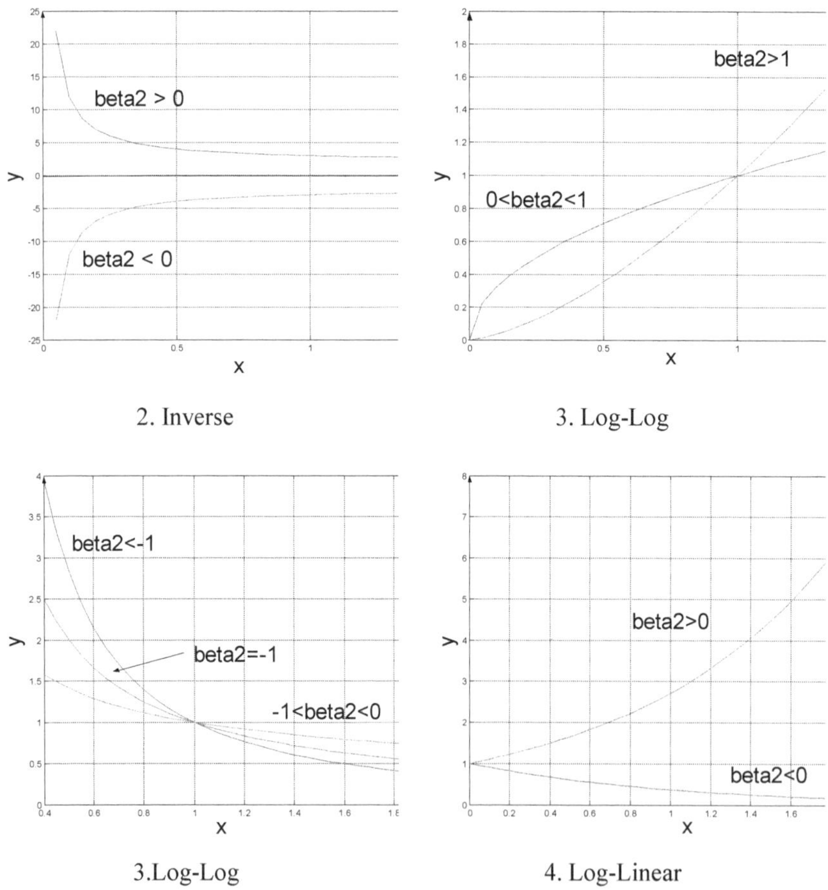

Figure 4.2: Graph of the functions for the transformations 2-6 in Table 4.1 (y_t on the vertical axis)

Type	Model: $f(y_t) = g(x_t, \beta) + e_t$	Slope*	Elasticity **
1. Linear	$y_t = \beta_1 + \beta_2 x_t + e_t$	β_2	$\beta_2 \frac{x_t}{y_t}$
2. Inverse	$y_t = \beta_1 + \beta_2 \frac{1}{x_t} + e_t$	$-\beta_2 \frac{1}{x_t^2}$	$-\beta_2 \frac{1}{x_t y_t}$
3. Log-Log	$\ln(y_t) = \beta_1 + \beta_2 \ln(x_t) + e_t$	$\beta_2 \frac{y_t}{x_t}$	β_2
4. Log-Linear	$\ln(y_t) = \beta_1 + \beta_2 x_t + e_t$	$\beta_2 y_t$	$\beta_2 x_t$
5. Linear-Log	$y_t = \beta_1 + \beta_2 \ln(x_t) + e_t$	$\beta_2 \frac{1}{x_t}$	$\beta_2 \frac{1}{y_t}$
6. Log-Inverse	$n(y_t) = \beta_1 - \beta_2 \frac{1}{x_t} + e_t$	$\beta_2 \frac{y_t}{x_t^2}$	$\beta_2 \frac{1}{x_t}$

**Slope* = $\partial f^{-1}(g(x_t, \beta))/\partial x_t$, *calculated as follows. We consider the expected value of* $f(y_t)$, *then we reverse the function f by isolating the* y_t *on the left, and then we compute the partial derivative with respect to* x_t *of the obtained composite function. Hence, we interpret it as* $\partial y_t / \partial x_t$ *when the error is 0 with probability 1.*

*** Elasticity: slope multiplied by* x_t / y_t. *As for the slope, it is the elasticity of* y_t *with respect to* x_t *when the error is 0 with probability 1.*

Table 4.1: Interpretation of the coefficients in models with transformed variables

4.8 An empirical analysis with simulated data

Using the EViews workfile "example_simul_chap4", we study the relationship between *Y1* and *X*. The graph of the two variables is shown in Figure 4.3.

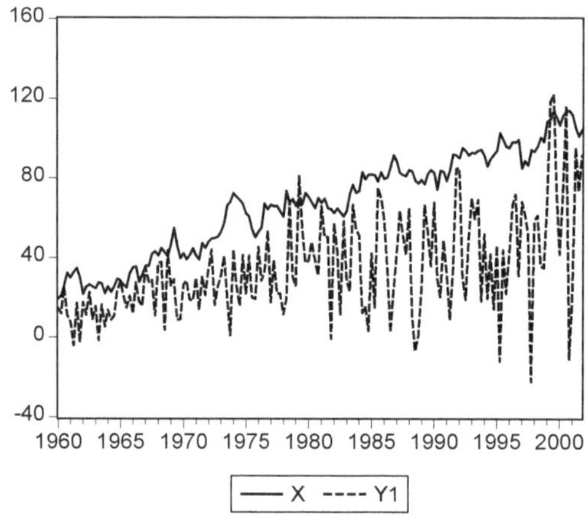

Figure 4.3: Graph of Y1 and X

The trough-peak structure of the variables is similar, indicating a positive correlation. However, there is a marked increase over time in the volatility of *Y1*, while that of *X* is rather stable. This may indicate that the variance of the errors grows with time, a possible case of heteroscedasticity.
If we regress *Y1* on a constant and *X*, we obtain the output in Table 4.2.

Dependent variable: Y1
Method: Least Squares
Sample: 1960:1 2001:4

Variable	Coefficient	Std. Error	*t*-Statistic	Prob.
C	1.303069	4.759673	0.273773	0.7846
X	0.498254	0.066273	7.518181	0.0000

R-squared	0.254010	Mean dependent var	34.75138
Adjusted R-squared	0.249516	S.D. dependent var	25.30699
S.E. of regression	21.92357	Akaike info criterion	9.024835
Sum squared resid	79786.73	Schwarz criterion	9.062025
Log likelihood	-756.0862	*F*-statistic	56.52305
Durbin-Watson stat	1.737912	Prob(*F*-statistic)	0.000000

Table 4.2: Output of the regression of Y1 on X

We get a rather low R^2, and the constant is not statistically significant. The estimated coefficient of X is significant and, it should be noted, in this case the F-statistic in the table is the square of the t-statistic. We interpret the coefficient of X as the marginal variation in the expected value of *Y1* following a marginal variation of X.

The graph of the residuals in Figure 4.4 seems to confirm our hypothesis of heteroscedasticity, as their variability increases (monotonically) with the sample size.

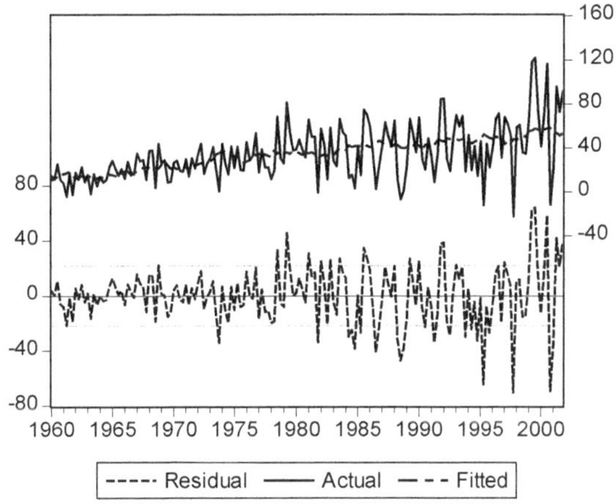

Figure 4.4: Actual and fitted values and residuals from the regression of Y1 on X

We then perform formal tests for homoscedasticity. The null hypothesis is $H_0 : \sigma^2 const$ and we formulate the alternative hypothesis as $H_1 : \sigma_i^2 = f(\gamma + \delta Z)$ with $Z = X, X^2$. White's test requires the following regression:

$$\hat{\varepsilon}_i^2 = \gamma + \alpha x + \beta x^2 + v$$

After running this regression, we compute the statistic $T * R^2 \underset{H_0}{\overset{a}{\sim}} \chi^2$, where there are 2 degrees of freedom since Z contains X and X^2.

The White test can be directly computed with EViews. From the OLS regression window, we click on "Residual Diagnostics" and choose "Heteroscedasticity tests" from the drop down menu, then we select the option "White". The EViews output is shown in Table 4.3.

White Heteroscedasticity Test:

F-statistic	31.39586	Probability	0.000000
Obs*R-squared	46.30989	Probability	0.000000

Test Equation:
Dependent variable: RESID^2
Method: Least Squares
Sample: 1960:1 2001:4
Included observations: 168

Variable	Coefficient	Std. Error	*t*-Statistic	Prob.
C	723.1487	355.3421	2.035077	0.0434
X	-31.02792	11.79412	-2.630795	0.0093
X^2	0.355704	0.089175	3.988840	0.0001

R-squared	0.275654	Mean dependent var	474.9210
Adjusted R-squared	0.266874	S.D. dependent var	856.4732
S.E. of regression	733.3361	Akaike info criterion	16.05078
Sum squared resid	88733992	Schwarz criterion	16.10657
Log likelihood	-1345.266	*F*-statistic	31.39586
Durbin-Watson stat	1.988219	Prob(*F*-statistic)	0.000000

Table 4.3: Output of the White test for homoscedasticity

The lower part of Table 4.3 reports the result of the regression underlying the construction of the White test; the upper part, second line, the White test statistic. If there is no correlation between (X, X^2) and the squared residuals, then T * R^2 should be close to 0. Instead we find a large value, with an associated very low *p*-value, so that we reject homoscedasticity. In the first line of Table 4.3 we find the *F*-test for the joint non-significance of α and β (associate with homoscedasticity). Again, we strongly reject the null hypothesis. Next, we apply the Goldfeld-Quandt Test for $H_0 : \sigma^2{}_i = \sigma^2$ vs. $H_1 : \sigma_i{}^2 = c * z_i$. This test is not programmed in EViews, but we can construct it. We have seen that the variance of $\hat{\varepsilon}$ tends to grow over time, and X also grows over time. We can then set $z_i = x_i$, and keep the observations in their original order. We must then exclude observations in the central part of the sample, and build two subsamples of approximately equal length. We consider the periods 1960:1-1979:4 and 1982:1-2001:4.

Now we regress *Y1* on *c* and *X* in each of the two subsamples, and calculate the residual sum of squares (RSS). To carry out these operations with EViews, we use the command:

```
smpl 1960:1 1979:4
ls y1 c x,
```

which yields RSS=13062.04.
Next we consider the second subsample:

```
smpl 1982:1 2001:4
ls y1 c x
```

and we get RSS= 63595.19.
Next, we compute the GQ statistic as (RSS2/RSS1)~ F(78,78). The computed value is 4.869 and we get the associate *p*-value with the command:

```
Scalar p=@fdist(4.869, 78,78)
```

The *p*-value is 0.0000, so that we again reject the null hypothesis of homoscedasticity.
Note that in general, to compute GQ, we should multiply (RSS2/RSS1) by $(T_1-K)/(T_2-K)$, while in our example it is $(T_1-K)/(T_2-K)=1$.
To address the problem of heteroscedasticity, we have three possibilities.
First, we can use OLS estimation combined with HAC standard errors. In EViews, we proceed as follows. First, we use the entire sample, with the command:

```
smpl @all
```

Next, we click on "Object", "New object", "Equation" and specify the model:

```
y1 c x
```

From the "Options" menu, we choose "Coefficients Covariance Matrix" and select "White". We get the output reported in Table 4.4.
Compared to the values in Table 4.2, the estimated parameters do not change, but the standard errors do, since they are now correctly estimated.
As a second possibility to handle homoscedasticity, we can use GLS estimation. If $y = \alpha + x\beta + \varepsilon$ and $\sigma_\varepsilon^{\ 2} = d \cdot x^2$, then:

$$\frac{y}{x} = \alpha\frac{1}{x} + \frac{x}{x}\beta + \frac{\varepsilon}{x}, \qquad u = \frac{\varepsilon}{x} \quad , \quad \sigma_u^{\ 2} = \frac{1}{x^2}\sigma_\varepsilon^{\ 2} = d\,.$$

Hence, the transformed model to be estimated by OLS in this example is:

$$\frac{y}{x} = \frac{\alpha}{x} + \beta + u\,.$$

We can do this easily in EViews, using the command

```
ls y1/x c 1/x
```

where now the constant refers to the coefficient of X in the original model, while the estimated coefficient of $1/X$ refers to the constant in the original model.

Dependent variable: Y1
Method: Least Squares
Sample: 1960:1 2001:4
Included observations: 168
Newey-West HAC Standard Errors & Covariance (lag truncation=4)

Variable	Coefficient	Std. Error	*t*-Statistic	Prob.
C	1.303069	4.213873	0.309233	0.7575
X	0.498254	0.078504	6.346889	0.0000

R-squared	0.254010	Mean dependent var	34.75138
Adjusted R-squared	0.249516	S.D. dependent var	25.30699
S.E. of regression	21.92357	Akaike info criterion	9.024835
Sum squared resid	79786.73	Schwarz criterion	9.062025
Log likelihood	-756.0862	*F*-statistic	56.52305
Durbin-Watson stat	1.737912	Prob(*F*-statistic)	0.000000

Table 4.4: OLS regression of Y1 on X with HAC standard errors

The output in Table 4.5 includes the GLS estimators of the parameters of the initial model. Compared to the Tables 4.2 and 4.4, the intercept is much more precisely estimated, its standard error is reduced, while the same does not hold for the slope parameter.

Dependent variable: Y1/X
Method: Least Squares
Sample: 1960:1 2001:4
Included observations: 168

Variable	Coefficient	Std. Error	*t*-Statistic	Prob.
C	0.489317	0.047861	10.22361	0.0000
1/X	1.899318	2.316769	0.819813	0.4135

R-squared	0.004032	Mean dependent var	0.523990
Adjusted R-squared	-0.001967	S.D. dependent var	0.290110
S.E. of regression	0.290396	Akaike info criterion	0.376689
Sum squared resid	13.99873	Schwarz criterion	0.413879
Log likelihood	-29.64187	*F*-statistic	0.672094
Durbin-Watson stat	1.879850	Prob(*F*-statistic)	0.413498

Table 4.5: Transformed model to eliminate heteroscedasticity

The residual graph in Figure 4.5 suggests that the errors in the transformed model are actually homoscedastic.

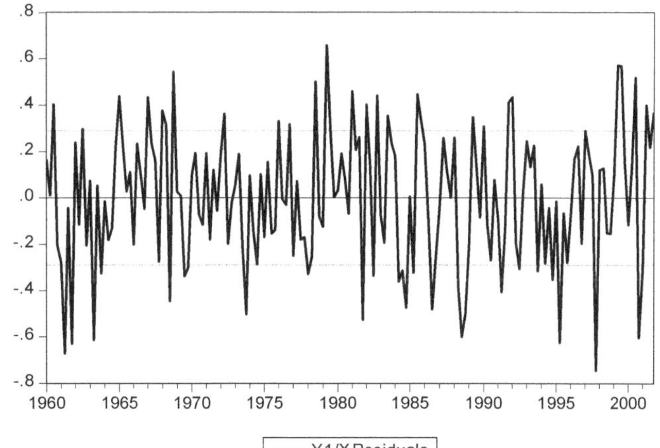

Figure 4.5: Residuals of the transformed model in Table 4.5

In order to more formally evaluate the hypothesis of homoscedasticity, we can calculate the White test in the transformed model. The test statistic has a value of 0.841976, with a *p*-value of 0.656, which leads us to not reject the null hypothesis of homoscedasticity.

For this example, we can also directly get the GLS estimator in EViews. We create a weight series with the command:

```
genr w=1/x
```

so that $w_t=1/x_t$. Next, we click on “Object”, “New object”, “Equation” and, after specifying the model

```
y1 c x
```

in the “Method” box we select “LS-Least Squares (NLS and ARMA)” and in the “Options” box we select “Weights”. Hence, in the “type” box we select “Inverse std. dev.” and in the “weight series” box we insert “w”. In practice, both the dependent variable and the regressors are multiplied by the weights w before applying OLS. The resulting estimated parameters and their standard errors, presented in Table 4.6, are the same as those in Table 4.5 (but reversed). The residuals computed by EViews are, however, the same as the OLS regression residuals, so that we do not know if we really solved the problem of heteroscedasticity – whereas when we manually construct the transformed regression, we can check (as we have done) if the residuals are actually homoscedastic.

As a third solution to the problem of heteroscedasticity if GLS does not work, in the sense that the residuals remain heteroscedastic in the transformed model, we have to respecify the model, for example by adding other variables whose omission might cause the heteroscedasticity.

The transformed model might, however, violate other OLS assumptions, such as that of no correlation of the errors. In the EViews output, the Durbin–Watson test for H_0: *ε, uncorrelated* versus $H_1: \rho \neq 0$ in $\varepsilon_t = \rho\varepsilon_{t-1} + u$ is readily available. In the model where we regressed $Y1/X$ on a constant and $1/X$, the value of the DW statistic is 1.879850. Since the value is very close to 2, we do not reject H_0.

Dependent variable: Y1
Method: Least Squares
Sample: 1960q1 2001q4
Included observations: 168
Weighting series: W
Weight type: Inverse standard deviation (EViews default scaling)

Variable	Coefficient	Std. Error	*t*-Statistic	Prob.
C	1.899318	2.316769	0.819813	0.4135
X	0.489317	0.047861	10.22361	0.0000

Weighted Statistics

R-squared	0.386372	Mean dependent var	28.70318
Adjusted R-squared	0.382676	S.D. dependent var	15.89171
S.E. of regression	15.90733	Akaike info criterion	8.383271
Sum squared resid	42005.17	Schwarz criterion	8.420461
Log likelihood	-702.1947	Hannan-Quinn criterion	8.398364
F-statistic	104.5223	Durbin-Watson stat	1.879850
Prob(*F*-statistic)	0.000000	Weighted mean dep.	22.82968

Unweighted Statistics

R-squared	0.253928	Mean dependent var	34.75138
Adjusted R-squared	0.249434	S.D. dependent var	25.30699
S.E. of regression	21.92477	Sum squared resid	79795.48
Durbin-Watson stat	1.736976		

Table 4.6: Weighted regression of Y1 on X

Alternatively, we can apply the LM test of Breusch-Godfrey for H_0: ε_t *uncorrelated* against H_1: ε_t *correlated of order m.*

With EViews, taking as an example the regression of Y1 on c and X, we can click on "View", "Residual Diagnostics", "Serial Correlation LM test ...", selecting the proper number of lags (*m*). With m=2, we obtain the values in Table 4.7, so that we do not reject the null hypothesis at the 5% level, while we reject it at the 10% level.

Breusch-Godfrey Serial Correlation LM Test:

F-statistic	2.513008	Probability	0.084140
Obs*R-squared	4.995507	Probability	0.082270

Table 4.7: Test of no correlation for the model in Table 4.2

If we repeat the same test in the transformed model in Table 4.5, the p-value becomes 0.77, a much higher value than that reported in Table 4.7, 0.08. This large difference is due to the fact that in the regression of $Y1$ on X, the presence of heteroscedasticity can distort the outcome of the test for no correlation (whose distribution is derived assuming that all the strong OLS assumptions hold).
Another violation of the OLS assumptions could be non-Normality of the errors. As an example, let us apply the Jarque-Bera test for Normality in the model where $Y1$ is regressed on c and X. This test is already programmed in EViews, and is accessible from the menu "View", "Residual Diagnostics", "Histogram-Normality test", after having estimated the model.

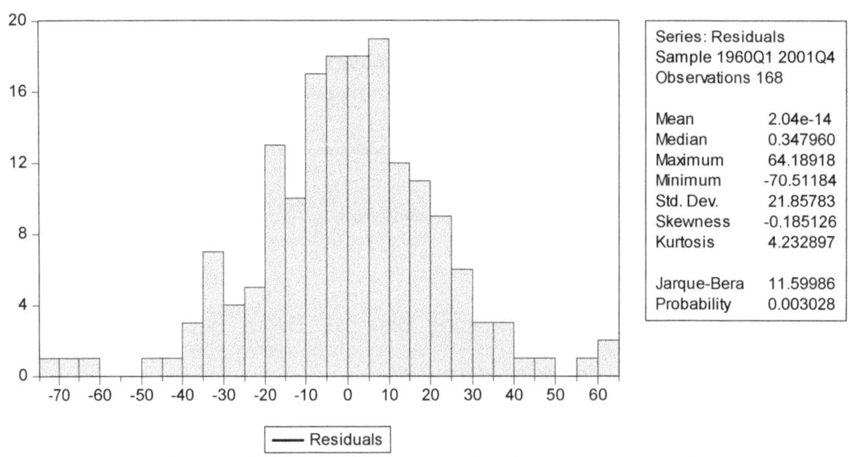

Table 4.8: Normality test in the original model with heteroscedastic errors

From Table 4.8, we can see the sample distribution of the residuals and the result of the Jarque-Bera test for H_0: $e \sim N(0, \sigma^2)$ against H_1: the errors are non Normal. Since the p-value is about 0.003, we reject the null hypothesis even at the 1% level. This is again a side effect of heteroscedasticity.

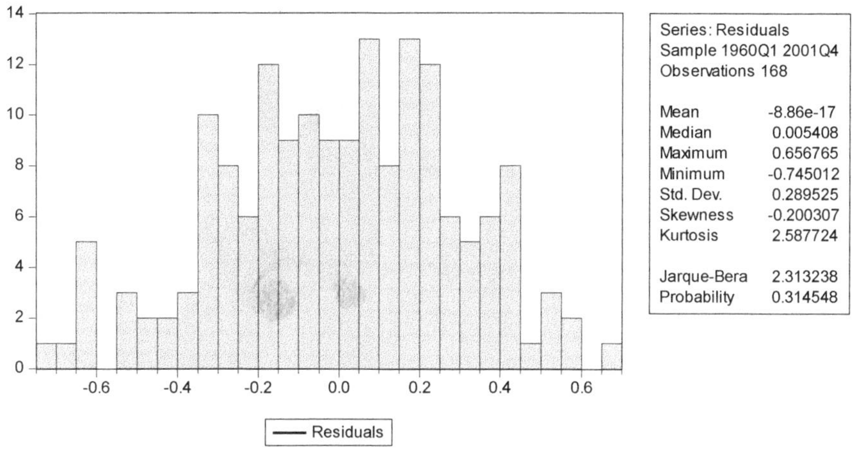

Table 4.9: Normality test in the transformed model

If we now run the Normality test in the transformed model of Table 4.5, the null hypothesis of Normality is not rejected at a 10% significance level, see Table 4.9.
We have seen that the Ramsey (RESET) test for linearity (in the parameters) adds powers of the fitted values ($\hat{y}1$ in our example) to the set of regressors, which should not be significant under the null hypothesis. In EViews, this test is implemented under the model "Stability" diagnostics, in the "View" menu of the "Equation" object. For example, for the model where *Y1* is regressed on *c* and *X*, and using $\hat{y}^2$ and $\hat{y}^3$ as additional regressors, we get a *p*-value of 0.15 for the *F* test for the significance of $\hat{y}^2$ and $\hat{y}^3$, whereby the linearity hypothesis is not rejected, see Table 4.10.

Ramsey RESET Test
Equation: EQ01
Specification: Y1 X C

	Value	Df	Probability
F-statistic	1.898410	(2, 164)	0.1531
Likelihood ratio	3.845086	2	0.1462

Table 4.10: Linearity test for the original model with heteroscedastic errors

We now try to model the *Y2* variable using *X* as explanatory variable, using the same EViews workfile.

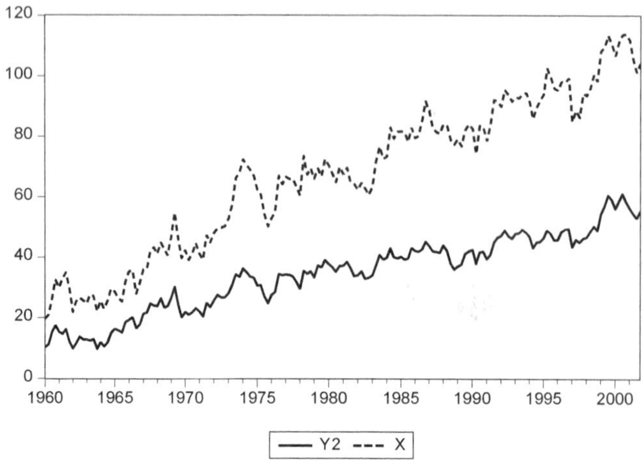

Figure 4.6: Graph of the Y2 and X variables

The evolution of the variables in Figure 4.6 seems to indicate a good explanatory capacity of *X* for *Y2*. This is confirmed by the high value of the determination coefficient in Table 4.11, and by the marked significance of the coefficients.

Dependent variable: Y2
Method: Least Squares
Sample: 1960:1 2001:4
Included observations: 168

Variable	Coefficient	Std. Error	*t*-Statistic	Prob.
C	1.087103	0.341813	3.180403	0.0018
X	0.498756	0.004759	104.7946	0.0000

R-squared	0.985109	Mean dependent var	34.56911
Adjusted R-squared	0.985020	S.D. dependent var	12.86355
S.E. of regression	1.574427	Akaike info criterion	3.757494
Sum squared resid	411.4843	Schwarz criterion	3.794684
Log likelihood	-313.6295	*F*-statistic	10981.90
Durbin-Watson stat	0.418693	Prob(*F*-statistic)	0.000000

Table 4.11: Output of the regression of Y2 on X

However, the Durbin-Watson statistic is very far from 2, which indicates rejection of the null hypothesis of non-correlated errors. In addition, we can notice a persistent behavior of the residuals, with negative values followed by several other negative values, and the same for positive values, see Figure 4.7.

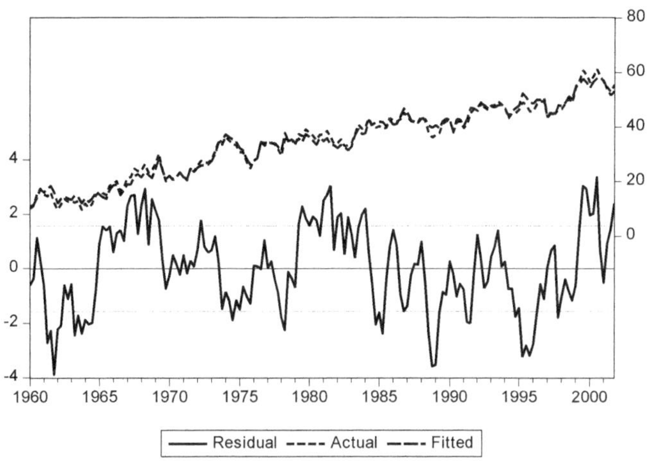

Figure 4.7: Actual and fitted values and residuals for the model in Table 4.11

Can this be due to error heteroscedasticity as in the previous example? From Figure 4.7, the variability of the dependent variable and of the residuals seems to be constant. The White test does not reject the null hypothesis of homoscedasticity at the 5% level, the value of the statistic is 5.8706, with a *p*-value of 0.0531. The *p*-value of the Goldfeld-Quandt test is even higher, 0.41788. Using the same sample breakdown as in the previous example, the test statistic is 174.4307/198.4508 = 0.87896, while

```
@fdist(0.87896,78,78)
```

gives a *p*-value of 0.41788 (the value appears in the lower left corner of the screen).

To gather more information about the error correlation, we run the Lagrange Multiplier test, for example with m=2 lags. It results in a *p*-value of 0.000, which leads us to strongly reject the hypothesis of no correlation.

Does error correlation lead us to reject (by mistake) also Normality and linearity? The output of the Jarque-Bera test gives a *p*-value of 0.276, while the Ramsey test with 2 terms returns a *p*-value of 0.000. Hence, we do not reject Normality but we strongly reject linearity.

Given that the main specification problem we have detected is error serial correlation, we try to address it by assuming that the errors follow an autoregressive process:

$$y_t = c + x_t\beta + \varepsilon_t \text{, with } \varepsilon_t = \rho\varepsilon_{t-1} + u_t,\; u_t \sim iid\left(0, \sigma_u^2\right)$$

As we shall see in Chapter 7, this model can be rewritten as:

$$y_t - \rho y_{t-1} = \widetilde{c} + x_t \beta - \rho \beta x_{t-1} + \underbrace{\varepsilon_t - \rho \varepsilon_{t-1}}_{u_t} , \ |\rho| < 1$$

Therefore, we extend the original model as follows:

$$y2_t = c + x_t \beta_1 + x_{t-1} \beta_2 + y_{t-1} \beta_3 + u_t$$

This model can now be estimated by OLS, since the errors u should be iid. The results are reported in Table 4.12.

Dependent variable: Y2

Variable	Coefficient	Std. Error	*t*-Statistic	Prob.
C	0.260970	0.219770	1.187466	0.2368
X	0.477493	0.017649	27.05571	0.0000
X(-1)	-0.378312	0.028342	-13.34818	0.0000
Y2(-1)	0.800594	0.048487	16.51163	0.0000

R-squared	0.994316	Mean dependent var	34.71447
Adjusted R-squared	0.994211	S.D. dependent var	12.76309
S.E. of regression	0.971083	Akaike info criterion	2.802851
Sum squared resid	153.7094	Schwarz criterion	2.877534
Log likelihood	-230.0381	*F*-statistic	9504.084
Durbin-Watson stat	1.849122	Prob(*F*-statistic)	0.000000

Table 4.12: Transformed model to eliminate error serial correlation

Note that we lose one sample observation because the regressors include the lagged dependent variable. We get significant parameter values, apart from the constant, both individually and jointly. The fit has improved and the DW statistic suggests that the correlation of the errors has been removed (but this test is not reliable in this dynamic model). The more appropriate LM test has a value of 1.34 with a *p*-value of 0.51, confirming the non-rejection of the hypothesis of no correlation.

Finally, you can verify that, in the transformed model, the null hypothesis of linearity is also not rejected.

4.9 An empirical analysis of the determinants of aggregate consumption

We now want to check if the assumptions underlying the linear regression model are satisfied in the regression of consumption on disposable income and wealth, using the data in the EViews workfile "example_cons_chap4.wf1".
By clicking on the "View" menu of the "Equation" object and choosing "Residual Tests", we have access to several diagnostic tests, some of which are reported in Table 4.13.

Heteroscedasticity Test: Breusch-Pagan-Godfrey

F-statistic	4.306076	Prob. F(3,85)	0.0071
Obs*R-squared	11.74166	Prob. Chi-Square(3)	0.0083
Scaled explained SS	18.30527	Prob. Chi-Square(3)	0.0004

Heteroscedasticity Test: White (with cross terms)

F-statistic	10.64282	Prob. F(8,80)	0.000000
Obs*R-squared	45.88574	Prob.Chi-Square(8)	0.000000
Scaled explained SS	71.53599	Prob.Chi-Square(8)	0.000000

Heteroscedasticity Test: White (without cross terms)

F-statistic	4.300474	Prob. F(3,85)	0.007100
Obs*R-squared	11.72840	Prob.Chi-Square(3)	0.008400
Scaled explained SS	18.28460	Prob.Chi-Square(3)	0.000400

Ramsey RESET Test

	Value	Df	Probability
t-statistic	3.269140	84.00000	0.001600
F-statistic	10.68728	(1, 84)	0.001600
Likelihood ratio	10.65889	1.000000	0.001100

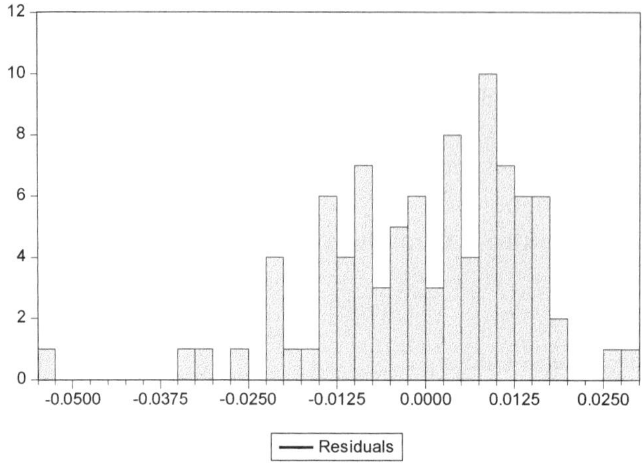

Series: Residuals	
Sample 1990Q1 2012Q1	
Observations 89	
Mean	-2.20e-15
Median	0.002563
Maximum	0.028593
Minimum	-0.053675
Std. Dev.	0.013889
Skewness	-0.874825
Kurtosis	4.418369
Jarque-Bera	18.81256
Probability	0.000082

Table 4.13: Diagnostic tests on the consumption model

The results are rather disappointing, with most null hypotheses rejected. Therefore, the empirical consumption model that we have used in Chapters 2 and 3 does not satisfy the assumptions underlying the linear regression model. This can be due to various factors. First, the sample period considered includes various events that could cause parameter instability, such as the devaluation of the Lira in 1992 as a result of Italy's exit from the EMS, the introduction of the Euro in 2002, and the global financial crisis. We will deal with these issues in Chapter 5. In addition, we will see in Chapter 7 that, in order to get a correctly specified model for consumption, we need to adopt a dynamic specification.

Similar considerations apply to the empirical models for investment and labor productivity, described in Chapters 2 and 3. You can apply the diagnostic tests to verify that most null hypotheses are rejected. To obtain good specifications, we need again to carefully define the dynamic component and to take care of possible parameter changes.

Finally, assuming that the estimated consumption model is correctly specified, let us compute the estimated marginal propensity to consume. Since the model is specified with variables in logarithms, the LRYD parameter corresponds to the consumption elasticity to income. Hence, to obtain the propensity to consume, we use the formula:

```
series Propmarg = 1.066816 * (rc/ryd)
```

The marginal propensity varies over time, and its average value is equal to about 0.84, in line with economic theory that predicts a value between 0 and 1.

The main concepts of this chapter

In this chapter we have seen that generalizing the hypothesis $V(\varepsilon) = \sigma^2 I_T$ by $V(\varepsilon) = \Omega$, to allow for heteroscedastic and correlated errors, the OLS estimator remains unbiased and consistent for the parameters of a static linear model, but it is no longer B.L.U.E.. In addition, the formulae of Chapter 3 to calculate interval estimators and statistics for hypothesis testing are no longer valid.
Since for every variance-covariance matrix Ω there exists a square invertible matrix H such that $H\Omega H' = I$, the errors of the model $HY = HX\beta + H\varepsilon$ have variance-covariance matrix equal to I_T. Hence, we can estimate the parameters of the transformed model by OLS. The resulting estimator is called the *Generalized Least Squares estimator* (GLS), $\hat{\beta}_{GLS} = (X'\Omega^{-1}X)^{-1}X'\Omega^{-1}Y$.
$\hat{\beta}_{GLS}$ is unbiased, $\text{var}(\hat{\beta}_{GLS}) = (X'\Omega^{-1}X)^{-1}$, $\hat{\beta}_{GLS}$ is consistent and if the errors are normally distributed then $\hat{\beta}_{GLS} \sim N(\beta, (X'\Omega^{-1}X)^{-1})$.
In practice, Ω is not known and has to be replaced by an estimator, $\hat{\Omega}$. The resulting parameter estimator is called *the Feasible Generalized Least Squares estimator*, (FGLS, feasible GLS), $\hat{\beta}_{FGLS} = (X'\hat{\Omega}^{-1}X)^{-1}X'\hat{\Omega}^{-1}Y$. Since, in general, $\hat{\Omega}$ and ε are correlated, the FGLS estimator is biased in finite samples, it is no longer a linear estimator (since Y is also used in the construction of $\hat{\Omega}$), and it has no longer necessarily the lowest variance. However, in general, $\hat{\beta}_{FGLS}$ remains consistent, $\hat{\beta}_{FGLS} \to \beta$, and the properties of $\hat{\beta}_{GLS}$ are also asymptotically valid for $\hat{\beta}_{FGLS}$, when $\hat{\Omega}$ is consistent.
To test for homoscedasticity, we saw the following tests:

Goldfeld-Quandt Test	H_0: $\sigma_i^2=\sigma^2$ H_1: $\sigma_i^2=cX_i^2$, $c>0$,	$GQ = \dfrac{RSS_2}{RSS_1} \underset{H_0,\text{ altre OLS}}{\sim}$	$F(p,p)$
Breusch-Pagan Test	H_0: $\sigma_i^2=\sigma^2$ H_1: $\sigma_i^2=f(\gamma+\delta Z)$, Z is $T\times q$	$\text{BP} = \dfrac{\sum_{i=1}^{T}\left(\hat{\gamma}+\hat{\delta}Z_i\right)^2}{2} \underset{H_0,altreOLS}{\sim}$	$\chi^2(q)$
White Test	H_0: $\sigma_i^2=\sigma^2$ H_1: $\sigma_i^2=f(\gamma+\delta Z)$, Z is $T\times q$	$\text{W} = \text{T}\cdot\text{R}^2 \underset{H_0,altreOLS}{\sim}$	$\chi^2(q)$

For the hypothesis of no correlation of the errors, we saw the following tests:

Durbin-Watson Test	H_0: ε_t uncorrelated H_1: $\varepsilon_t = \varrho\varepsilon_{t-1} + u_t$, $\varrho \neq 0$	$DW = \frac{\sum_{t=2}^{T}(\hat{\varepsilon}_t - \hat{\varepsilon}_{t-1})^2}{\sum_{t=1}^{T}\hat{\varepsilon}_t^2} \approx 2 - 2\hat{\rho}$
LM Test	H_0: ε_t uncorrelated H_1: ε_t correlated of order m	$LM = T\left(\sum_{j=1}^{m}\hat{r}_j^2\right) \overset{a}{\sim} \chi^2(m)$ H_0, othe OLS

For the hypothesis of Normal errors:

Jarque-Bera Test	H_0: ε Normal H_1: ε non-Normal	$JB = T\left(\frac{\hat{\mu}_3^2}{6\hat{\sigma}^6} + \frac{(\hat{\mu}_4 - 3\hat{\sigma}^4)^2}{24\hat{\sigma}^4}\right) \overset{a}{\sim} \chi^2(2)$ H_0, othe OLS

Finally, if $E(Y) = f(X;\beta)$, where f is a generic function, the OLS estimators $\hat{\beta}$ and $\hat{\sigma}^2$, derived under the assumption of a linear model (in the parameters), are biased and inconsistent. To test for linearity, we can use:

Ramsey RESET Tests	or	$Y = X\beta + W\gamma + Z\delta + u$ H_0: $\gamma=0$, $\delta=0$ H_1: $\gamma\neq 0$, $\delta\neq 0$	*F-test*, $F(q, T-p)$, where q is the number of parameters in γ and δ and p that in γ, δ e β .

The linearity of the model can be improved by appropriate transformations of the variables, which however can complicate the interpretation of the parameters. The logarithmic transformation is common for economic data. When applied to both the dependent variable and the regressors, the model parameters are interpretable as elasticities.

Exercises

1)
Using the workfile "example_inv_chap4", estimate the model for aggregate investments of Section 3.13 and calculate:
- the Breusch Godfrey test for no correlation,
- the White test for homoscedasticity,
- the Jarque-Bera test for Normality,
- the Ramsey test for linearity.
Then estimate the same models with correlated errors (of order 1 or 2) and evaluate if the answers of the specification tests change.

2)
Consider the model

$$Y_t = X\beta + u_t$$

$$VAR(u_t) = \sigma_1^2 \qquad \text{for t=1....T}_1$$

$$VAR(u_t) = \sigma_2^2 \qquad \text{for t=T}_2\text{....T}$$

with known σ_1^2 and σ_2^2.
- How would you transform the model in order to have a constant error variance?
- What is the variance of the OLS estimator in the original model?
- What is the variance of the OLS estimator in the transformed model?

3)
Given the model:

$$Y_t = X_t\beta + u_t,$$

show that the variance of Y can be decomposed into the sum of the variances of $\hat{y}_t$ and $\hat{u}_t$ when the parameters are estimated by OLS. Is it possible to run a similar decomposition when the estimator is GLS?

4)
Consider the model

$$Y_t = X_t\beta + \varepsilon_t$$

a) Define the null and alternative hypotheses of the Breusch-Pagan test, the test statistic, and its distribution under the null hypothesis.
b) If you have estimated the model by OLS but the Breusch-Pagan test rejects the null hypothesis, how would you construct a robust test for the hypothesis β = 0?

5)
Consider the model

$$y_t = \alpha + \beta x_t + e_t, \qquad e_t \sim \text{iid}(0, \sigma^2)$$

a) Define a test for Normality of e_t.
b) On the basis of the following dataset, compute the Normality test and and its *p*-value.

obs	Y	X_1
1	2	1
2	9	3
3	10	3
4	15	4

6)
Compute the variance-covariance matrix of the OLS and GLS estimators when the regressors are deterministic, $E(\varepsilon)=0$ but $\text{Var}(\varepsilon)=\Sigma$, where Σ is a positive definite matrix.

7)
Show that under the GLS assumption the OLS estimator is unbiased but inefficient.

8)
Using the EViews workfile "example_simul2_chap4", regress Y on a constant, X and Z. Then, using the statistics implemented in EViews, test the null hypotheses of homoscedasticity, no correlation and Normality of the errors. Given the results you have obtained (and in particular the outcome of the White test), how would you transform the model? Do the residuals of the transformed model satisfy the OLS assumptions?

5. Parameter Instability in the Linear Regression Model

5.1 Structural breaks and tests for parameter stability

One of the main problems for the empirical analysis of economic data is the presence of institutional or policy changes, which could lead to changes in the relationships between economic variables, and therefore in the parameters of econometric models. Infrequent events such as wars, terrorist attacks, jumps in commodity prices or stock market collapses might have similar, although temporary, effects.

There could be parameter changes also in cross-sectional models for very heterogeneous units, such as families or firms in different geographical areas or economic conditions.

In this chapter, we study the effects of parameter changes on the OLS estimators, how to incorporate changes in the model, and how to test for parameter stability.

Let us suppose the dependent variable y is explained by k regressors grouped in X, and the event that potentially changes the parameters of the linear model takes place in period T_1. We can write the econometric model as:

$$Y_t = X_t \beta_1 + \varepsilon_{1t} \quad t = 1,\ldots,T_1 \qquad o \quad t \in T_1 \tag{5.1.1}$$

$$Y_t = X_t \beta_2 + \varepsilon_{2t} \quad t = T_1 + 1,\ldots,T \qquad o \quad t \in T_2 \tag{5.1.2}$$

$$\begin{pmatrix} \varepsilon_{1t} \\ \varepsilon_{2t} \end{pmatrix} \sim N\left(\begin{pmatrix} 0 \\ 0 \end{pmatrix}, \begin{pmatrix} \sigma_1^2 I_{T_1} & 0 \\ 0 & \sigma_2^2 I_{T_2} \end{pmatrix} \right) \tag{5.1.3}$$

If we assume by mistake that the parameters are stable, namely, the model is:

$$Y_t = X_t\beta + \varepsilon_t \quad t \in T \tag{5.1.4}$$

$$\varepsilon_t \sim N(0, \sigma^2 I_T) \tag{5.1.5}$$

then $\hat{\beta}$ and $\hat{\sigma}^2$, the OLS estimators, are biased and inconsistent. Therefore, it is important to test for parameter stability.
The hypotheses of interest relate to the stability of the parameters of the regression function and/or of the error variance:

$$H_0': \beta_1 = \beta_2, \quad H_0'': \sigma_1^2 = \sigma_2^2 \tag{5.1.6}$$

$$H_1': \beta_1 \neq \beta_2, \quad H_1'': \sigma_1^2 \neq \sigma_2^2 \tag{5.1.7}$$

To build a test for the hypotheses of interest, we distinguish two cases based on whether the number of observations in the second subperiod, T_2, is larger or smaller than the number of regressors, *k*.

T >₂k

In this case we can estimate both the model in (5.1.4) over the whole sample, and the models in (5.1.1) and (5.1.2) on the subsamples T_1 and T_2. We get three sets of residuals; the associated sums of squared residuals (RSS) are indicated by, respectively, RSS_T, RSS_{T_1} and RSS_{T_2}. The *Chow test* for $H_0': \beta_1 = \beta_2 \mid H_0''$ is:

$$CH_1 : \left(\frac{RSS_T - RSS_{T_1} - RSS_{T_2}}{RSS_{T_1} + RSS_{T_2}}\right)\frac{T-2k}{k} \underset{H_0}{\sim} F(k, T-2k) \tag{5.1.8}$$

,

with $T=T_1+T_2$. The test is one sided and the rejection region is in the right tail of the distribution. This test is called "Chow Breakpoint" in EViews.
The intuition underlying the Chow test is that if the parameters β are stable, then the residual sum of squares resulting from the model estimated over the whole sample, RSS_T, should be similar to the sum of the RSS resulting from the models estimated over the two subsamples, $RSS_{T_1} + RSS_{T_2}$. Otherwise, RSS_T should be larger than $RSS_{T_1} + RSS_{T_2}$.

The Chow test in (5.1.8) is expressed in terms of RSS. In Chapter 3 we saw that the F-test for the significance of (all or subsets of) the coefficients of the linear model can be expressed in terms of RSS. We will see in Section 5.3 that we can rewrite the models in (5.1.1) and (5.1.2) into a single equation, and that an F-test for the significance of selected regressors in this equation is equivalent to the CH_1 test.

Note that the null hypothesis of CH_1 is slightly different from what we would like to test, i.e. $H_0': \beta_1 = \beta_2$. In particular, with CH_1 we test whether $\beta_1 = \beta_2$ under the assumption that the error variance is unchanged. This is necessary because if $\beta_1 = \beta_2$ but the variance changes in the two subsamples, then RSS_T should be greater than $RSS_{T_1} + RSS_{T_2}$ and the test should incorrectly reject the null hypothesis $\beta_1 = \beta_2$.

Consequently, it is also important to verify that the variance remains constant in the two subsamples. Assuming that it is:

$$H_0'': \sigma_1^2 = \sigma_2^2, \text{ versus } H_1'': \sigma_1^2 < \sigma_2^2, \tag{5.1.9}$$

the test is:

$$CH_2 : \frac{\hat{\sigma}_2^2}{\hat{\sigma}_1^2} = \frac{RSS_{T_2}}{RSS_{T_1}} \frac{T_1 - k}{T_2 - k} \underset{H_0}{\sim} F(T_2 - k, T_1 - k) \tag{5.1.10}$$

The CH_1 and CH_2 statistics are independent, so we can first apply CH_2 to check the stability of the variance, and then CH_1 for stability of the parameters of the regression function. For more details, see Judge *et al.* (1988, p. 363).

Note that the assumption of constant error variance is the same as that of homoscedasticity, analyzed in Chapter 4. It follows that the tests we have considered in that context can be applied here too.

Furthermore, if variance stability is rejected, we can try to transform the model so that the new errors have constant variance, similarly to what we did to derive the GLS estimators. For example, in the case of the model in (5.1.1) and (5.1.2), we can simply divide the variables by σ_1 in the first subsample and by σ_2 in the second subsample (in practice, we would use the OLS estimators of σ_1 and σ_2). The CH_1 test can be then applied in the transformed model, whose errors have constant variance.

$T_2<k$

The number of observations in the second period is now lower than that of regressors, so that it is not possible to estimate the model in (5.1.2). Hence, to verify $H_0': \beta_1 = \beta_2 \mid H_0$" we use a different test:

$$CH_3 : \left(\underset{T_2x1}{Y_2 - X_2\hat{\beta}_1} \right)' \frac{\left[I + X_2(X_1'X_1)^{-1} X_2' \right]^{-1}}{\underset{T_1x1}{\hat{u}_1'\hat{u}_1}} \left(Y_2 - X_2\hat{\beta}_1 \right) \frac{T_1 - k}{T_2} \underset{H_0}{\sim} F(T_2, T_1 - k) \qquad (5.1.11)$$

The test is one sided and the rejection region is in the right tail of the distribution. This test is called "Chow Forecast" in EViews.

As we will see later, in Section 5.5, $Y_2 - X_2\hat{\beta}_1$ is the forecast error when predicting Y_2 as $X_2\hat{\beta}_1$, the best predictor when X_2 is known and the model parameters are unknown. The term $\left(\left[I + X_2(X'_1 X_1)^{-1} X'_2 \right]^{-1} / \hat{u}'_1 \hat{u}_1 \right)(T_1 - k)$ is an estimator of the variance-covariance matrix of the forecast error. If the null hypothesis is satisfied, the forecast errors should be small and the CH_3 statistic should take low values. The opposite holds if the parameters change in T_2.

The CH_3 test can be also rewritten in terms of sum of squares of the residuals as:

$$CH_3 = \left(\frac{RSS_T - RSS_{T_1}}{RSS_{T_1}} \right) \frac{T_1 - k}{T_2} \underset{H_0}{\sim} F(T_2, T_1 - k) \qquad (5.1.12)$$

The numerator of the statistic is an estimate of RSS_2. Hence, the numerator and denominator should be very similar if the parameters of the regression function in the first subsample are equal to those in the second subsample, assuming again that the variance remains constant.

As discussed above, it is important to assess whether the error variance is stable. In this case, the test for $H_0'': \sigma_1^2 = \sigma_2^2$ versus $H_1'': \sigma_1^2 < \sigma_2^2$ is:

$$CH_4 : \frac{\left(Y_2 - X_2\hat{\beta}_1\right)'\left(Y_2 - X_2\hat{\beta}_1\right)}{\hat{\sigma}_1^2} \overset{a}{\sim} \chi^2(T_2) \qquad (5.1.13)$$

,

The intuition is that the numerator is an estimator for $T_2\sigma_2^2$ assuming that β is constant. Hence, under H_0, the CH_4 statistics should take values close to the number of observations in the second period, T_2.
Note that CH_4 is asymptotically χ^2, rather than F-distributed. This is because the numerator and denominator in (5.1.13) are not independent, since they both rely on the parameter estimator in the first period, $\hat{\beta}_1$.
Finally, all the Chow tests we have discussed require the specification of the date of the parameter change. If this is not known with certainty, we can compute the tests for a set of possible candidate break dates, and then take the maximum (or supremum) of the resulting set of statistics. Unfortunately, the max-Chow test does not have a standard distribution, so that the proper critical values must be tabulated using simulation methods.

5.2 Recursive estimation methods

Recursive methods can provide a useful tool for assessing parameter stability. The recursive estimators of the parameters are:

$$\hat{\beta}_t = \left(\underset{kxt}{\overline{X}_t' \overline{X}_t}\right)^{-1} \underset{tx1}{\overline{X}_t' \overline{Y}_t}, \quad t{=}T_0, T_0{+}1,\ldots,T \tag{5.2.1}$$

where $T_0>k$ and $\overline{X}_t'$ and $\overline{Y}_t$ group the first t observations on the regressors X and the dependent variable y. The corresponding formula for the variance of $\hat{\beta}_t$ is:

$$\text{var}\left(\hat{\beta}_t\right) = \hat{\sigma}_t^2 \left(\underset{kxt}{\overline{X}_t' \overline{X}_t}\right)^{-1} \tag{5.2.2}$$

with

$$\hat{\sigma}_t^2 = \hat{e}_t{}'\hat{e}_t / t - k. \tag{5.2.3}$$

Combining the formulae in Chapter 3 with those in (5.2.1) and (5.2.2), we can also construct recursive interval estimators (confidence intervals) for β.
A graph of the recursive parameter estimators, $\hat{\beta}_t$, $t=T_0, T_0+1,\ldots,T$, possibly with their respective confidence intervals at a given significance level, can

provide useful information on the stability of the parameters. For example, if the model is as in (5.1.1) and (5.1.2), then we should observe a progressive change in $\hat{\beta}_t$ starting from $t=T_1+1$, with initial values close to β_1 and then getting closer to β_2 as the number of observations from the second subperiod increases.

The combined use of punctual values and confidence intervals for $\hat{\beta}_t$ allows to distinguish random variations in $\hat{\beta}_t$ from changes due to a structural break. In particular, after a certain number of periods from the break date, the new punctual estimates $\hat{\beta}_t$ should fall outside the pre-break confidence intervals. This observation implies that small variations in the parameters, or variations that take place at the beginning or at the end of the sample, are very difficult to identify.

The time when $\hat{\beta}_t$ begins to change also provides information on the break date, when the latter is not known. Formal tests for the break date are also available, but they are too complex for an introductory textbook.

Another useful tool to detect parameter changes are the *one-step ahead forecast errors*, also called *recursive residuals*. They are defined as:

$$\tilde{\varepsilon}_t = \left(Y_t - X_t\hat{\beta}_{t-1}\right) = \varepsilon_t + X_t(\beta_t - \hat{\beta}_{t-1}), \tag{5.2.4}$$

with

$$\hat{\beta}_{t-1} = \left(\underset{kx(t-1)}{\bar{X}'_{t-1}\,\bar{X}_{t-1}}\right)\bar{X}'_{t-1}\,\underset{(t-1)x1}{\bar{Y}_{t-1}}, \tag{5.2.5}$$

and

$$\tilde{\sigma}_t^2 = Var(\tilde{\varepsilon}_t) = (1 + X_t(\bar{X}'_{t-1}\bar{X}_{t-1})X'_t)\sigma^2. \tag{5.2.6}$$

We shall return in Section 5.5 on the interpretation of $\tilde{\varepsilon}_t$. Here, it is interesting to observe that, by defining

$$\omega = \left(\tilde{\varepsilon}_{k+1}/\tilde{\sigma}_{k+1},\ldots,\tilde{\varepsilon}_T/\tilde{\sigma}_T\right)', \tag{5.2.7}$$

under $H_0: \beta_1 = \beta_2$, $\sigma_1 = \sigma_2$, it is:

$$\omega \sim N(0, I_{T-k}), \qquad (5.2.8)$$

Therefore, the fact that the recursive standardized residuals ω exceed the critical reference value, such as ±1.96, provides evidence against the hypothesis of parameter stability. The dates in which this occurs provide indication on the likely timing of the structural changes.

To conclude, similar problems to those caused by parameter changes arise when there are abnormal observations, known as *outliers*, i.e. values of the dependent variable or of the regressors very different from the averages. The outliers can be caused by statistical problems, such as measurement errors, or by the occurrence of exceptional events, such as wars, natural disasters or sudden stock market collapses. Figure 5.1 illustrates how a single anomalous observation can significantly distort the OLS estimators.

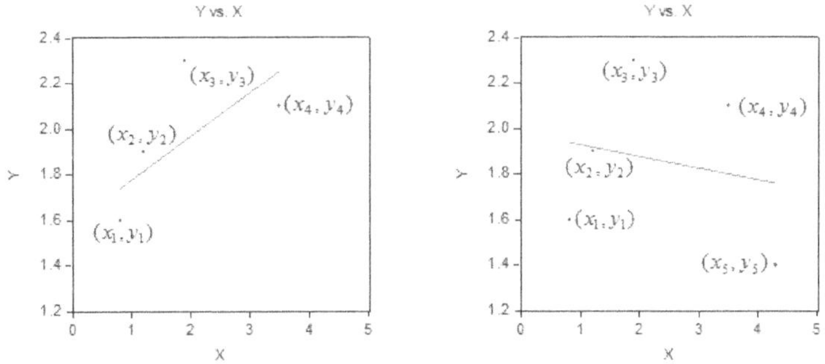

Figure 5.1: The distortionary effects of outliers

Outliers can be graphically identified, as in Figure 5.1, but this is harder with multiple regressors. Recursive methods can also be useful in this case, since an unusual observation should lead to a change in the recursive estimator of the parameters and, above all, to a large value of the corresponding recursive residual.

5.3 Remedies for parameter instability

If the Chow tests of Section 5.1 or the recursive methods of Section 5.2 indicate instability in the linear model, a simple remedy is to increase the set of explanatory variables by adding appropriate so-called *dummy variables.* These are binary variables, taking values of 0 or 1, which typically assume the value 1 during the period of changing parameters.

To illustrate the use of dummy variables, let us consider the effects of a war on aggregate consumption, due to rationing, forced saving, or simply a worsening of expectations. We define a dummy variable, D_t, taking the value 1 in times of war and 0 elsewhere, and we insert it in the linear model. We need to evaluate various cases:

i) $C_t = \beta_1 + \beta_2 Y_t + \varepsilon_t$	Standard model
ii) $C_t = \beta_1 + \alpha D_t + \beta_2 Y_t + \varepsilon_t$	Changing autonomous consumption model
iii) $C_t = \beta_1 + \beta_2 Y_t + \gamma D_t Y_t + \varepsilon_t$	Changing marginal propensity to consume model
iv) $C_t = \beta_1 + \alpha D_t + \beta_2 Y_t + \gamma D_t Y_t + \varepsilon_t$	Changing autonomous consumption and marginal propensity to consume model

Let us focus on the more general model, iv), which allows for change in both autonomous consumption and the marginal propensity to consume. The model in iv) is equivalent to

$$\begin{cases} C_t = \beta_1 + \beta_2 Y_t + \varepsilon_t, & t \notin war \\ C_t = \gamma_1 + \gamma_2 Y_t + \varepsilon_t, & t \in war \end{cases} \tag{5.3.1}$$

with $\gamma_1 = \beta_1 + \alpha$ and $\gamma_2 = \beta_2 + \gamma$. Hence, the inclusion of dummy variables as additional regressors in the original model allows the parameters of the latter to change over time.

The parameter change can be not only temporarily, as in i)-(iv)) but also permanent. For example, if during the period t_0 there is a permanent change in the marginal propensity to consume, we can adopt a specification such as:

$$C_t = \beta_1 + \beta_2 Y_t + \beta_3 (Y_t - Y_{t_0}) D_t + \varepsilon_t \quad , \tag{5.3.2}$$

$$D_t = \begin{cases} 1 & t \geq t_o \\ 0 & t < t_o \end{cases} \quad , \tag{5.3.3}$$

Note that there are no breaks in the expected value of consumption, since in t_0 it is:

$$\beta_1 + \beta_2 Y_{t_0} = \beta_1 - \beta_3 Y_{t_0} + \beta_2 Y_{t_0} + \beta_3 Y_{t_0} \quad , \tag{5.3.4}$$

The models for the two subsamples in (5.1.1) and (5.1.2) can be also rewritten as a single equation by using proper dummy variables. Assuming that the error variance is stable, it is:

$$y_t = \beta_1 X_t + \gamma_2 D_t X_t + \varepsilon_t \tag{5.3.5}$$

with $\beta_2 = \beta_1 + \gamma_2$ and

$$D_t = \begin{cases} 1 & t > T_1 \\ 0 & t \leq T_1 \end{cases} \tag{5.3.6}$$

It can be shown that an *F-test* for $\gamma_2 = 0$ in (5.3.5) is equivalent to the CH_1 statistic in (5.1.8). Under the null hypothesis (of no parameter change) the distribution of the *F-test* is $F(k, T-2k)$, since the model in (5.3.5) has *2k* regressors and we are testing that only *k* of them are not significant.
If in the two subperiods there is also a change in the error variance,

$$\begin{cases} V(\varepsilon_t) = \sigma_1^2, & t \leq T_1 \\ V(\varepsilon_t) = \sigma_2^2, & t > T_1 \end{cases} \tag{5.3.7}$$

then the model in (5.3.5) can be written as:

$$Y_t/(D_t\sigma_1+(1-D_t)\sigma_2)=\beta_1 X_t/(D_t\sigma_1+(1-D_t)\sigma_2)+\gamma_2 D_t X_t/(D_t\sigma_1+(1-D_t)\sigma_2)+u_t \qquad (5.3.8)$$

with

$$u_t=\varepsilon_t/(D_t\sigma_1+(1-D_t)\sigma_2), \qquad \mathrm{var}(u_t)=1 \qquad (5.3.9)$$

Hence, the model in (5.3.8) allows changes in the variance while having homoscedastic errors.

Outliers can be also treated by means of dummy variables. Specifically, their effects can be removed by inserting as additional regressors as many dummies as abnormal observations, where each dummy takes the value 1 in correspondence of each outlier (and 0 otherwise). For example, if only the observation in t_0 is an outlier, we need the single dummy:

$$D_t=\begin{cases}1 & t=t_0\\ 0 & t\neq t_0\end{cases} \qquad (5.3.10)$$

We insert it into a model such as:

$$y_t=\beta_1+\alpha D_t+\beta_2 X_t+\varepsilon_t \qquad (5.3.11)$$

The dummy variables we have used so far are binary. This implies sudden parameter changes, for example from β_1 to β_2. We can also allow for gradual change, e.g., by using:

$$D_t=\begin{cases}1/(1+\exp(\mu t)) & t\geq t_0\\ 0 & t<t_0\end{cases} \qquad (5.3.12)$$

In more complex models, parameter time variation can be also driven by the behavior of a certain variable, such as:

$$D_t=\begin{cases}1/(1+\exp(\mu Z_t)) & Z_t\geq Z_0\\ 0 & Z_t<Z_0\end{cases} \qquad (5.3.13)$$

In these cases, however, parameter estimation is more complex, since the model becomes nonlinear. Hence, we will only consider binary dummy variables.

Finally, we mention that dummy variables are also useful for analyzing qualitative or categorical data. For example, if a certain good, Y, can be produced by two different production techniques (or plants), A and B, we can write the model as

$$Y_i = \beta_1 + \beta_2 X_i + \varepsilon_i \tag{5.3.14}$$

$X_i = 1$ if A is active;

$X_i = 0$ if B is active.

To construct a test for the null hypothesis of "same productive capacity", we can use a *t*-test for $\beta_2 = 0$.

When the dependent variable is binary (rather than a regressor as in the case of dummy variables), the model becomes much more complex, as we will see in Chapter 9.

5.4 Forecasting with the linear regression model

Structural changes are one of the main sources of forecast errors. To examine this phenomenon, we start by considering the construction of forecasts of future values of the dependent variable based on the linear regression model. Let us assume that y depends on a single explanatory variable:

$$y_t = \alpha + \beta x_t + \varepsilon_t, \qquad \varepsilon_t \sim iid\ N(0,\sigma^2) \tag{5.4.1}$$

with $t=1,\dots,T$. We want to forecast y_{T+1} and the forecast should minimize the *predictive loss function*, which we suppose to be equal to the second moment of the *forecast error*. In formulae, the forecast error is:

$$e_{T+1} = \hat{y}_{T+1} - y_{T+1} \tag{5.4.2}$$

and we want to find $\hat{y}_{T+1}$ such that the loss function:

$$L = E(e_{T+1}^2) \tag{5.4.3}$$

is minimized. Note that this loss function is similar to the one we used for estimation, except that now we use the forecast error rather than the model errors.

The optimal forecast is a function of all the available information. If x_{T+1} and the α and β parameters are known, it can be easily shown that the optimal point forecast of y_{T+1} is:

$$\hat{y}_{T+1} = E(y_{T+1}|I_T) = \alpha + \beta x_{T+1} \tag{5.4.4}$$

where I_T indicates the available information set in period T, when we formulate the forecast. Hence, the optimal forecast coincides with the expectation of the future value of y conditional on all the available information.

The resulting forecast error is:

$$e_{T+1} = \alpha + \beta x_{T+1} - \alpha - \beta x_{T+1} - \varepsilon_{T+1} = -\varepsilon_{T+1}\,. \tag{5.4.5}$$

Hence, the expected value of the forecast error is 0,

$$E(e_{T+1}) = 0 \tag{5.4.6}$$

and the optimal forecast is an *unbiased forecast.* Moreover,

$$V(e_{T+1}) = E(e_{T+1}^2) = \sigma^2 \tag{5.4.7}$$

Hence, the optimal one-step ahead forecast error has 0 mean and is uncorrelated and homoscedastic.

Combining these results with the hypothesis $\varepsilon_T \sim iid\ \mathrm{N}(0,\sigma^2)$ and with $e_{T+1} = -\varepsilon_{T+1}$, we get:

$$\lambda = \frac{e_{T+1}}{\sigma} = \frac{\hat{y}_{T+1} - y_{T+1}}{\sigma} \sim N(0,1) \tag{5.4.8}$$

Using critical values from the $N(0,1)$, we can find the interval:

$$\Pr\left(-t_c \le \frac{\hat{y}_{T+1} - y_{T+1}}{\sigma} \le t_c\right) = 1-\alpha \quad (5.4.9)$$

The expression in (5.4.9) can be rewritten as:

$$\hat{y}_{T+1} - t_c\sigma \le y_{T+1} \le \hat{y}_{T+1} + t_c\sigma \quad (5.4.10)$$

which is an *optimal forecast interval* (or interval forecast) for y_{T+1}, at the $(1-\alpha)\%$ significance level. The interval forecast is graphed in Figure 5.2.

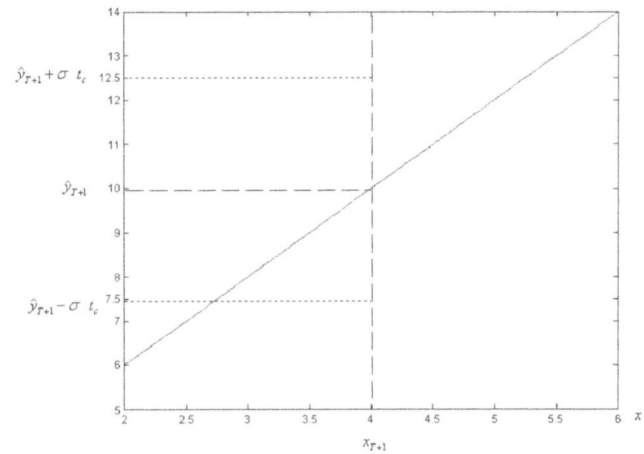

Figure 5.2: An example of optimal point and interval forecast of y_{T+1}

If in period *T+1* the model parameters become α^* and β^*, i.e., there is a structural break, but we still use the forecast in (5.4.4), the forecast error is:

$$e_{T+1} = \alpha + \beta x_{T+1} - \alpha^* - \beta^* x_{T+1} - \varepsilon_{T+1} = (\alpha - \alpha^*) + (\beta - \beta^*)x_{T+1} - \varepsilon_{T+1} \quad (5.4.11)$$

with

$$E(e_{T+1}) = (\alpha - \alpha^*) + (\beta - \beta^*)x_{T+1}. \quad (5.4.12)$$

Hence, the forecast is no longer unbiased, and the forecast error becomes heteroscedastic and possibly correlated over time.

Note that it was not possible in period T to predict the parameter change, so that the optimal forecast remains $\hat{y}_{T+1} = E(y_{T+1}|I_T) = \alpha + \beta x_{T+1}$. The use of dummy variables does not help either, since they can only capture in-sample known parameter changes. A possible remedy is the adoption of more sophisticated models, such as that in (5.3.13), where parameter changes are modelled and can therefore be predicted. However, as we said, specification and estimation of these models is complex. Besides, it is still unclear whether they lead to any significant and systematic forecasting gains.

To conclude, we make a few important comments. First, if we change the forecast loss function, the formula for the optimal forecast also changes, as well as the properties of the optimal forecast. In general, the optimal forecast can no longer be computed analytically, but simulation methods are needed. Second, forecasting is different from parameter estimation: parameters are unknown numbers, while in forecasting the target is the realization of a random variable. Hence, for example, it is not possible to speak of consistency of a forecast. Finally, about forecast intervals, as in the case of interval estimation, we should keep in mind that the significance level is associated with the notion of repeated samples. Hence, for example, a 95% forecast interval will contain the actual future realization of the dependent variable in 95% of the samples while, strictly speaking, we cannot say that it contains the actual value of the dependent variable with a 95% probability in a single sample.

5.5 Forecasting with unknown parameters

In practice the model parameters are unknown, but can be estimated by OLS, assuming that the OLS assumptions are satisfied. The OLS estimators are then used in the expression of the optimal forecast.

For the model considered in the previous Section, we have:

$$\hat{y}_{T+1} = \hat{\alpha} + \hat{\beta} \ x_{T+1} \tag{5.5.1}$$

Hence, the forecast error becomes:

$$e_{T+1} = \hat{y}_{T+1} - y_{T+1} = (\hat{\alpha} - \alpha) + (\hat{\beta} - \beta) x_{T+1} - \varepsilon_{T+1} \tag{5.5.2}$$

If we assume that the model errors are Normal, then also $\hat{\alpha}$, $\hat{\beta}$ are Normal, and therefore the forecast error is also Normal, $e_{T+1} \sim N$. Moreover, since the OLS estimators are unbiased, then:

$$E(e_{T+1}) = E(\hat{\alpha} - \alpha) + E(\hat{\beta} - \beta)x_{T+1} - E(\varepsilon_{T+1}) = 0 \tag{5.5.3}$$

and the optimal forecast remains unbiased.
The variance of the forecast error increases, due to parameter estimation, to:

$$V(e_{T+1}) = V(\hat{\alpha}) + x_{T+1}^2 V(\hat{\beta}) + 2x_{T+1}\,\mathrm{cov}(\hat{\alpha}, \hat{\beta}) + \sigma^2, \tag{5.5.4}$$

since the expected values of the other cross products are equal to 0 because $\hat{\alpha}, \hat{\beta}$ depend on $\varepsilon_1, \ldots, \varepsilon_T$ that are independent from ε_{T+1}. From Chapter 2 we know that:

$$V(\hat{\alpha}) = \sigma^2 \sum_t x_t^2 \Big/ \left(T \sum_t (x_t - \bar{x})^2 \right) \tag{5.5.5}$$

$$V(\hat{\beta}) = \sigma^2 \Big/ \sum_t (x_t - \bar{x})^2 \tag{5.5.6}$$

$$Cov(\hat{\alpha}, \hat{\beta}) = -\bar{x}\sigma^2 \Big/ \sum_t (x_t - \bar{x})^2 \tag{5.5.7}$$

By replacing these values in (5.5.4), we get:

$$V(e_{T+1}) = \sigma_f^2 = \sigma^2 \left[1 + \frac{1}{T} + \frac{(x_{T+1} - \bar{x})^2}{\sum_t (x_t - \bar{x})^2} \right] \tag{5.5.8}$$

From (5.5.8) we see that the variance of e_{T+1} decreases when:

- The sample size, T, increases (this reduces the variance of the OLS estimators). We also know that when T diverges to infinity the OLS estimators are consistent and in this case the formula in (5.5.8) reduces to that in (5.4.7) for the case of known parameters.

- The variability of x increases (we know from Chapter 2 that this also reduces the variance of the OLS estimators)
- x_{T+1} is close to $\bar{x}$ (it is easier to predict y in the case of realizations close to the average than in the case of extreme values)

Since $e_{T+1} \sim N\left(0, \sigma_f^2\right)$, we can construct a forecast interval starting from:

$$\lambda = \frac{e_{T+1}}{\sigma_f} = \frac{\hat{y}_{T+1} - y_{T+1}}{\sigma_f} \sim N\left(0,1\right) \tag{5.5.9}$$

However, σ_f depends on σ that is unknown but can be replaced by its OLS estimator:

$$\hat{\sigma}^2 = \sum_t \hat{\varepsilon}_t^2 \Big/ (T-2) \tag{5.5.10}$$

It follows that:

$$\lambda = \frac{\hat{y}_{T+1} - y_{T+1}}{\hat{\sigma}_f} \sim t\,(\mathrm{T}-2) \tag{5.5.11}$$

and a $(1-\alpha)\%$ forecast interval for y_{T+1} is:

$$\hat{y}_{T+1} - t_c \hat{\sigma}_f \le y_{T+1} \le \hat{y}_{T+1} + t_c \hat{\sigma}_f \tag{5.5.12}$$

where t_c is the proper critical value from the Student-t-distribution with $T-2$ degrees of freedom and:

$$\hat{\sigma}_f^2 = \hat{\sigma}^2 \left[1 + \frac{1}{T} + \frac{\left(x_{T+1} - \bar{x}\right)^2}{\sum_t \left(x_t - \bar{x}\right)^2} \right]. \tag{5.5.13}$$

In practice, the future value of the explanatory variable x_{T+1} is also typically not known and should be replaced by a predicted value. In this case, the forecast error becomes;

$$e_{T+1} = \hat{y}_{T+1} - y_{T+1} = (\hat{\alpha} - \alpha) + \hat{\beta}\hat{x}_{T+1} - \beta x_{T+1} - \varepsilon_{T+1} \tag{5.5.14}$$

The analysis of the properties of the forecast error is more complex, due to the multiplicative term $\hat{\beta}\hat{x}_{T+1}$ in (5.5.14). For this reason, we assume for simplicity that x_{T+1} is known.

5.6 Multi-step ahead forecasting

We now consider the case where the model has multiple regressors and we are interested in predicting $y_{T+1}, y_{T+2}, \ldots, y_{T+p}$.
Let us write the model as

$$\underset{T\times 1}{y_1} = \underset{T\times k}{x_1}\ \underset{k\times 1}{\beta} + \underset{T\times 1}{\varepsilon_1} \qquad \text{estimation period,} \tag{5.6.1}$$

$$\underset{p\times 1}{y_2} = \underset{p\times k}{x_2}\ \underset{k\times 1}{\beta} + \underset{p\times 1}{\varepsilon_2} \qquad \text{forecast period,} \tag{5.6.2}$$

$$\begin{pmatrix} \varepsilon_1 \\ \varepsilon_2 \end{pmatrix} \sim \text{iid } N\left(\begin{pmatrix} 0 \\ 0 \end{pmatrix}, \begin{pmatrix} \sigma^2 I_T & 0 \\ 0 & \sigma^2 I_p \end{pmatrix} \right) \tag{5.6.3}$$

The optimal one- up to *p*-step ahead forecasts are grouped in the vector:

$$\hat{y}_2 = x_2\hat{\beta}, \qquad \hat{\beta} = (x'_1 x_1)^{-1} x'_1 y_1 \tag{5.6.4}$$

with associated forecast errors:

$$e_2 = \hat{y}_2 - y_2 = x_2\hat{\beta} - x_2\beta - \varepsilon_2 = x_2(\hat{\beta} - \beta) - \varepsilon_2 \tag{5.6.5}$$

Since $\hat{\beta} - \beta = (x'_1 x_1)^{-1} x'_1 \varepsilon_1$, it is also:

$$e_2 = x_2 (x'_1 x_1)^{-1} x'_1 \varepsilon_1 - \varepsilon_2$$

Wherefrom

$$E(e_2)=0 \qquad (5.6.6)$$

This shows that all the elements of $\hat{y}_2$ are unbiased, and we say that $\hat{y}_2$ is an unbiased forecast.
Next, we have:

$$V(e_2)=E(e_2e'_2)=E\left[x_2(x'_1x_1)^{-1}x'_1\varepsilon_1\varepsilon'_1x_1(x'_1x_1)^{-1}x'_2\right]+E(\varepsilon_2\varepsilon'_2)-2x_2(x'_1x_1)^{-1}x'_1\underbrace{E(\varepsilon_1\varepsilon'_2)}_{=0}=$$

$$=\sigma^2\underbrace{\left[x_2(x'_1x_1)^{-1}x'_2\right]}_{p\times p}+\sigma^2I_p=\sigma^2\left[I_p+x_2(x'_1x_1)^{-1}x'_2\right] \qquad (5.6.7)$$

If σ^2 is not known, we can replace it by $\hat{\sigma}^2=\hat{\varepsilon}_1'\hat{\varepsilon}_1/(T-k)$, so that:

$$\hat{V}(e_2)=\hat{\sigma}^2\left[I_p+x_2(x'_1x_1)^{-1}x'_2\right] \qquad (5.6.8)$$

The elements on the main diagonal of $\hat{V}(e_2)$ are the variances of the one- up to *p*-step ahead forecast errors $\hat{e}_{T+1},\hat{e}_{T+2},\ldots,\hat{e}_{T+p}$:

$$\hat{V}(e_2)=\begin{bmatrix}\hat{\sigma}^2_{f,T+1} & \hat{\text{cov}}(\hat{e}_{T+1},\hat{e}_{T+2}) & \cdots & \hat{\text{cov}}(\hat{e}_{T+1},\hat{e}_{T+p})\\ & \hat{\sigma}^2_{f,T+2} & & \\ \vdots & & \ddots & \\ & & & \hat{\sigma}^2_{f,T+p}\end{bmatrix} \qquad (5.6.9)$$

Hence, a $(1-\alpha)\%$ forecast interval for y_{t+i} is:

$$\hat{y}_{T+i}-t_c\cdot\hat{\sigma}_{f,T+i}\le y_{T+i}\le\hat{y}_{T+i}+t_c\cdot\hat{\sigma}_{f,T+i},\quad \text{i=1,2,...,p}, \qquad (5.6.10)$$

where t_c is the proper critical value form the Student-*t* distribution with $(T-k)$ degrees of freedom.
Finally, it is worth mentioning that a model with a good in-sample performance (over the period *1, ..., T*) does not produce necessarily good

predictions. This is especially true when the model parameters change during the forecast period ($T+1, ..., T+p$). On the other hand, an incorrectly specified model in-sample hardly produces good forecasts: a careful specification analysis is generally a prerequisite for using the model to predict the future.

5.7 An empirical analysis with simulated data

The dataset for this example is contained in the EViews workfile "example_simul1_chap5.wf1". It consists of the series *Y*, *X1* and *X2*, sampled on a quarterly basis over the period 1951q1-2003q4.
We want to study, to begin with, the forecasting ability of the linear model. Specifically, we forecast the values of *Y* for the last two quarters of 2003, given the observations from 1951q1 to 2003q2. Using the notation in Section 5.6, we calculate $\hat{y}_2 = x_2\hat{\beta}$, where the subscript 2 indicates the second subsample in which we carry out the forecast, in our case 2003q3-2003q4.
In EViews, we first run a regression of *Y* on a constant, *X1* and *X2* over the sample 1951q1-2003q2 (which is set by writing "smpl 2003q2 1951q1" in the command line). The results are presented in Table 5.1.

Dependent variable: Y
Method: Least Squares
Sample: 1951q1 2003q2
Included observations: 210

Variable	Coefficient	Std. Error	*t*-Statistic	Prob.
C	-0.526532	0.040584	-12.97376	0.0000
X1	1.528189	0.063166	24.19318	0.0000
X2	-0.985975	0.019699	-50.05242	0.0000

R-squared	0.933660	Mean dependent var	-2.223227
Adjusted R-squared	0.933019	S.D. dependent var	0.395824
S.E. of regression	0.102442	Akaike info criterion	-1.704853
Sum squared resid	2.172342	Schwarz criterion	-1.657037
Log likelihood	182.0096	*F*-statistic	1456.637
Durbin-Watson stat	1.943011	Prob(*F*-statistic)	0.000000

Table 5.1: Output of the regression of Y on X1 and X2

All parameters are highly significant. From the estimation window, we click on "Forecast" and run a static forecast one- and two-step ahead. In the "Forecast Sample" field we indicate

```
2003q3 2003q4
```

and click "OK". We obtain Figure 5.3, and the series "yf" (the forecast of *y*) is added to the workfile.

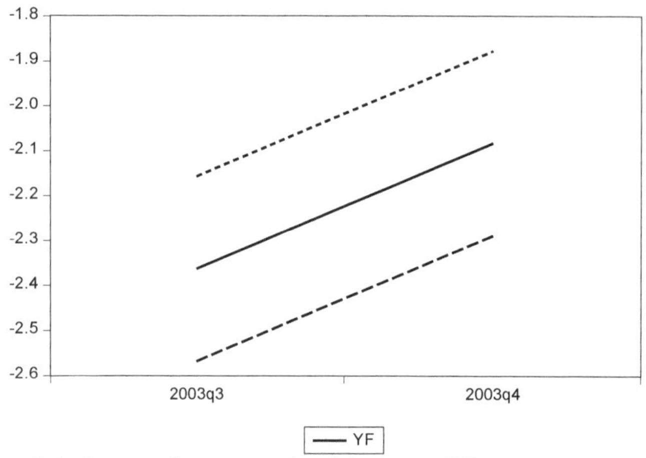

Figure 5.3: One- and two-step ahead forecast of Y with confidence bands

Figure 5.3 shows the two expected values based on the model estimated by OLS over 1951q1-2003q2. The dashed lines are confidence bands computed as $\pm 2*$ standard error. The point forecasts can be read in the "Spreadsheet" view of "yf". If we also want to save the forecast standard error, we can indicate in the forecast window a name, such as "se", in the proper box. This automatically creates a series in the workfile whose elements are the standard errors for the two forecasts. In our case, they are very close and the three lines in Figure 5.3 appear parallel. We can compare the forecasts with the actual values of *Y* by typing the command lines:

```
smpl 2003q2 2003q4
plot y yf
smpl @all
```

We obtain Figure 5.4, while Figure 5.5 reports the forecast errors. The first value in Figure 5.4 is equal for *Y* and *YF* by construction (since *Y* is not forecast in 2003q2), while the two prediction errors in Figure 5.5 for 2003q3 and 2003q4 are small compared to the scale of the *Y* variable. This outcome, in addition to providing evidence in favor of the estimated model, suggests that the parameters are stable over the forecast sample.

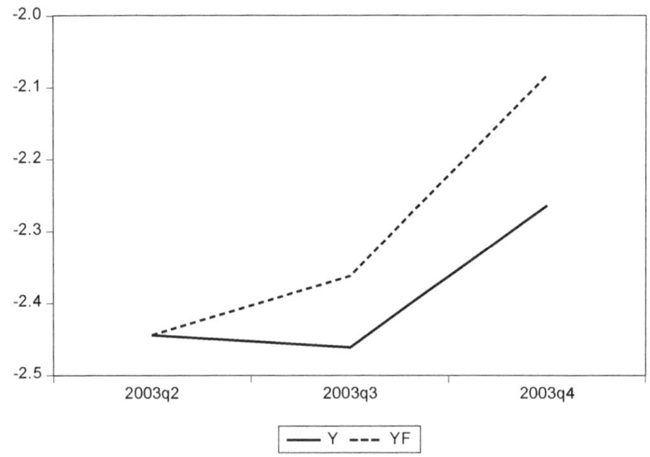

Figure 5.4: Actual and forecast values

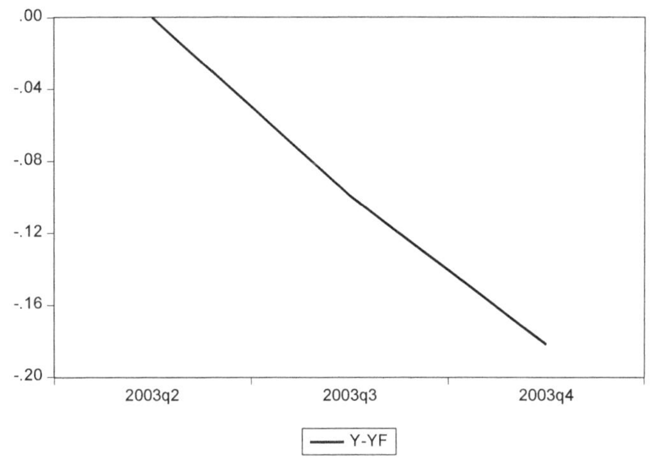

Figure 5.5: Forecast errors

In fact, a Chow forecast test for the null hypothesis "the parameters are stable" in 2003q4 against the alternative hypothesis of unstable parameters does not to reject the null hypothesis; the result is shown in Table 5.2.

Chow Forecast Test
Equation: EQ01
Specification: Y C X1 X2
Test predictions for observations from 2003Q3 to 2003Q4

	Value	df	Probability
F-statistic	2.012818	(2, 207)	0.1362
Likelihood ratio	4.083296	2	0.1298

Table 5.2 Chow forecast test

This test is valid under the assumption that the variance is stable. To verify (at the 5% level of significance) whether this is the case, we can use the statistic:

$$CH_4 : \frac{\left(Y_2 - X_2\hat{\beta}_1\right)'\left(Y_2 - X_2\hat{\beta}_1\right)}{\hat{\sigma}_1^2} \overset{a}{\sim} \chi^2(T_2) \cdot$$

We can compute it in EViews using the commands:

```
smpl 2003q3 2003q4
vector ch4=@transpose(y-yf)*(y-yf)/( 0.102442^2)
smpl @all
=@chisq(ch4(1),2)
```

The last command returns the *p*-value of the test statistic (in our case it is $T_2 = 2$). The result, reported in the lower-left corner of the screen, is 0.1294, whereby we do not to reject the null hypothesis of constant variance. The CH_3 test is valid and therefore the overall hypothesis of parameter stability in the linear model is not rejected.

Let us now look at another example, based on the EViews workfile "example_simul2_chap5.wf1". We begin by considering the Y and X series, whose behavior is reported in Figure 5.6.

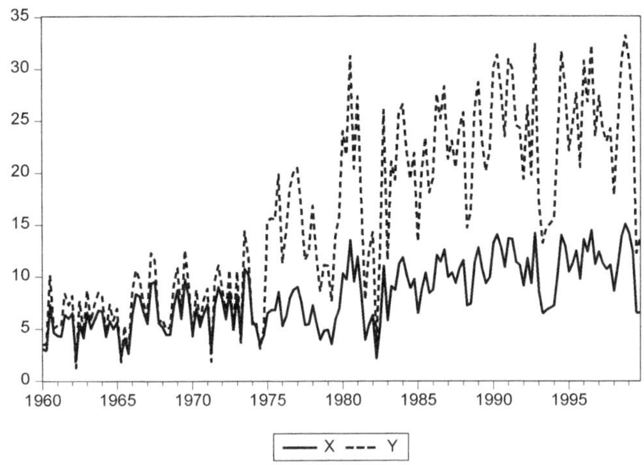

Figure 5.6: Graph of the dependent and explanatory variables

Figure 5.6 shows that *X* explains very well the behavior of *Y* until about 1975. Later, the peak structure of the two variables remains similar, indicating that *X* still has a strong explanatory ability for *Y*, but the relationship between *Y* and *X* has evidently changed.
Let us run a standard OLS regression over the entire sample (e.g., using the command "ls y c x"). The output is shown in Table 5.3.

Dependent variable: Y
Method: Least Squares

Variable	Coefficient	Std. Error	*t*-Statistic	Prob.
C	-5.692651	0.654182	-8.701934	0.0000
X	2.627614	0.074925	35.06972	0.0000

R-squared	0.886158	Mean dependent var	15.74488
Adjusted R-squared	0.885437	S.D. dependent var	8.707361
S.E. of regression	2.947192	Akaike info criterion	5.012004
Sum squared resid	1372.378	Schwarz criterion	5.050443
Log likelihood	-398.9603	*F*-statistic	1229.885
Durbin-Watson stat	0.339089	Prob(*F*-statistic)	0.000000

Table 5.3: Output of the regression of Y on X over the entire sample

The R^2 coefficient is high enough and both the constant and the *X* regressor are strongly statistically significant (*p*-value is 0.0000 for both). However, Figure 5.7 clearly shows that there is a change in the parameters of the model.

In fact, the fitted values are systematically higher than the actual values of y in the first part of the sample, and systematically lower in the second part. A side effect of the change in the parameters is that many of the other OLS assumptions are rejected when tested. For example, the value of the DW-test is close to 0, suggesting a rejection of the null hypothesis of no correlation of the errors.

Finally, we have seen that in the presence of structural change the OLS estimators are biased and inconsistent, so that the estimated parameter values in Table 5.3 are likely unreliable.

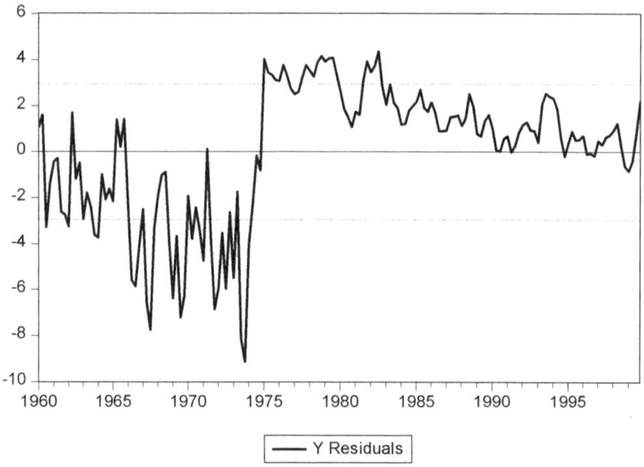

Figure 5.7: Residuals of the regression in Table 5.3

To verify the stability of the parameters, let us run a Chow breakpoint test, in the case $T_2 > k$, denoted by CH_1 in Section 5.1. It tests the null hypothesis H_0: β constant, given $\sigma_1^2 = \sigma_2^2$, against $H_1: \beta_1 \neq \beta_2$.

The Figures above suggest taking as breakpoint t^* = 1975q1. We choose a significance level, for example, of 5%. To perform the test using EViews, we select "View" from the estimation window, choose "Stability Diagnostics", "Chow Breakpoint Test" and we indicate 1975q1 as the break date. We get Table 5.4.

Chow Breakpoint Test: 1975q1
Null Hypothesis: No breaks at specified breakpoints
Varying regressors: All equation variables
Equation Sample: 1960q1 1999q4

F-statistic	1410.380	Prob. F(2,156)	0.0000
Log likelihood ratio	471.7975	Prob. Chi-Square(2)	0.0000
Wald Statistic	2820.760	Prob. Chi-Square(2)	0.0000

Table 5.4: Chow breakpoint (CH_1) test

The low *p*-value of the *F*-statistic leads us to reject the null hypothesis of parameter stability.
Since the CH_1 test assumes $\sigma_1^2 = \sigma_2^2$, we need to also test this hypothesis, e.g., at the 5% confidence level. We can use

$$CH_2 : \frac{\hat{\sigma}_2^{\,2}}{\hat{\sigma}_1^{\,2}} = \frac{RSS_2}{RSS_1} \frac{T_1 - K}{T_2 - K} \underset{H_0}{\sim} F(T_2 - K, T_1 - K).$$

We use 1960q1-1974q4 and 1975q1-1999q4) as subsamples. To get RSS_1 and RSS_2 in EViews, we type and run (one by one) the command lines:

```
smpl 60:1 74:4
ls y c x
scalar rss1=@inner(resid,resid)
```

Then, we repeat the procedure for the second subsample:

```
smpl 75:1 99:4
ls y c x
scalar rss2=@inner(resid,resid)
```

We get the required values: RSS_1=20.74 and RSS_2 =51.17. Moreover, T_1=60, K=2, T_2 =100. Hence, the realization of the test statistic is:

$$CH_2 : \frac{51.17}{20.74} * \frac{58}{98} = 1.46 \underset{H_0}{\sim} F(98,58)$$

and to compute its *p*-value we write

```
@fdist(1.46,98,58)
```

which yields 0.059. Since $\alpha = 0.05$ < 0.059, we do not reject the null hypothesis. Finally, we type

```
smpl @all
```

to return to the full sample.

Additional evidence on the presence and timing of structural change is provided by recursive methods, studied in Section 5.2. In EViews, we can get the recursive estimators by clicking, in the equation object, on "View", "Stability Diagnostics", "Recursive Estimates (OLS)". We get a window in which we can choose "Recursive Coefficients", and select the parameters we are interested in (both *c(1)* and *c(2)* in our example, which are the constant and the coefficient of *X*). We get Figure 5.8, which shows that the estimates of α decrease when the sample size increases, with a decided drop from the mid-1970s onwards, while β grows from the same period. The estimates are clearly unstable and we must correct the model.

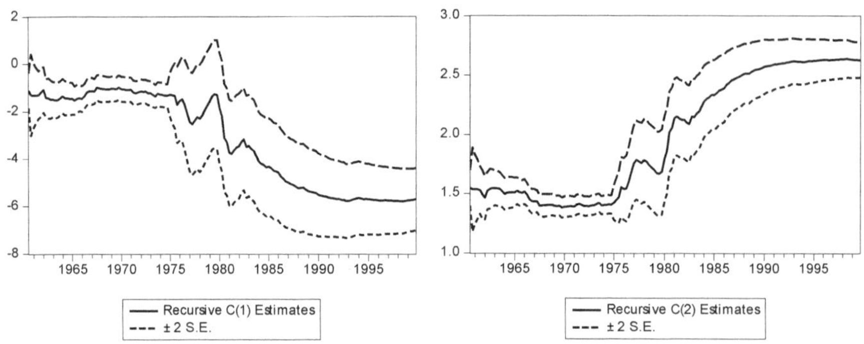

Figure 5.8: Recursive estimates for the model in Table 5.3

In order to take into account the parameter change, we can use binary or dummy variables, as we saw in Section 5.3. Assuming that the breakpoint is at 1975q1, we create a variable whose value is 0 until 1974q4 and 1 afterwards. We use the commands:

```
genr d744=0
smpl 75:1 99:4
genr d744=1
smpl @all
```

We have thus created a series, d744, which is equal to 0 on the whole sample; then changing its values to 1 only in the second subsample. To check if we have correctly generated the dummy variable, we graph it in Figure 5.9.

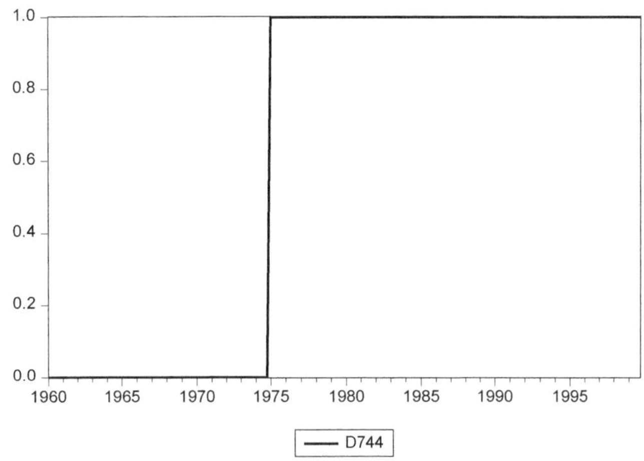

Figure 5.9: An example of a dummy variable

By performing a regression of *Y* on *X*, the constant, and the dummy variable, we allow the intercept to change over time (in 1975q1). The results are summarized in Table 5.5.

Dependent variable: Y
Method: Least Squares
Sample: 1960:1 1999:4
Included observations: 160

Variable	Coefficient	Std. Error	*t*-Statistic	Prob.
C	-5.287685	0.256893	-20.58324	0.0000
X	2.071070	0.034915	59.31803	0.0000
D744	6.617012	0.224270	29.50466	0.0000

R-squared	0.982606	Mean dependent var	15.74488
Adjusted R-squared	0.982384	S.D. dependent var	8.707361
S.E. of regression	1.155688	Akaike info criterion	3.145842
Sum squared resid	209.6917	Schwarz criterion	3.203501
Log likelihood	-248.6673	*F*-statistic	4434.431
Durbin-Watson stat	1.279125	Prob(*F*-statistic)	0.000000

Table 5.5: Regression with changing intercept

The coefficient associated with the dummy variable is highly statistically significant, and the inclusion of d744 does significantly increase the adjusted R^2, confirming the significance of this regressor. Did we solve the problem of the structural change? To address this issue, we can look at the other diagnostic tests, starting with the Durbin-Watson statistic. From Table 5.5, it is still rather low, about 1.28, suggesting some positive correlation in the errors. Looking at the graph of the residuals in Figure 5.10, indeed it seems that there is some persistence in the first subsample, with fewer problems in the second subsample. This could be due to a form of serial correlation but, looking again at Figure 5.6, it could be also due to a change in the relationship between *Y* and *X*. In fact, the variability of the *Y* values with respect to the *X* values increases after 1975, suggesting that the coefficient of *X* could have also changed. If this is true and our model does not take it into account, we could still get biased and inconsistent estimators and tests.

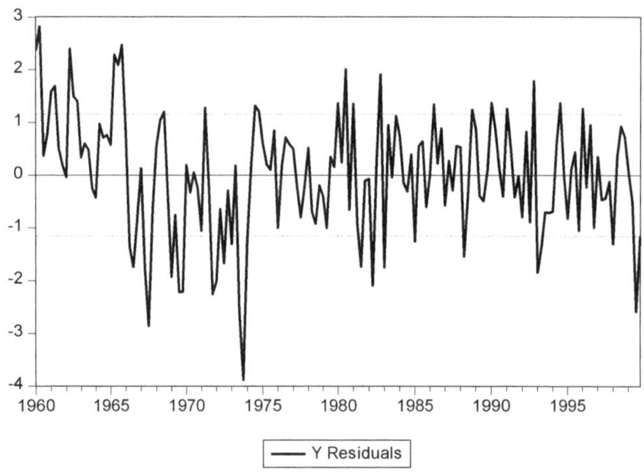

Figure 5.10: Residuals of the regression in Table 5.5

To consider this additional source of parameter change, we can run a regression of *Y* on a constant, *X*, the dummy variable and the dummy multiplied by *X*:

$$d744*x=\begin{pmatrix} 0*x_1 \\ \vdots \\ 0*x_{t^*-1} \\ \hline 1*x_{t^*} \\ \vdots \\ 1*x_T \end{pmatrix}=\begin{pmatrix} 0 \\ \vdots \\ 0 \\ \hline x_{t^*} \\ \vdots \\ x_T \end{pmatrix}.$$

We write:

```
ls y c x d744 d744*x
```

and the output is reported in Table 5.6. Each regressor is statistically significant (individually and jointly, looking at the *t*-statistics and *F*-statistic). The adjusted R^2 is higher than in Table 5.5, and the graph of the residuals greatly improves. Note also that the *F*-test for the joint significance of the two dummies has a value of 1410.38, exactly equal to the CH_1 Chow test in Table 5.4.

Dependent variable: Y
Method: Least Squares
Sample: 1960:1 1999:4
Included observations: 160

Variable	Coefficient	Std. Error	*t*-Statistic	Prob.
C	-1.319952	0.274702	-4.805029	0.0000
X	1.408636	0.043465	32.40826	0.0000
D744	0.853570	0.358495	2.380982	0.0185
D744*X	0.852257	0.049301	17.28676	0.0000

R-squared	0.994034	Mean dependent var	15.74488
Adjusted R-squared	0.993919	S.D. dependent var	8.707361
S.E. of regression	0.678993	Akaike info criterion	2.088269
Sum squared resid	71.92083	Schwarz criterion	2.165148
Log likelihood	-163.0615	*F*-statistic	8664.034
Durbin-Watson stat	1.666360	Prob(*F*-statistic)	0.000000

Table 5.6: Regression with changing coefficients

Breusch-Godfrey Serial Correlation LM Test:

F-statistic	7.968794	Probability	0.000509
Obs*R-squared	15.00559	Probability	0.000552

Table 5.7: Test for no correlation of the residuals of the model in Table 5.6

Note that the likelihood of serially correlated errors diminishes according to the Durbin-Watson test. The LM test, however, still rejects the null hypothesis of no correlation. Since the model in Table 5.6 coincides with the Data Generating Process (DGP) and there is no correlation in the errors in the DGP, the LM test is committing a type I error.

How do we interpret the coefficients obtained from the regression with the dummy variables? It is:

$$\frac{\partial E(y_t)}{\partial x_t} = \begin{cases} 1.40 & \textit{before 1975} \\ 1.40 + 0.85 & \textit{after 1975} \end{cases}$$

We now use this augmented model to test the hypothesis of a parameter change in the last quarter of 1999, with a 5% significance level. Now $T_2 < K$ (1<4), and therefore we use the Chow forecast test (CH_3 in section 5.1):

$$CH_3 = \left(\frac{RSS_T - RSS_{T_1}}{RSS_{T_1}} \right) \frac{T_1 - k}{T_2} \underset{H_0}{\sim} F(T_2, T_1 - k).$$

As we saw in the previous example, we can run this test in EViews by clicking on "View", in the equation window, and selecting "Stability Tests", "Chow Forecast" and finally selecting "1999q4" as the break date. The outcome, reported in Table 5.8, implies that we do not reject the null hypothesis of parameter stability at the last observation of the sample.

Chow Forecast Test
Equation: EQ03
Specification: Y C X D744 D744*X
Test predictions for observations from 1999q4 to 1999q4

	Value	df	Probability
t-statistic	0.887013	155	0.3764
F-statistic	0.786793	(1, 155)	0.3764
Likelihood ratio	0.810119	1	0.3681

Table 5.8: Chow Forecast Test for the model in Table 5.6

5.8 An empirical analysis of the determinants of aggregate consumption

Permanent income theory predicts that consumption depends on the expected future value of average disposable income. For the sake of simplicity, let us suppose permanent income in period t, y_{pt}, is a simple average of current disposable incomes, y_{dt}, and expected income next period, $E(y_{dt+1})$.
We can replace $E(y_{dt+1})$ with the best forecast of y_{dt+1}, $\hat{y}_{dt+1}$, computed as in Section 5.4. Hence, permanent income is defined as:

$$y_{pt} = \frac{1}{2} yd_t + \frac{1}{2} \hat{yd}_{t+1}.$$

To compute $\hat{yd}_{t+1}$, let us suppose that the model for y_d is:

$$yd_t = \alpha + \beta yd_{t-1} + \varepsilon_t \tag{5.8.1}$$

where y_d denotes the log of real disposable income. This is a dynamic model, as we will see in Chapter 7, but to compute the optimal forecast we can still use the formula in Section 5.4: $\hat{yd}_{t+1} = \hat{\alpha} + \hat{\beta}\, yd_t$, where $\hat{\alpha}$ and $\hat{\beta}$ are the OLS estimators of a and β.
The output of the regression in (5.8.1), using data for Italy, is shown in Table 5.9.

Dependent variable: LRYD
Method: Least Squares
Sample (adjusted): 1990q2 2012q1
Included observations: 88 after adjustments

Variable	Coefficient	Std. Error	*t*-Statistic	Prob.
C	0.432123	0.251645	1.717190	0.0895
LRYD(-1)	0.945808	0.031488	30.03692	0.0000

R-squared	0.912975	Mean dependent var	7.990439
Adjusted R-squared	0.911963	S.D. dependent var	0.074802
S.E. of regression	0.022194	Akaike info criterion	-4.755482
Sum squared resid	0.042363	Schwarz criterion	-4.699179
Log likelihood	211.2412	Hannan-Quinn criterion	-4.732799
F-statistic	902.2167	Durbin-Watson stat	1.734234
Prob(*F*-statistic)	0.000000		

Table 5.9: Model of disposable income

Given the estimated parameters in Table 5.9, it is $\hat{yd}_{t+1} = 0.4321123 + 0.945808\, yd_t$.
As we have seen in the examples with simulated data, there is an alternative way to compute $\hat{yd}_{t+1}$ in EViews. We click on "Forecast" in the equation window, which produces Figure 5.11, and select "Static Forecast".

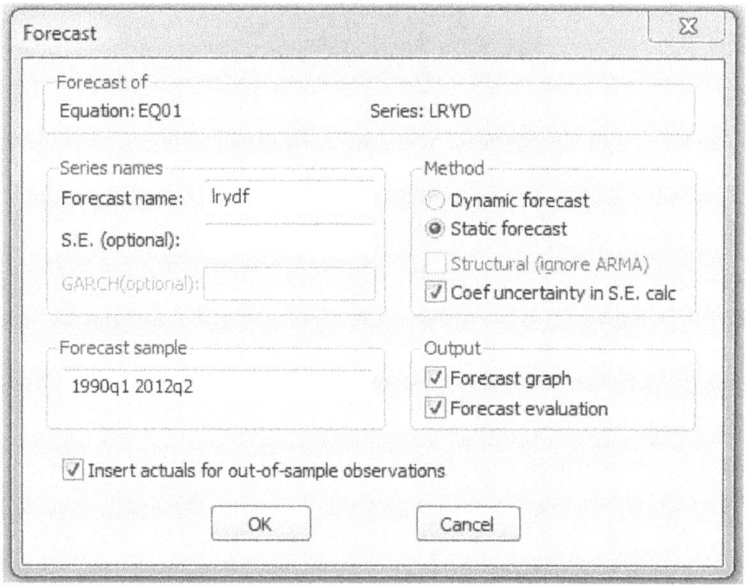

Figure 5.11: EViews window to compute forecasts

The series of forecast values is saved in the workfile, with the name "lrydf" . Then, we generate the series of disposable income (lyp) using the command:
genr lyp=0.5*lryd+0.5*lrydf,
and we regress consumption on lyp.
The results are illustrated in Table 5.10. They suggest that lyp is strongly significant, but the Durbin-Watson statistic indicates possible serial correlation that, as we have seen in Chapter 4, could lead to biased estimators.
To evaluate whether income expectations are really important for consumer decisions, we then regress consumption on current income and on future expected income. The results are shown in Table 5.11 (and similar values are obtained when using LRYDF as a regressor).
The coefficient of the expected future income is statistically significant at the 5% level. However, the hypothesis that this coefficient is equal to that of LRYD is rejected (the *p*-value is 0.00). Hence, according to this simple model,

we should reject our idea of computing disposable income as a simple average of LRYD and its predicted value for the next period. On the other hand, the Durbin–Watson statistic still indicates potential correlation in the errors, and we did not consider the possibility of structural breaks in the relationship.

Dependent variable: LRC
Method: Least Squares
Sample (adjusted): 1990q2 2012q1
Included observations: 88 after adjustments

Variable	Coefficient	Std. Error	*t*-Statistic	Prob.
C	-1.489178	0.243441	-6.117211	0.0000
LYP	1.157816	0.030465	38.00448	0.0000

R-squared	0.943803	Mean dependent var	7.762283
Adjusted R-squared	0.943150	S.D. dependent var	0.086189
S.E. of regression	0.020550	Akaike info criterion	-4.909414
Sum squared resid	0.036319	Schwarz criterion	-4.853111
Log likelihood	218.0142	Hannan-Quinn criterion	-4.886731
F-statistic	1444.341	Durbin-Watson stat	0.430446
Prob(*F*-statistic)	0.000000		

Table 5.10: Regression of consumption on permanent income

Dependent variable: LRC
Method: Least Squares
Sample (adjusted): 1990q3 2012q1
Included observations: 87 after adjustments

Variable	Coefficient	Std. Error	*t*-Statistic	Prob.
C	-1.447444	0.237231	-6.101407	0.0000
LRYD	0.940922	0.065188	14.43406	0.0000
LRYDF(-1)	0.211701	0.067766	3.123996	0.0024

R-squared	0.949829	Mean dependent var	7.761457
Adjusted R-squared	0.948634	S.D. dependent var	0.086338
S.E. of regression	0.019568	Akaike info criterion	-4.996004
Sum squared resid	0.032163	Schwarz criterion	-4.910973
Log likelihood	220.3262	Hannan-Quinn criterion	-4.961765
F-statistic	795.1309	Durbin-Watson stat	0.341432
Prob(*F*-statistic)	0.000000		

Table 5.11: Regression of consumption on current and future expected income

A first possible source of structural change is the introduction of the euro, which could have changed the consumption habits. Hence, we want to test for a break in 2002q1, when the euro was introduced. We create a proper dummy variable with the following commands:

```
series eurodu
eurodu=0
smpl 02:1 @last
eurodu=1
smpl @all
```

Then we run the regression in Table 5.12.

Dependent variable: LRC
Sample (adjusted): 1990q1 2012q1

Variable	Coefficient	Std. Error	*t*-Statistic	Prob.
C	-0.726053	0.235799	-3.079121	0.0028
LRYD	1.061113	0.029639	35.80116	0.0000
EURODU	4.004151	1.018063	3.933108	0.0002
EURODU*LRYD	-0.495889	0.126784	-3.911292	0.0002

R-squared	0.961126	Mean dependent var	7.763188
Adjusted R-squared	0.959754	S.D. dependent var	0.086122
S.E. of regression	0.017277	Akaike info criterion	-5.234943
Sum squared resid	0.025373	Schwarz criterion	-5.123094
Log likelihood	236.9550	Hannan-Quinn criterion	-5.189860
F-statistic	700.5149	Durbin-Watson stat	0.342236
Prob(*F*-statistic)	0.000000		

Table 5.12: Effects of the introduction of the euro on consumption

The dummies are significant both individually and jointly, on the basis of the *F*-test in Table 5.13. Note also that the *F*-test for the significance of the dummies is identical to the *F*-version of the Chow breakpoint test, also shown in table 5.13.

Based on these results, it seems that autonomous consumption has increased with the introduction of the euro, while the marginal propensity to consume has substantially decreased. Similar conclusions are obtained using the permanent income instead of the current one. This result should be interpreted with caution, however, since other assumptions on the model could not hold. For example, the DW statistic still suggests that the errors are correlated. In addition, we have not considered the possible effects of the financial crisis, which started around 2007. We will re-evaluate the effects of the introduction of the euro in a more appropriate dynamic model in Chapter 7, while now we try to understand the effects of the crisis on consumption.

Wald Test:
Equation: EQEURO

Test Statistic	Value	df	Probability
F-statistic	20.90163	(2, 85)	0.0000
Chi-square	41.80326	2	0.0000

Null Hypothesis: C(3)=C(4)=0
Null Hypothesis Summary:

Normalized Restriction (= 0)	Value	Std. Err.
C(3)	4.004151	1.018063
C(4)	-0.495889	0.126784

Restrictions are linear in coefficients.

Chow Breakpoint Test: 2002q1
Null Hypothesis: No breaks at specified breakpoints

F-statistic	20.90163	Prob. F(2,85)	0.000000
Log likelihood ratio	35.59871	Prob. Chi-Square(2)	0.000000
Wald Statistic	41.80326	Prob. Chi-Square(2)	0.000000

Table 5.13: Tests for a structural break associated with the introduction of the euro

Let us create the dummy for the crisis as we did previously, namely, by executing the following commands:

```
series crisis
crisis=0
smpl 07:3 @last
crisis=1
smpl @all
```

When we add this dummy to the consumption function, we obtain the output shown in Table 5.14.

Dependent variable: LRC
Method: Least Squares
Sample (adjusted): 1990q1 2012q1
Included observations: 89 after adjustments

Variable	Coefficient	Std. Error	*t*-Statistic	Prob.
C	-0.780582	0.130945	-5.961134	0.0000
LRYD	1.068078	0.016408	65.09678	0.0000
CRISIS	7.006581	0.928609	7.545244	0.0000
CRISIS*LRYD	-0.867501	0.115616	-7.503280	0.0000

R-squared	0.984334	Mean dependent var	7.763188
Adjusted R-squared	0.983781	S.D. dependent var	0.086122
S.E. of regression	0.010968	Akaike info criterion	-6.143768
Sum squared resid	0.010225	Schwarz criterion	-6.031919
Log likelihood	277.3977	Hannan-Quinn criterion	-6.098685
F-statistic	1780.234	Durbin-Watson stat	0.884805
Prob(*F*-statistic)	0.000000		

Table 5.14: Effects of the financial crisis on consumption

Both dummies are strongly significant, and the goodness of fit of the model is slightly larger than with the euro dummies, even though the Durbin-Watson statistic remains low (DW = 0.88). The main effect of the crisis seems to be an increase in autonomous consumption (typically related to primary goods) and a decrease in the marginal propensity to consume (with a corresponding increase in the propensity to save).

Since both the introduction of the euro and the financial crisis seem to have affected the consumption function, we may want to account for both events. Actually, a regression of consumption on a constant and income generates regressors (plotted in Figure 5.12) that show marked changes around both 2002 and 2007.

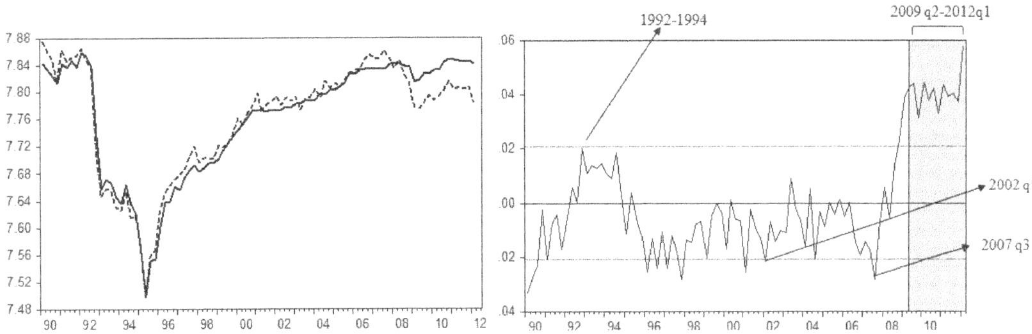

Figure 5.12: Residual of the model $c_t = \alpha + \beta y_t^d$

From the graph of actual and fitted values (Figure 5.12, left panel), after 1995 there seems to be a quadratic trend in the data. Moreover, the graph of the residuals (Figure 5.12, right panel) suggests to introduce the euro dummy and two dummies for the crisis, generated as follows:

```
genr crisisdu1=0
smpl 2007q3 2007q3
crisisdu1=1
smpl @all
genr crisisdu2=0
smpl 2009q2 @last
crisisdu2=1
smpl @all
```

For simplicity, we exclude from the sample the turbulent years that began with the exit of the Italian Lira from the EMS in 1992, and we estimate a model with a quadratic trend and the dummies we have discussed. The results are presented in Table 5.15. All the regressors are statistically significant at the 10% level, and we find additional evidence of changes both in the autonomous consumption and in the marginal propensity to consume.

Dependent variable: LRC

Variable	Coefficient	Std. Error	t-Statistic	Prob.
C	-0.265121	0.148702	-1.782899	0.0795
@TREND^2	1.43E-05	1.23E-06	11.54001	0.0000
CRISISDU1	-0.012700	0.006361	-1.996491	0.0503
LRYD	1.000150	0.018834	53.10304	0.0000
EURODU	3.971558	0.459055	8.651601	0.0000
EURODU*LRYD	-0.496420	0.057314	-8.661444	0.0000
CRISISDU2*LRYD	-0.000799	0.000590	-1.354581	0.1805

R-squared	0.995509	Mean dependent var	7.764666
Adjusted R-squared	0.995074	S.D. dependent var	0.084126
S.E. of regression	0.005904	Akaike info criterion	-7.330397
Sum squared resid	0.002161	Schwarz criterion	-7.103749
Log likelihood	259.8987	Hannan-Quinn criterion	-7.240478
F-statistic	2290.580	Durbin-Watson stat	1.990265
Prob(F-statistic)	0.000000		

Table 5.15: Effects of the euro introduction and of the financial crisis on consumption

Finally, in order to provide additional evidence on structural change and possibly identify other problematic periods, we recursively estimate the model. The graph of the recursive estimators in Figure 5.13 highlights substantial changes in the first part of the sample. These could be either spurious, due to the very small number of available observations for parameter estimation, or related to the aforementioned problematic exit of the Lira from the EMS. If we re-compute the recursive estimators starting the estimation sample in 1995, the resulting graphs confirm that there have been substantial changes around 2002 and 2007.

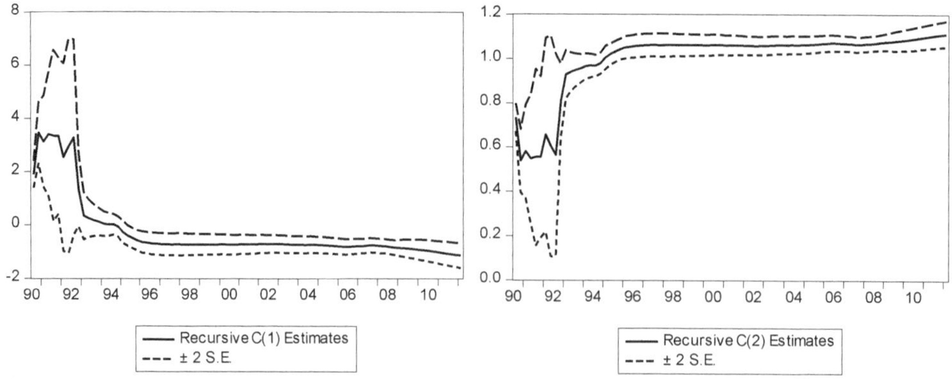

Figure 5.13: Recursive estimation of the model LRC=C(1)+C(2)*LRYD

The main concepts of this Chapter

In this chapter we have dealt with the problem of parameter instability. If this happens, the $\hat{\beta}$ and $\hat{\sigma}^2$ OLS estimators are biased and inconsistent. It is important, therefore, to test the hypothesis of parameter stability. Given a break date, we split the sample into two subperiods. If the number of observations in the second subperiod, T_2, is greater than the number of regressors, the model can be estimated both in the first and in the second subsample. The (one-sided) *Chow test* for $H_0': \beta_1 = \beta_2 \mid H_0'': \sigma_1^2 = \sigma_2^2$ is:

$$CH_1 : \left(\frac{RSS_T - RSS_{T_1} - RSS_{T_2}}{RSS_{T_1} + RSS_{T_2}} \right) \frac{T-2k}{k} \underset{H_0}{\sim} F(k, T-2k),$$

It is also important to test whether the error variance is stable. The test for $H_0'': \sigma_1^2 = \sigma_2^2$ versus $H_1'': \sigma_1^2 < \sigma_2^2$ is:

$$CH_2 : \frac{\hat{\sigma}_2^2}{\hat{\sigma}_1^2} = \frac{RSS_{T_2}}{RSS_{T_1}} \frac{T_1 - k}{T_2 - k} \underset{H_0}{\sim} F(T_2 - k, T_1 - k).$$

The CH_1 and CH_2 statistics are independent, so we can first apply CH_2 to check the stability of the variance, and then CH_1 for the stability of the parameters of the regression function.

If the sample size in the second subperiod is smaller than the number of regressors, the (one-sided) test statistic for $H_0': \beta_1 = \beta_2 \mid H_0'': \sigma_1^2 = \sigma_2^2$ is:

$$CH_3 : \left(\underset{T_2 x 1}{Y_2} - X_2 \hat{\beta}_1 \right)' \frac{\left[I + X_2 (X_1' X_1)^{-1} X_2' \right]^{-1}}{\underset{T_1 x 1}{\hat{u}_1' \hat{u}_1}} \left(Y_2 - X_2 \hat{\beta}_1 \right) \frac{T_1 - k}{T_2} \underset{H_0}{\sim} F(T_2, T_1 - k),$$

The CH_3 statistic can be also written as:

$$CH_3 = \left(\frac{RSS_T - RSS_{T_1}}{RSS_{T_1}} \right) \frac{T_1 - k}{T_2} \underset{H_0}{\sim} F(T_2, T_1 - k).$$

The statistic for testing $H_0'': \sigma_1^2 = \sigma_2^2$ versus $H_1'': \sigma_1^2 < \sigma_2^2$ becomes:

$$CH_4 : \frac{\left(Y_2 - X_2\hat{\beta}_1\right)'\left(Y_2 - X_2\hat{\beta}_1\right)}{\hat{\sigma}_1^2} \overset{a}{\sim} \chi^2(T_2).$$

The Chow tests require to know the break date. When the latter is unknown, we could compute a set of tests for different break dates, and take their maximum. However, the resulting statistic has a non-standard distribution.
Recursive estimation methods provide an additional useful tool to assess parameter stability and, possibly, identify the break dates. The recursive OLS estimators are:

$$\hat{\beta}_t = \left(\underset{kxt}{\overline{X}_t'}\,\overline{X}_t\right)^{-1} \overline{X}_t'\underset{tx1}{\overline{Y}_t}, \quad \text{t=T}_0\text{, T}_0\text{+1,....,T}$$

The corresponding formula for the variance of $\hat{\beta}_t$ is:

$$\text{var}\left(\hat{\beta}_t\right) = \hat{\sigma}_t^2\left(\underset{kxt}{\overline{X}_t'}\,\overline{X}_t\right)^{-1} \text{ with } \hat{\sigma}_t^2 = \hat{e}_t{}'\hat{e}_t\,/t-k.$$

By combining these formulae with those in Chapter 3, we can also build recursive confidence intervals for the parameters. A few periods after a break, the recursive point estimators should fall outside the recursive intervals computed before the break.
We have then defined the recursive residuals as

$$\tilde{\varepsilon}_t = \left(Y_t - X_t\hat{\beta}_{t-1}\right) = \varepsilon_t + X_t(\beta_t - \hat{\beta}_{t-1}),$$

and mentioned that they correspond to one-step ahead forecast errors. If we define $\omega = \left(\tilde{\varepsilon}_{k+1}/\tilde{\sigma}_{k+1},\ldots,\tilde{\varepsilon}_T/\tilde{\sigma}_T\right)'$, then under H_0 : $\beta_1 = \beta_2$, $\sigma_1 = \sigma_2$, it should be $\omega \sim N(0, I_{T-k})$. Hence, if the standardized recursive residuals are larger (in absolute value) than the proper critical value, e.g., 1.96, there is evidence against parameter stability.
The use of dummy variables can alleviate the problems due to parameter instability Dummy variables are binary variables, usually taking the values of 0and 1, that can be added to the model regressors to allow for (temporary or permanent) changes in their coefficients. For example, assuming that the error variance is constant, in the model $y_t = \beta_1 X_t + \gamma_2 D_t X_t + \varepsilon_t$ where

$$D_t = \begin{cases} 1 & t > T_1 \\ 0 & t \le T_1 \end{cases}$$

the coefficient of X can differ before and after the period T_1. In this model we can also construct an alternative and equivalent version of the *CH1* statistic: an F-test for the non-significance of the dummy ($\gamma_2 = 0$), with an $F(k, T-2k)$ distribution under the null hypothesis.

Parameter change has important effects for forecasting. Given the loss function $L = E(e_{T+1}^2)$ where $e_{T+1} = \hat{y}_{T+1} - y_{T+1}$ is the forecast error, the best forecast of y_{T+1} in the simple linear model with known parameters and future values of the explanatory variable is $\hat{y}_{T+1} = E(y_{T+1} | I_T) = \alpha + \beta x_{T+1}$. This forecast is unbiased, in the sense that the associated forecast error has 0 expected value. Using the critical values for the standard Normal, we can determine the interval $\Pr\left(-t_c \le \frac{\hat{y}_{T+1} - y_{T+1}}{\sigma} \le t_c\right) = 1 - \alpha$, so that $\hat{y}_{T+1} - t_c\sigma \le y_{T+1} \le \hat{y}_{T+1} + t_c\sigma$ is a $(1-\alpha)\%$ forecast interval for y_{T+1}.

If the model parameters change in period $T + 1$, the optimal forecast is no longer unbiased and the forecast error can be heteroscedastic and serially correlated.

When the model parameters are unknown, we replace them by the OLS estimators, so that the optimal forecast becomes $\hat{y}_{T+1} = \hat{\alpha} + \hat{\beta}\ x_{T+1}$, with forecast error $e_{T+1} = \hat{y}_{T+1} - y_{T+1} = (\hat{\alpha} - \alpha) + (\hat{\beta} - \beta)x_{T+1} - \varepsilon_{T+1} = 0$. The optimal forecast remains unbiased, and it is:

$$V(e_{T+1}) = \sigma_f^2 = \sigma^2 \left[1 + \frac{1}{T} + \frac{(x_{T+1} - \bar{x})^2}{\sum_t (x_t - \bar{x})^2} \right].$$

The $(1-\alpha)\%$ forecast interval for y_{T+1} becomes $\hat{y}_{T+1} - t_c\hat{\sigma}_f \le y_{T+1} \le \hat{y}_{T+1} + t_c\hat{\sigma}_f$, where now t_c is the proper critical value for the Student-t distribution with $T - 2$ degrees of freedom.

In the case of multiple regressors, the optimal (one- to p-step ahead) forecast is $\hat{y}_2 = x_2\hat{\beta}$, $\hat{\beta} = (x'_1 x_1)^{-1} x'_1 y_1$, with associated forecast error $e_2 = \hat{y}_2 - y_2 = x_2\hat{\beta} - x_2\beta - \varepsilon_2 = x_2(\hat{\beta} - \beta) - \varepsilon_2$. The optimal forecast remains unbiased. Moreover, it is $V(e_2)=\sigma^2\left[I_p + x_2(x'_1 x_1)^{-1} x'_2\right]$, where p is the length of the forecast period. If σ^2 is unknown, we replace it by its OLS estimator, $\hat{\sigma}^2 = \hat{\varepsilon}_1'\hat{\varepsilon}_1/(T-k)$ whereby $\hat{V}(e_2)=\hat{\sigma}^2\left[I_p + x_2(x'_1 x_1)^{-1} x'_2\right]$. As a $(1-\alpha)\%$ forecast interval for y_{t+i}, we get:

$$\hat{y}_{T+i} - t_c \cdot \hat{\sigma}_{f,T+i} \leq y_{T+i} \leq \hat{y}_{T+i} + t_c \cdot \hat{\sigma}_{f,T+i}, \quad \text{i=1,2,...,p}$$

where t_c is the proper critical value for the Student-t distribution with $T-k$ degrees of freedom.

Exercises

1)
Estimate the investment function using the data in the EViews workfile "example_inv_chap3 workfile". Then apply an appropriate Chow test to verify parameter stability after 2007q1. In case of rejection of the null hypothesis, generate an appropriate dummy variable and check for its significance.

2)
Using the EViews workfile "example_simul2_chap5", regress *Y1* on a constant and *X*. By looking at the graph of *Y1* and the actual and fitted values and the regression residuals, you may have noticed the presence of a "break" in the second quarter of 1985. Apply an appropriate Chow test to check parameter stability and, in case of rejection of the null hypothesis, generate an appropriate dummy variable and check for its significance.

3)
Consider the model

$$y_t = a + bx_t + \varepsilon_t$$

Assuming you know $y_1,\ldots, y_t$, and $x_1,\ldots, x_{t+1}$, construct the optimal predictor for y_{t+1} and derive its variance.
If t=1,2,3 with ($x_1 = 0, y_1 = 1$), ($x_2 = -1, y_2 = 0$), and $x_3 = 1$, compute the best predictor of y_3 and its variance.
If t=1,2,3,4 with ($x_1 = 0, y_1 = 1$), ($x_2 = -1, y_2 = 0$), ($x_3 = 1, y_3 = 1$) and $x_4 = 2$, compute the best predictor of y_4 and its variance.

4)
Consider the model:

$$y_t = a + bx_t + \varepsilon_t$$

$\varepsilon_t \sim$ iid $N(0,\sigma^2)$ t=1,...,T_1,...,
Suppose that the residuals generated by separate regressions for the full sample and for the subsamples 1,...,T_1 and T_1+1,...,T, with T_1=4 and T=8, are the following:

$$\hat{\varepsilon}_{1,T} = \begin{bmatrix} -3.58 \\ 0.03 \\ -1.03 \\ 4.33 \\ 1.64 \\ 0.33 \\ -3.74 \\ -0-05 \end{bmatrix}; \quad \hat{\varepsilon}_{1,T_1} = \begin{bmatrix} 0.36 \\ -1.05 \\ -0.05 \\ 0.73 \end{bmatrix}; \quad \hat{\varepsilon}_{T_1,T} = \begin{bmatrix} 1.4 \\ -1.3 \\ -1.2 \\ 1.1 \end{bmatrix}$$

Are the model parameters stable? Motivate your answer using an appropriate parameter stability test.

5)
Consider the model:

$$y_t = x_t\beta + e_t$$
$$e_t \sim i.i.d.\ \mathrm{N}(0, \sigma_1^2) \quad \mathrm{t} = 1, \ldots, \mathrm{T}_1$$
$$e_t \sim i.i.d.\ \mathrm{N}(0, \sigma_2^2) \quad \mathrm{t} = \mathrm{T}_1 + 1, \ldots, \mathrm{T};$$

a) Construct the Chow test for H_0: $\sigma_1 = \sigma_2$ versus H_1: $\sigma_1 > \sigma_2$.
b) Suppose the null hypothesis in (a) is not rejected. Indicate two tests for the null hypothesis that β is stable over the two subsamples $1, \ldots, T_1$ and $T_1+1, \ldots .T$

6)
Consider the model

$$y_t = x_t\beta + e_t$$
$$e_t \sim i.i.d.\ \mathrm{N}(0, \sigma^2)$$

and suppose, x_{t+1}, β and σ^2 are known.

a) Indicate the best predictor for y_{t+1} given information available in period t, the associate forecast error and its variance.
b) Using the results in (a), construct the forecast interval for y_{t+1}, for a generic significance level.

7)
Consider the model

$$y_t = x_t\beta + e_t$$
$$e_t \sim i.i.d.\ \mathrm{N}(0, \sigma_1^2) \quad \mathrm{t} = 1, \ldots, \mathrm{T}_1$$
$$e_t \sim i.i.d.\ \mathrm{N}(0, \sigma_2^2) \quad \mathrm{t} = \mathrm{T}_1 + 1, \ldots, \mathrm{T};$$

with σ_1 and σ_2 known and different from each other.

a) Describe a test for the null hypothesis that β is stable over the two subperiods $1,\ldots,T_1$ e $T_1+1,\ldots T$ and indicate its distribution, assuming that in the second subperiod there are enough observations to re-estimate the model.

b) Describe a test for the null hypothesis that β is stable over the two subperiods $1,\ldots,T_1$ e $T_1+1,\ldots T$ and indicate its distribution, assuming that in the second subperiod there are not enough observations to re-estimate the model.

6. Stochastic Regressors

6.1 Stochastic regressors, independent of the error term

We have assumed thus far that explanatory variables, in contrast to the dependent one, are deterministic. This is convenient for a first analysis of the linear model but it is not realistic, especially in economics, where the nature of the variable to be explained and of the regressors are typically the same. We therefore now study what changes occur in the econometric theory of the linear regression model when the hypothesis of deterministic regressors is relaxed.
Let us consider the model:

$$\underset{T\times 1}{y} = \underset{T\times k}{X}\ \underset{k\times 1}{\beta} + \underset{T\times 1}{\varepsilon}, \tag{6.1.1}$$

and suppose that the elements of the matrix X are stochastic but *independently distributed of* the errors in the vector ε, thus relaxing hypothesis (1a) in Chapter 4. The hypotheses on the model become:

1) $E(\varepsilon \mid X)=E(\varepsilon)=E(y \mid X)-X\beta=X\beta-X\beta=0$.
2) $V(\varepsilon \mid X)= V(\varepsilon)=\sigma^2 I_T$.
3) $\varepsilon \mid X \sim N(0, \sigma^2 I_T)$.

The formula for the OLS estimator of the parameters β is unchanged:

$$\hat{\beta}_{OLS} = (X'X)^{-1}X'y = \beta + (X'X)^{-1}X'\varepsilon \tag{6.1.2}$$

Let us see what the properties of $\hat{\beta}_{OLS}$ are in this context. We have:

$$E(\hat{\beta}_{OLS}) = \beta + E_{f(\varepsilon,X)}\left[(X'X)^{-1}X'\varepsilon\right] =$$
$$= \beta + \underbrace{E_{f(x)}\left[(X'X)^{-1}X'\right]E_{f(\varepsilon|X)=f(\varepsilon)}(\varepsilon \mid X)}_{=0} = \beta \tag{6.1.3}$$

where $f(\varepsilon,X), f(\varepsilon|X), f(X), f(\varepsilon)$ indicate, respectively, the joint density of ε and *X*, that of ε given *X*, and the marginal densities of X and ε. Therefore, under assumption (1), the OLS estimator remains unbiased.

Next we have:

$$V\left(\hat{\beta}_{OLS}\right) = E\left[\left(\beta-\hat{\beta}\right)\left(\beta-\hat{\beta}\right)'\right] = E_{f(e,X)}\left[(X'X)^{-1}X'ee'X\left((X'X)^{-1}\right)'\right]$$
$$= E_{f(X)}\left[(X'X)^{-1}X'\sigma^2 IX(X'X)^{-1}\right] = \sigma^2 E_{f(X)}\left[(X'X)^{-1}\right] \tag{6.1.4}$$

This expression requires hypotheses (1) and (2). As $V\left(\hat{\beta}_{OLS}\right)$ converges to 0 when the sample size diverges to infinity, and $\hat{\beta}_{OLS}$ is unbiased, the OLS estimator is also consistent.

Strictly speaking, $\hat{\beta}_{OLS}$ is no longer the B.L.U.E. estimator. In fact, when *X* is stochastic, from (6.1.2) we see that $\hat{\beta}_{OLS}$ is no longer a linear function of *y*, it becomes a complicated product of random variables. However, if we consider the estimator conditional on *X*, then $\hat{\beta}_{OLS}\,|X$ is again a linear function of *y* and the linear, unbiased, minimum variance (B.L.U.E.) estimator of β.

Adding hypothesis (3) to (1) and (2), and continuing to consider $\hat{\beta}_{OLS}$ given *X*, it is:

$$\hat{\beta} \mid X \sim N\left(\beta, \sigma^2 (X'X)^{-1}\right). \tag{6.1.5}$$

Therefore, the inferential theory we developed in Chapter 3 still applies in this context. Without hypothesis (3), $\hat{\beta} \mid X$ has a Normal distribution in large samples and tests based on it have an asymptotic justification.
It can then be shown that for the estimator of the error variance:

$$\hat{\sigma}^2 = \hat{e}'\hat{e}/T - k \tag{6.1.6}$$

it is:

$$E(\hat{\sigma}^2) = \sigma^2 .$$

Moreover, $\hat{\sigma}^2$ is consistent.

It is also useful to note that the Maximum Likelihood Estimators (MLE) of the parameters can be obtained from $f(y, X) = f(y|X) f(X)$, considered as a function of the parameters. As the parameters only appear in $f(y|X)$, if *f(X)* does not contain any information on the parameters, then we can obtain the MLE estimators using $f(y|X)$ only. As it is $f(y|X) = f(\varepsilon \mid X) = f(\varepsilon)$, as for the case with deterministic regressors, it is $\hat{\beta}_{MLE} = \hat{\beta}_{OLS}$.

In conclusion, with stochastic regressors but independence of ε and X (meaning that each element of the former is independent of each element of the latter), and if $X\beta$ is interpreted as $E(y|X)$, then there are no substantial variations in the analysis of the linear regression model with respect to the case with deterministic regressors.

6.2 Stochastic regressors, asymptotically uncorrelated with the error term

Unfortunately, in many cases the hypothesis of independence of regressors and error is not satisfied. An important example is that of dynamic models, which we will study in Chapter 7. So let us replace the assumption of independence with less restrictive conditions and evaluate the consequences for estimation and inference in the linear model with stochastic regressors.

Assumptions (1) –(2) are replaced by:

1*) $E(\varepsilon)=0$
2*) $V(\varepsilon)=\sigma^2 I_T$.

and we need to add hypotheses on the asymptotic behavior of particular moments:

4*) plim $X'\varepsilon/T = 0$
5*) plim $X'X/T = \Sigma_{XX}$.
6*) plim $\varepsilon'\varepsilon/T = \sigma^2$

where, as usual, *plim* indicates the limit in probability of the sample moments. The key assumption is (4^*), which requires the regressors to be *asymptotically uncorrelated* with the error. Hypotheses (5^*) and (6^*) assume the existence of the limit of the sample second moments of the regressors and the error.

Let us now consider the OLS estimator of the parameters β. It is:

$$\hat{\beta} = \beta + (X'X)^{-1}X'\varepsilon = \beta + \left(\frac{X'X}{T}\right)^{-1}\frac{X'e}{T} \overset{T\to\infty}{\to} \beta + \Sigma_{XX}^{-1} * 0 = \beta\,. \qquad (6.2.1)$$

Hence, $\hat{\beta}$ is consistent. Note that with deterministic regressors it is true that $\frac{X'e}{T} \to 0$, which is an alternative proof of consistency of $\hat{\beta}_{OLS}$ in that context (in Chapter 2 for the proof we used instead the unbiasedness of $\hat{\beta}_{OLS}$ combined with $Var(\hat{\beta}) \to 0$).

Moreover, it is:

$$\hat{\sigma}^2 = \frac{\hat{\varepsilon}'\hat{\varepsilon}}{T-k} = \frac{\varepsilon'\left[I - X(X'X)^{-1}X'\right]\varepsilon}{T-k} =$$
$$= \left[\frac{\varepsilon'\varepsilon}{T} + \frac{\varepsilon'X}{T}\left(\frac{X'X}{T}\right)^{-1}\frac{X'\varepsilon}{T}\right]\frac{T}{T-k} \overset{T\to\infty}{\to} \sigma^2 + 0 * \Sigma_{XX}^{-1} * 0 = \sigma^2 \qquad (6.2.2)$$

so that $\hat{\sigma}^2$ is also consistent.

Finally, it can be shown that

$$\sqrt{T}(\hat{\beta}_{OLS} - \beta) = (X'X/T)^{-1}(X'\varepsilon/\sqrt{T}) \overset{T\to\infty}{\to} N(0, \sigma^2\Sigma_{XX}^{-1}) \tag{6.2.3}$$

so that $\hat{\beta}_{OLS}$ has a limiting Normal distribution. The term $\sigma^2\Sigma_{XX}^{-1}$ can be estimated by $\hat{\sigma}^2(X'X/T)^{-1}$.

Using the result in (6.2.3), we can also build interval estimators and test statistics, which will have an asymptotic justification. Note that, since all results are asymptotic, the hypothesis of Normality of the errors is not very relevant in this context.
In summary, if the stochastic regressors and the error are not independent but at least asymptotically uncorrelated, the toolbox to analyze the linear model is still applicable, as long as the sample is sufficiently large to justify the reference to asymptotic results.

6.3 Stochastic regressors, correlated with the error term

Let us now consider what happens if hypothesis (4*) does not apply, i.e., if the regressors are correlated with the error and, a fortiori, non-independent of the error (and $E(e|X)\neq 0$), and look at cases where this can happen.
From the derivation of $E(\hat{\beta})$ in (6.1.3) it follows that, when $E(e|X)\neq 0$, then $E(\hat{\beta}) \neq \beta$. Hence, the OLS estimator is biased. From (6.2.1), it instead follows that, if plim $X'\varepsilon/T \neq 0$, then $\hat{\beta}_{OLS}$ is not consistent, and the same holds for $\widehat{\sigma^2}$, given (6.2.2).
The hypothesis of no correlation between the explanatory variables and the error typically does not apply when the regressors are subject to measurement error. This happens frequently in economics, where there is often not a one-to-one correspondence between the theoretical definition of a variable and its observable counterpart. Examples include the natural rate of unemployment, the level of potential output, the desired amount of a certain variable or expectations about future realizations of an event. Even clearly defined economic concepts, such as gross domestic product or the rate of inflation, can be hard to measure without error.
To illustrate the consequences of measurement error, let us consider the model:

$$y = X\beta + \varepsilon \tag{6.3.1}$$

with

$$\text{cov}(X,\varepsilon)=0, \qquad 6.3.2)$$

and assume that the observed variables are:

$$X^* = X + v \qquad (6.3.3)$$

where v represents the measurement error, with

$$E(v) = 0, Var(v) = \sigma_v^2 I, Cov(X, v) = 0, \text{cov}(\varepsilon, v) = 0 \qquad (6.3.4)$$

The regression model in terms of observable variables is:

$$Y = X\beta + \varepsilon = X^*\beta + \varepsilon - v\beta \qquad (6.3.5)$$

and the OLS estimator of β is:

$$\hat{\beta}_{OLS} = \left(X^{*'}X^*\right)^{-1} X^{*'}y = \beta + \left(X^{*'}X^*\right)^{-1} X^{*'}(\varepsilon - v\beta) = \beta + \left(X^{*'}X^*\right)^{-1}(X + v)'(\varepsilon - v\beta) \qquad (6.3.6)$$

Assumptions (6.3.2) and (6.3.4) are insufficient to guarantee consistency of $\hat{\beta}_{OLS}$, as:

$$\hat{\beta}_{OLS} \xrightarrow{T\to\infty} \beta - \Psi^{-1}\beta\sigma_v^2 \text{ with } \left(\frac{X^{*'}X^*}{T}\right) \to \Psi \qquad (6.3.7)$$

What happens instead when the dependent variable, *y*, is measured with error? Let us assume it is:

$$y^* = y + u \qquad (6.3.7)$$

With

$$E(u) = 0, Var(u) = \sigma_u^2, \text{cov}(X, u) = 0, \text{cov}(\varepsilon, u) = 0 \qquad (6.3.8)$$

The regression model in terms of observable variables is:

$$y = y^* - u = X\beta + \varepsilon \text{ o } \; y^* = X\beta + \varepsilon + u, \tag{6.3.9}$$

with OLS estimator:

$$\hat{\beta}_{OLS} = \left(X'X\right)^{-1} X'y^* = \beta + \left(X'X\right)^{-1} X'(\varepsilon + u) \tag{6.3.10}$$

so that

$$\hat{\beta}_{OLS} \xrightarrow{T \to \infty} \beta \tag{6.3.11}$$

as X is uncorrelated with ε (by the assumption in (6.3.2)) and with u (by the assumption in (6.3.8)).
Hence, the OLS estimator $\hat{\beta}_{OLS}$ remains consistent (and unbiased), but its variance increases to $(\sigma_u^2 + \sigma_v^2)E(X'X)^{-1}$, so that its efficiency decreases.

6.4 Instrumental Variables (IV) and IV estimator

We now propose an alternative to the OLS estimator, called Instrumental Variable (IV) estimator, which is consistent even when the regressors are asymptotically (and in finite samples) correlated with the error, namely, (4*) does not hold and plim $X'\varepsilon/T$ is different from 0.
Let us assume there exist q variables grouped in the matrix Z with the following properties:

- plim $Z'\varepsilon/T = 0$ (6.4.1)
- plim $Z'X/T = \Sigma_{ZX}$ (6.4.2)
- plim $Z'Z/T = \Sigma_{ZZ}$ (6.4.3)

Thus, the variables Z must be asymptotically uncorrelated with the error, asymptotically correlated with the k explanatory variables X, and have a well-defined matrix of asymptotic second moments. The Z variables are labeled *Instrumental Variables (IV)*. We will see later that their number, q, must be

greater than or equal to be the number of regressors correlated with the error term. Let us assume for the moment that $q = k$.
The Instrumental Variable (IV) estimator is defined as:

$$\hat{\beta}_{IV} = (Z'X)^{-1} Z'y \tag{6.4.4}$$

It has the following properties:

$$\begin{aligned} - \hat{\beta}_{IV} &= (Z'X)^{-1} Z'y = \beta + (Z'X)^{-1} Z'\varepsilon \\ &= \beta + \left(\frac{Z'X}{T}\right)^{-1} \left(\frac{Z'\varepsilon}{T}\right) \xrightarrow{T\to\infty} \beta + \Sigma_{ZX}^{-1} \cdot 0 = \beta \end{aligned} \tag{6.4.5}$$

so that $\hat{\beta}_{IV}$ is consistent. Moreover,

$$\begin{aligned} &- Var\left(\hat{\beta}_{IV}\right) = E\left[\left(\hat{\beta}_{IV} - \beta\right)\left(\hat{\beta}_{IV} - \beta\right)'\right] = E\left[(Z'X)^{-1} Z'\varepsilon\varepsilon' Z (Z'X)^{-1}\right] \\ &\lim_{T\to\infty} Var\left(\sqrt{T}\hat{\beta}_{IV}\right) = \sigma^2 \Sigma_{ZX}^{-1} \Sigma_{ZZ} \Sigma_{ZX}^{-1} \end{aligned} \tag{6.4.6}$$

– The asymptotic distribution of $\hat{\beta}_{IV}$ is:

$$\hat{\beta}_{IV} \overset{a}{\sim} N(\beta, Var(\hat{\beta}_{IV})) \tag{6.4.7}$$

Note that $Var(\hat{\beta}_{IV})$ "decreases" when Σ_{ZX} "increases", where we use the quotation marks as we are talking of matrices so that the statement is not formally precise (on purpose). Therefore, we would like the IV to be highly correlated with the explanatory variables X (and uncorrelated with the error term ε), otherwise the IV estimator loses precision. The practical problem is to find instruments with these properties, which is not easy.
To derive the $\hat{\beta}_{IV}$ estimator we can proceed as follows. Let us pre-multiply both sides of the model

$$y = X\beta + e \tag{6.4.8}$$

by Z', which gives:

$$Z'y = Z'X\beta + Z'e \tag{6.4.9}$$

Then, imposing that the condition in (6.4.1) holds also in finite samples, namely,

$$Z'e=0 \tag{6.4.10}$$

we get:

$$Z'y = Z'X\hat{\beta}_{IV} \Rightarrow \hat{\beta}_{IV} = (Z'X)^{-1} Z'y \tag{6.4.11}$$

Note that, if we pre-multiply (6.4.8) by X' and impose no correlation of regressors and errors:

$$X'e=0 \tag{6.4.12}$$

we get:

$$X'y = X'X\hat{\beta}_{OLS} \Rightarrow \hat{\beta}_{OLS} = (X'X)^{-1} X'y \tag{6.4.13}$$

This estimator generation procedure is quite general and is called *Method of Moments estimation*. It is based on imposing the theoretical restrictions derived from economic or econometric theory on the empirical moments.
To conclude, we mention that we have by now seen five alternative procedures to derive the OLS estimator, $\hat{\beta}_{OLS}$:

- $min \sum \varepsilon_t^2$;
- $\min Var(\tilde{\beta})$ in the class of linear and unbiased estimators;
- such that $\hat{y}$ and $\hat{e}$ are orthogonal;
- as the MLE estimator;
- as the Method of Moments estimator.

6.5 Two-Stage Least Squares (TSLS) estimator and the over-identification test

Let us consider the linear model with a single explanatory variable:

$$y_i = \beta_1 X_{1i} + \varepsilon_i, \qquad i = 1,...T \tag{6.5.1}$$

and assume that X_1 is correlated with the error ε but there exists a valid instrumental variable, Z_1.

A particular IV estimator is obtained when using the fitted values in a regression of X_1 on the instrument Z_1, $\hat{X}_1$ as instruments. The first stage is therefore:

$$X_{1i} = \gamma_1 Z_{1i} + u_i \;\Rightarrow\; X_{1i} = Z_{1i}\hat{\gamma}_1 + \hat{u}_i = \hat{X}_{1i} + \hat{u}_i \tag{6.5.2}$$

where $\hat{\gamma}_1$ is the OLS estimator of γ_1. The second stage is:

$$\hat{\beta}_{1,TSLS} = \left(\hat{X}_1'\hat{X}_1\right)^{-1}\hat{X}_1'y \tag{6.5.3}$$

so that $\hat{\beta}_{TSLS}$ is the estimator of β_1 in:

$$y_i = \beta_1 \hat{X}_{1i} + v_i. \tag{6.5.4}$$

$\hat{\beta}_{TSLS}$ is called Two-Stage Least Squares (TSLS) estimator and it has the same properties as the IV estimators. In addition, we will see that the TSLS approach is convenient to compute IV estimators when the number of instruments is greater than that of explanatory variables correlated with the error, and therefore the formula for calculating $\hat{\beta}_{IV}$ in (6.4.4) does not apply.
The correlation between the instrument and the regressor (hypothesis 6.4.2) can be assessed by testing for the significance of Z_1 in (6.5.2). The better the explanatory capacity of Z_1 for X_1, the greater the precision of the estimator $\hat{\beta}_{1,TSLS}$.
Hypothesis (6.4.1), no correlation between instrument and error (also said *instrument validity*), instead cannot be tested, as the error is not observable.

However, suppose we have a second instrument, Z_2. Using the estimator $\hat{\beta}_{1,TSLS}$, the residuals of model (6.5.1) are:

$$\hat{\varepsilon}_i = y_i - \hat{\beta}_{1,TSLS} X_{1i}, \qquad i = 1,...T \tag{6.5.5}$$

Hence, we can now test for no correlation of Z_2 and the error (validity of Z_2) by replacing the error with the residual $\hat{\varepsilon}$. However, this procedure requires the presence of an a *priori* valid instrument, Z_1. We will see shortly how to formally test the hypothesis of no correlation in a more general case.
Let us now add a second regressor to the model in (6.5.1), X_2, assuming that it is uncorrelated with the error term

$$y_i = \beta_1 X_{1i} + \beta_2 X_{2i} + \varepsilon_i = \beta \ X_i + \varepsilon_i, \qquad i = 1,...T \tag{6.5.6}$$

In this case,

$$\hat{X} = (\hat{X}_1, \hat{X}_2) \tag{6.5.7}$$

And

$$\hat{\beta}_{TSLS} = \left(\hat{X}\ '\hat{X}\right)^{-1} \hat{X}\ 'y \tag{6.5.8}$$

Note that to compute $\hat{\beta}_{TSLS}$ in (6.5.8), the instrument Z_1 must be different from X_2 otherwise, since $\hat{X}_1 = Z_1 \hat{\gamma}_1$, the two vectors in $\hat{X}$ become perfectly collinear and the matrix $\left(\hat{X}\ '\hat{X}\right)$ is non invertible. As we have seen in Chapter 2, in this case the model parameters are not identified.
More generally, a necessary condition for identifiability of the coefficients of the linear model when some regressors are correlated with the error is that the number of instruments is at least equal to the number of regressors that are correlated with the error (assuming implicitly that no instrument is also an explantory variable for y). This condition is often referred to as *order condition.*

The regressors that are correlated with the error are also called *endogenous variables,* the others (those not correlated with the error) are the *exogenous variables.*

Let us now consider an example that shows how the stated order condition is not sufficient for the identifiability of the model parameters. In model (6.5.6), let us assume that X_2 is also correlated with the error but there is a second valid instrument, Z_2. Moreover, it is:

$$\begin{aligned} X_{1i} &= \gamma_{11} Z_{1i} + \gamma_{12} Z_{2i} + u_{1i} \\ X_{2i} &= \gamma_{21} Z_{1i} + \gamma_{22} Z_{2i} + u_{2i} \end{aligned} \tag{6.5.9}$$

The order condition is clearly satisfied, as there are two regressors correlated with the error term and two instruments (that are uncorrelated with the error and do not affect y). However, if $\gamma_{12} = \gamma_{22} = 0$, then both X_1 and X_2 depend on Z_1 only, so that $\hat{X}_1$ and $\hat{X}_2$ are perfectly collinear, the matrix $\left(\hat{X}'\hat{X}\right)^{-1}$ cannot be inverted, and the TSLS estimator $\hat{\beta}_{TSLS}$ cannot be computed.

Let us now look at the general case. We suppose that the dependent variable y depends on k regressors correlated with the error, X, and p regressors uncorrelated with the residual, W:

$$\underset{T\times 1}{y} = \underset{T\times k}{X}\;\underset{k\times 1}{\beta} + \underset{T\times p}{W}\;\underset{p\times 1}{\gamma} + \underset{T\times 1}{\varepsilon} \tag{6.5.10}$$

There are q valid instruments available, Z, with

$$q \geq k \tag{6.5.11}$$

$$\underset{T\times k}{X} = \underset{T\times q}{Z}\;\underset{q\times k}{\Gamma} + \underset{T\times k}{u} = \underset{T\times q}{Z}\;\underset{q\times k}{\hat{\Gamma}} + \underset{T\times k}{\hat{u}} = \underset{T\times k}{\hat{X}} + \underset{T\times k}{\hat{u}} \tag{6.5.12}$$

Where

$$\underset{q\times k}{\hat{\Gamma}} = \underset{q\times q}{\left(Z'Z\right)^{-1}}\;\underset{q\times k}{\left(Z'X\right)} \tag{6.5.13}$$

And

$$\text{cov}(W,\varepsilon)=0,\ \text{cov}(Z,\varepsilon)=0,\ \text{cov}(W,u)=0,\ \text{cov}(Z,u)=0. \tag{6.5.14}$$

To compute the TSLS estimator of the parameters of the model in (6.5.10), let us re-write it as:

$$\underset{T\times 1}{y} = \underset{T\times k}{\hat{X}}\ \underset{k\times 1}{\beta} + \underset{T\times p}{W}\ \underset{p\times 1}{\gamma} + \underset{T\times 1}{\varepsilon} + \underset{T\times 1}{\hat{u}\beta} =$$
$$= \begin{pmatrix} \underset{T\times k}{\hat{X}} & \underset{T\times p}{W} \end{pmatrix} \begin{pmatrix} \underset{k\times 1}{\beta} & \underset{p\times 1}{\gamma} \end{pmatrix} + \underset{T\times 1}{\xi} = \underset{T\times k+p}{R}\ \underset{k+p\times 1}{\eta} + \underset{T\times 1}{\xi}. \tag{6.5.15}$$

Given the assumptions in (6.5.14), the regressors in R are uncorrelated with the error ξ and therefore the parameters of (6.5.15) can be consistently estimated by OLS. The resulting OLS estimator is the TSLS estimator for the parameters of the original model in (6.5.10):

$$\underset{k+p\times 1}{\hat{\eta}_{OLS}} = \underset{k+p\times 1}{\hat{\eta}_{TSLS}} = \underset{k+p\times k+p}{(R'R)^{-1}}\ \underset{k+p\times 1}{(R'y)}. \tag{6.5.16}$$

The necessary (order) condition for the identifiability of parameters is respected, as $q \geq k$ from (6.5.11). The sufficient condition requires the rank of $(R'R)$ to be equal to $k + p$. This is often called *rank condition*.

When $q > k$, we can test that q-k instruments are uncorrelated with the error. Since when $q > k$ there are additional instruments besides those strictly necessary for identification, this case is labeled *over-identification*, and the test for the validity of the q-k additional instruments is called *over-identification test*.

The over-identification test can be built as follows. First, we compute the residuals of the regression in (6.5.15), $\hat{\xi}$, as:

$$\hat{\xi} = y - R\hat{\eta}_{TSLS} \tag{6.5.17}$$

Next, we regress $\hat{\xi}$ on all the variables that are uncorrelated with the error term ε:

$$\hat{\xi} = Z\alpha + W\phi + v \tag{6.5.18}$$

Finally, if we indicate by R_ζ^2 the coefficient of determination in (6.5.18), and as usual with T the sample size, the test statistic for the null hypothesis of no correlation between all the instruments Z and the error ε is:

$$TR_\zeta^2 \overset{a}{\sim} \chi^2(s) \qquad (6.5.19)$$

where $\overset{a}{\sim}$ indicates the asymptotic distribution, $s = q\text{-}k$, and the rejection region is one sided, so that the null hypothesis is rejected when $TR_\zeta^2 > \chi_c^2$, where χ_c^2 is the proper critical value given the chosen significance level of the test.

The rationale of this procedure is that if the instruments are uncorrelated with the errors, the regressors (6.5.18) should not be significant, and then R_ζ^2 should be low. Note that the residuals $\hat{\xi}$ are uncorrelated by construction with $\hat{X}$ and W. If $q = k$, then $\hat{X}$ is a simple rotation of Z, $\hat{X} = Z\hat{\Gamma}$, and Z and $\hat{\xi}$ are uncorrelated by construction. For this reason, it is not possible to test the hypothesis of no correlation of the instruments with the errors in the case of exact identification, and even in the case of over-identification the hypothesis can be only tested for q-k instruments (assuming *a priori* that it holds for the remaining k instruments).

6.6 The Hausman Test

Let us consider again the model

$$y = X\beta + \varepsilon \qquad (6.6.1)$$

and search for a statistic to test the null hypothesis of no correlation between the stochastic regressors and the error term against the alternative of correlation, namely, $H_0 : \text{cov}(X,\varepsilon) = 0,\;\; H_1 : \text{cov}(X,\varepsilon) \neq 0$.

To build the test, let us consider the OLS and IV estimators of the parameters β in (6.6.1). Under the null hypothesis, we know that the OLS estimator is consistent and efficient, while the IV estimator, though consistent, is inefficient. Under the alternative, the IV estimator is instead consistent, but OLS becomes inconsistent, as we saw in Section 6.3. Hence, let us consider the statistic:

$$H = \left(\hat{\beta}_{IV} - \hat{\beta}_{OLS}\right)' \left[V\left(\hat{\beta}_{IV}\right) - V\left(\hat{\beta}_{OLS}\right)\right]^{-1} \left(\hat{\beta}_{IV} - \hat{\beta}_{OLS}\right) \underset{H_0}{\sim} \chi^2(k) \qquad (6.6.2)$$

which is known as *Hausman Test.* Under H_0, since both $\hat{\beta}_{OLS}$ and $\hat{\beta}_{IV}$ are consistent, the H statistic will be close to 0, while it will be large under H_1 due to inconsistency of $\hat{\beta}_{OLS}$. Therefore, the rejection region is one-sided, $[\chi_c^2, \infty)$, where χ_c^2 is the critical value from the $\chi^2(k)$ distribution, with k the number of regressors and for a given significance level α.

Note that the statistic in (6.6.2) is standardized by $V\left(\hat{\beta}_{IV}\right) - V\left(\hat{\beta}_{OLS}\right)$ rather than $V(\hat{\beta}_{IV} - \hat{\beta}_{OLS})$. Actually, these two variance matrices are equal. In fact:

$$V\left(\hat{\beta}_{IV} - \hat{\beta}_{OLS}\right) = V\left(\hat{\beta}_{IV}\right) + V\left(\hat{\beta}_{OLS}\right) - 2\operatorname{cov}\left(\hat{\beta}_{OLS}, \hat{\beta}_{IV}\right) \qquad (6.6.3)$$

but as the OLS estimator is the most efficient in the linear class, it must be

$$\operatorname{cov}\left(\hat{\beta}_{OLS}, \hat{\beta}_{OLS} - \hat{\beta}_{IV}\right) = 0; \qquad (6.6.4)$$

otherwise, it can be shown that is possible to build an alternative estimator with lower variance than OLS by taking a linear combination of $\hat{\beta}_{OLS}$ e $\hat{\beta}_{OLS} - \hat{\beta}_{IV}$. From (6.6.4) it follows that:

$$\operatorname{cov}\left(\hat{\beta}_{OLS}, \hat{\beta}_{OLS} - \hat{\beta}_{IV}\right) = V\left(\hat{\beta}_{OLS}\right) - \operatorname{cov}\left(\hat{\beta}_{OLS}, \hat{\beta}_{IV}\right) = 0 \qquad (6.6.5)$$

F
rom which

$$\operatorname{cov}\left(\hat{\beta}_{OLS}, \hat{\beta}_{IV}\right) = V\left(\hat{\beta}_{OLS}\right) \qquad (6.6.6)$$

and, by replacing (6.6.6) in (6.6.3),

$$V\left(\hat{\beta}_{IV} - \hat{\beta}_{OLS}\right) = V(\hat{\beta}_{IV}) - V(\hat{\beta}_{OLS}) \qquad (6.6.7)$$

An alternative procedure to test $H_0 : \text{cov}(X, \varepsilon) = 0$ is the following. First, we regress by OLS each regressor in X on all the instruments Z:

$$X_i = Z\gamma_i + v_i \qquad i=1,...,k, \tag{6.6.8}$$

With

$$X_i = Z\hat{\gamma}_i + \hat{v}_i \tag{6.6.9}$$

Next, we group the k residuals from these regressions in the vector $\hat{v}$ and we treat them as additional regressors in the model in (6.6.1):

$$y = X\beta + \hat{v}\delta + u \tag{6.6.10}$$

Finally, we run an F-test for $\delta = 0$ in (6.6.10). It can be shown that this F-test is equal to the H-statistic. The rationale is that under H_0 the residuals $\hat{v}$ should have no explanatory power for y, once conditioning on X as in (6.6.10). As an example of an application of the alternative version of the Hausman test, we reconsider the situation analyzed in Section 6.3, where the regressors may be subject to measurement error. If:

$$\begin{aligned} y &= X\beta + \varepsilon \\ X^* &= X + v \end{aligned} \tag{6.6.11}$$

then X^* is likely correlated with the error u in the model:

$$y = X^*\beta + u \tag{6.6.12}$$

If there are k valid instruments, Z, we regress each component of X on Z, which gives:

$$X^* = Z\Gamma + \omega = Z\hat{\Gamma}_{OLS} + \hat{\omega} \tag{6.6.13}$$

Finally, we run the regression:

$$y = X^*\beta + \hat{\omega}\delta + u \tag{6.6.14}$$

and test the significance of $\hat{\omega}$ by means of an *F*-test, namely, we test for $\delta = 0$ that is equivalent to $corr(X^*, u) = 0$ and therefore implies consistency of the OLS estimator for the β parameters.

6.7 An empirical analysis based on simulated data

Data and output for this analysis are contained in the "example_simul_chap6.wf1" workfile. Let us consider a small economy, A, whose real interest rate, *y*, is determined by *x*, the corresponding rate of a large foreign economy, B. Given that A's debt-to-GDP ratio is lower than that of B, we can expect the rate in A to be lower than that in B, due to the lower default risk. Assuming the average differential between the rates is of two points, we would expect the relationship

$$y = -2 + x \quad (6.7.1)$$

To verify if this relationship is satisfied from the data, we can run an *F*-test on the coefficients of a regression of *y* on *x* and an intercept. The estimated values of the parameters are shown in Table 6.1, and the equation is labeled OLS for future use.

Dependent variable: Y
Method: Least Squares
Sample: 1951q1 2003q4
Included observations: 212

Variable	Coefficient	Std. Error	*t*-Statistic	Prob.
C	-1.921902	0.043808	-43.87106	0.0000
X	1.503399	0.032343	46.48347	0.0000

R-squared	0.911419	Mean dependent var	-2.025141
Adjusted R-squared	0.910997	S.D. dependent var	2.135308
S.E. of regression	0.637033	Akaike info criterion	1.945400
Sum squared resid	85.22043	Schwarz criterion	1.977066
Log likelihood	-204.2124	*F*-statistic	2160.713
Durbin-Watson stat	1.661925	Prob(*F*-statistic)	0.000000

Table 6.1: OLS regression of y on x and intercept

The estimated value of the intercept is close enough to that suggested in (6.7.1), but that of the *x* variable is instead about 50% larger. The *F* -test for the null hypothesis that the coefficients are equal to -2 and 1 is equal to 121.62, with a *p*-value of 0.000, so that the hypothesis is strongly rejected. Also, the value of the Durbin-Watson statistic indicates to further investigate the hypothesis of no serial correlation of the errors, and the LM test with two lags gives a *p*-value of 0.01, leading to a rejection of the hypothesis of no correlation for a significance level of 5%.
The source of these results could be the correlation between *x* and the error, due for example to the difficulty of measuring the real interest rate, which depends on unobservable expected future inflation. We then need a good instrument for *x*, and we can try with the nominal interest rate in country B, called *z*, which is independently decided by the Central Bank of B.
The real rate *x* is positively correlated with *z* (the sample correlation between *x* and *z*, calculated by using the command "=@cor(x,z)" is approximately 0.72), and *z* is highly significant in the regression of *x* on *z* and a constant. In addition, the fact that *z* is decided independently by the Central Bank of B suggests that it is exogenous and hence uncorrelated with the error in the regression of *y* on *x*, Then *z* is a valid instrument for *x*.
In summary, we are facing a potential case of stochastic regressors measured with errors. We also have a valid instrument and we can use it to construct an IV estimator for the model parameters. In particular, EViews makes the TSLS estimator directly available.
From the main toolbar, we choose "Quick" and then "Estimate Equation". We select "TSLS – Two Stage Least Squares (TSLS and ARMA)" and indicate in the first field the model:

```
y c x
```

and in the "Instrument list" field:

```
c z
```

Note that EViews requires to always insert the intercept in the instrument list if it appears in the model. Clicking "Ok" we get Table 6.2, and the underlying equation is labeled TSLS for future use.

Dependent variable: Y
Method: Two-Stage Least Squares
Sample: 1951q1 2003q4
Included observations: 212
Instrument specification: C Z

Variable	Coefficient	Std. Error	*t*-Statistic	Prob.
C	-1.952470	0.060490	-32.27772	0.0000
X	1.058252	0.061899	17.09631	0.0000

R-squared	0.831514	Mean dependent var	-2.025141
Adjusted R-squared	0.830711	S.D. dependent var	2.135308
S.E. of regression	0.878566	Sum squared resid	162.0946
F-statistic	292.2837	Durbin-Watson stat	2.033008
Prob(*F*-statistic)	0.000000	Second-Stage SSR	736.4550
J-statistic	0.000000	Instrument rank	2

Table 6.2: TSLS regression of y on x and intercept

The estimated coefficient of x is now very close to 1. In addition, the value of the F-test for the joint hypothesis $\alpha = -2, \beta = 1$ against the alternative hypothesis "at least one of the two equalities is not true" is 0.703, with a p-value of 0.496, allowing us not to reject the null hypothesis. In addition, the DW and LM tests for no residual correlation do not reject the null hypothesis. Is there correlation between x and the errors? We can test this hypothesis with the Hausman test:

$$\left(\hat{\beta}_{IV} - \hat{\beta}_{OLS}\right)' \left[V\left(\hat{\beta}_{IV}\right) - V\left(\hat{\beta}_{OLS}\right)\right]^{-1} \left(\hat{\beta}_{IV} - \hat{\beta}_{OLS}\right) \underset{H_0}{\overset{d}{\rightarrow}} \chi^2(2)$$

as in this example we have two regressors. This statistic is not directly available in EViews, but we can compute it in a few steps. First, we get $V\left(\hat{\beta}_{OLS}\right)$ with the command:

```
matrix vols=ols.@cov
```

which creates a matrix called "vols" that is equal to the variance-covariance matrix of the estimators ("@cov") of the equation we labeled ols ("ols."). Next, and similarly, for the tsls equation we type:

```
matrix vts=tsls.@cov
```

Further, the command:

```
matrix ih=@inverse(vts-vols)
```

computes $\left[V\left(\hat{\beta}_{IV}\right)-V\left(\hat{\beta}_{OLS}\right)\right]^{-1}$.

Next, we need to define a vector that contains the difference between the OLS and IV estimators. We type:

```
vector diff=tsls.@coef-ols.@coef
```

where "@coef" tells EViews to take the estimated coefficients of the relevant equations (further explanations and options are available, as usual, from the "Help" menu).

Finally, the value of the Hausman statistic is computed with:

```
=@transpose(diff)*ih*diff
```

If we want to save the value of the statistic, we use the command:

```
vector th=@transpose(diff)*ih*diff"
```

and then we can recover the value with the command "th(1)", which returns the first (and unique) component of the vector "th".

To compute the *p*-value associated with the Hausman test, we use the command:

```
=@chisq(@transpose(diff)*ih*diff,2)
```

wich in our example returns a very low value, $3.33 \cdot 10^{-16}$. Therefore, we strongly reject the null hypothesis of exogeneity of x (no correlation between regressor and error). This implies that we should trust the IV rather than OLS estimators and, in turn, this leads to the acceptance of the hypothesis that the real rates in country A are equal to those of B minus a risk premium.

The alternative version of the Hausman test we have considered in Section 6.6 requires in this example to implement a *t*-test for the significance of $\hat{u}$ (the residuals in the regression of x on z) in the OLS regression of y on x and $\hat{u}$. Hence, we first run the regression of x on z and save the residuals, using the commands:

```
ls x c z
series uhat=resid
```

Next, we regress y on x and $\hat{u}$ and report the results in Table 6.3. It turns out that $\hat{u}$ is strongly significant, in line with the rejection of the null hypothesis by the Hausman statistic.

Dependent variable: Y

Variable	Coefficient	Std. Error	t-Statistic	Prob.
C	-1.952470	0.007031	-277.6945	0.0000
X	1.058252	0.007195	147.0844	0.0000
UHAT	0.926003	0.010377	89.23512	0.0000

R-squared	0.997735	Mean dependent var	-2.025141
Adjusted R-squared	0.997713	S.D. dependent var	2.135308
S.E. of regression	0.102120	Akaike info criterion	-1.711289
Sum squared resid	2.179549	Schwarz criterion	-1.663790
Log likelihood	184.3967	Hannan-Quinn criterion	-1.692091
F-statistic	46022.26	Durbin-Watson stat	2.070081
Prob(F-statistic)	0.000000		

Table 6.3: An alternative version of the Hausman test

6.8 An empirical analysis of aggregate consumption

This time, let us try to explain the growth rate of aggregate private consumption with that of income, using data in the "example_cons_chap6.wf1" workfile. We also add to the model a dummy variable labeled "CRISIS", whose value is 0 up to 2007q2 (inclusive) and 1 afterwards. The dummy, interacted with income growth, aims to capture a possible structural change in the consumption-income relationship associated with the financial crisis. The resulting estimated model is called "OLS" and reported in Table 6.4.

Table 6.4 shows that income growth and the dummy are strongly significant, while the intercept is not. Specifically, it seems that after the crisis the change in consumption reacts much less to a change in income. It has to be noted that a graph of the residuals suggests there are substantial violations of the OLS assumptions, but here we focus on evaluating whether there are endogeneity problems, as in the previous example.

What are the appropriate instruments in this context? A common choice in macroeconomic analyses is lagged values of the explanatory variables. This option typically works when the regression errors are not correlated over time while the explanatory variables are. In our example we could therefore use as instruments income growth in periods t-1, t-2, etc. Let us just use the first lag so that the model is exactly identified. We also include the dummy in the list of instruments (as we do for the intercept)[1]. In EViews, we use the TSLS estimation option, with "c d(lryd(-1)) crisis" as instruments. The outcome is shown in Table 6.5.

[1] Using DLRYD(-1)*CRISIS(-1) instead of CRISIS in the instrument list produces very similar results.

Dependent variable: DLRC
Method: Least Squares
Sample (adjusted): 1990q2 2012q1
Included observations: 88 after adjustments

Variable	Coefficient	Std. Error	*t*-Statistic	Prob.
C	0.000420	0.000999	0.420652	0.6751
DLRYD	0.870766	0.046159	18.86465	0.0000
CRISIS*DLRYD	-0.559093	0.170955	-3.270403	0.0016

R-squared	0.808714	Mean dependent var	-1.65E-05
Adjusted R-squared	0.804213	S.D. dependent var	0.021035
S.E. of regression	0.009307	Akaike info criterion	-6.482524
Sum squared resid	0.007363	Schwarz criterion	-6.398069
Log likelihood	288.2310	Hannan-Quinn criterion	-6.448499
F-statistic	179.6806	Durbin-Watson stat	2.720654
Prob(*F*-statistic)	0.000000		

Table 6.4: OLS regression of consumption growth on income growth

Dependent variable: DLRC
Method: Two-Stage Least Squares
Sample (adjusted): 1990q3 2012q1
Included observations: 87 after adjustments
Instrument specification: C DLRYD(-1) CRISIS

Variable	Coefficient	Std. Error	*t*-Statistic	Prob.
C	0.000310	0.003803	0.081526	0.9352
DLRYD	2.228595	1.381005	1.613748	0.1103
CRISIS*DLRYD	-2.258306	2.847987	-0.792948	0.4300

R-squared	-1.143272	Mean dependent var	8.25E-05
Adjusted R-squared	-1.194303	S.D. dependent var	0.021136
S.E. of regression	0.031309	Sum squared resid	0.082342
F-statistic	1.304667	Durbin-Watson stat	2.242869
Prob(*F*-statistic)	0.276702	Second-Stage SSR	0.035861
J-statistic	0.000000	Instrument rank	3

Table 6.5: TSLS regression of consumption growth on income growth, with lagged income growth as instrument

In the IV regression, the estimated coefficient of DLRYD is much larger than with OLS estimation, but its standard error is also larger, so much so that the coefficient is no longer statistically significant. We saw that if the instrument is not highly correlated with the endogenous regressors, then the variance of the

IV estimator increases. This is just what happens in our case, since a regression of DLRYD on DLRYD (-1) shows that the explanatory variable is not significant
Using instead DLRC (-1) as an instrument, as it explains well DLRYD, Table 6.6 shows that the estimated values of the parameters are similar to those obtained with OLS. We call this equation "IV" and save the output for subsequent processing.

Dependent variable: DLRC
Method: Two-Stage Least Squares
Sample (adjusted): 1990q3 2012q1
Included observations: 87 after adjustments
Instrument specification: C DLRC(-1) CRISIS

Variable	Coefficient	Std. Error	*t*-Statistic	Prob.
C	0.000157	0.001396	0.112694	0.9105
DLRYD	1.156891	0.230249	5.024529	0.0000
CRISIS*DLRYD	-1.234387	0.983304	-1.255347	0.2128

R-squared	0.709911	Mean dependent var	8.25E-05
Adjusted R-squared	0.703004	S.D. dependent var	0.021136
S.E. of regression	0.011519	Sum squared resid	0.011145
F-statistic	12.70621	Durbin-Watson stat	2.976491
Prob(*F*-statistic)	0.000015	Second-Stage SSR	0.035047
J-statistic	6.45E-46	Instrument rank	3

Table 6.6: TSLS regression of consumption growth on income growth, with lagged consumption growth as instrument

To evaluate whether IV estimation really is useful, we now apply the Hausman test for the null hypothesis that the covariance between errors and regressors is 0 against the alternative hypothesis that the covariance is different from 0. The EViews commands are the following:

```
vector(2) bols=ols.@coefs
matrix(2,2) vols=ols.@coefcov
vector(2) biv=iv.@coefs
matrix(2,2) viv=iv.@coefcov
vector(1) haus1 = @transpose(bols-biv)*@inverse((viv-vols))*(bols-biv)
vector(1) pvhaus1=1-@cchisq(haus1(1),2)
```

The resulting *p*-value is well above 0.10, so that the null hypothesis is not rejected and OLS estimation is preferable.

We can now replicate the analysis adding DLRC(-2) and CRISIS(-1) as instruments, which gives an over-identified model as the number of instruments is larger than that of endogenous variables. Table 6.7 presents the estimation results, which are rather similar to those in Table 6.6.

Dependent variable: DLRC
Method: Two-Stage Least Squares
Sample (adjusted): 1990q4 2012q1
Included observations: 86 after adjustments
Instrument specification: C DLRC(-1) DLRC(-2) CRISIS CRISIS(-1)

Variable	Coefficient	Std. Error	*t*-Statistic	Prob.
C	0.000134	0.001319	0.101917	0.9191
DLRYD	1.159242	0.232553	4.984855	0.0000
CRISIS*DLRYD	-1.193206	0.673927	-1.770526	0.0803

R-squared	0.710904	Mean dependent var	0.000202
Adjusted R-squared	0.703938	S.D. dependent var	0.021230
S.E. of regression	0.011552	Sum squared resid	0.011076
F-statistic	12.48152	Durbin-Watson stat	2.968538
Prob(*F*-statistic)	0.000018	Second-Stage SSR	0.034981
J-statistic	0.010331	Instrument rank	5
Prob(J-statistic)	0.994848		

Table 6.7: TSLS regression of consumption growth on income growth, with consumption growth in t-1 and t-2 as instruments

Also in this case, the Hausman test supports OLS estimation, as it does not reject the null hypothesis so that OLS is consistent and efficient. But, as we are now considering the case of over-identification, we can test whether one of the two instruments is valid by applying the test that we have introduced in Section 6.5. Hence, in the first stage we regress DLRYD on the instruments and save the fitted values. Next, in the second stage, we regress DLRC on a constant and the fitted values of the first stage regression. Finally, we regress the residuals of the second stage regression on a constant and the entire set of instruments. Specifically, the EViews commands are:

```
equation sovraid1.ls d(lryd) c d(lrc(-1)) d(lrc(-2)) crisis crisis(-1)
series resid01=res
genr dlrydf=dlryd-resid01
equation sovraid2.ls d(lrc) c dlrydf
```

```
series res=resid
equation sovraid3.ls res c d(lryd(-1)) d(lryd(-2)) crisis crisis(-1)
```

The last step is to use the R^2 of this regression (sovraid3. @r2) multiplied by the number of observations (with the command sovraid3. @regobs) to get the test statistic, which is distributed as Chi-square with as many degrees of freedom as the over-identification conditions (two in this case). The *p*-value is calculated using the following command:

```
scalar pvsovraid=@chisq((sovraid3.@r2)*(sovraid3.@regobs),2)
```

and the result is 0.608. Hence, the null hypothesis of no correlation (validity) of the additional instruments with errors is not rejected.

To conclude, we should point out that the use of the lagged endogenous variable as an instrument requires the error term to be uncorrelated over time (which is unlikely in this example so we should search for better instruments, which we leave as an exercise). Actually, if ε_t is correlated with ε_{t-1}, since y_t is correlated with ε_t by definition and therefore y_{t-1} is also correlated with ε_{t-1}, it follows that y_{t-1} is also correlated with ε_t, and this makes it an invalid instrument.

The main concepts of this chapter

In this chapter we have seen that if we have stochastic regressors but they are independent of the error term, then the OLS estimator remains unbiased and consistent. If it is also $V(\varepsilon|X)= V(\varepsilon)=\sigma^2 I_T$, then $V(\hat{\beta}_{OLS})= \sigma^2 E_{f(X)}\left[(X'X)^{-1}\right]$ and moreover $\hat{\beta}_{OLS}|X$ is a linear function of y and it is B.L.U.E. for β. With the additional assumption that $\varepsilon|X=\varepsilon \sim N(0, \sigma^2 I_T)$, then $\hat{\beta}\,|\,X \sim N\left(\beta, \sigma^2 (X'X)^{-1}\right)$. Without Normal errors, the OLS estimator is only asymptotically Normal, as in the case with fixed regressors, and tests based on a Normal distribution for the OLS estimator have an asymptotic justification. The OLS estimator $\hat{\sigma}^2$ is also unbiased and consistent. Overall, there are no major changes with respect to the case with fixed regressors, once we interpreted all the estimators as conditioned on X.

When instead the regressors are only asymptotically uncorrelated with the error term, and $E(\varepsilon)=0$, $V(\varepsilon)=\sigma^2 I_T$, $plim\ X'X/T = \Sigma_{XX}$, $plim\ \varepsilon'\varepsilon/T = \sigma^2$, $\hat{\beta}_{OLS}$ and $\hat{\sigma}_{OLS}$ are consistent (not necessarily unbiased in finite samples), and $\hat{\beta}_{OLS}$ is asymptotically normally distributed. Using these results, it is possible to construct interval estimators and test statistics, which have an asymptotic justification.

If instead $E(e|X) \neq 0$, then the explanatory variables are endogenous and $\hat{\beta}_{OLS}$ is biased, while if $plim\ X'\varepsilon/T \neq 0$, then $\hat{\beta}_{OLS}$ is not consistent. Measurement error is a typical cause of correlation between the regressors and the errors, while if only the dependent variable is measured with error, then $\hat{\beta}_{OLS}$ remains consistent but its variance increases. To obtain a consistent estimator for the parameters in the presence of endogenous regressors, we introduced the Instrumental Variable (IV) estimator. If there are $q=k$ variables (the instruments) grouped in Z such that $plim\ Z'\varepsilon/T = 0$, $plim\ Z'X/T = \Sigma_{ZX}$, $plim\ Z'Z/T = \Sigma_{ZZ}$, then we can use them to form the estimator $\hat{\beta}_{IV} = (Z'X)^{-1} Z'y$. This estimator is consistent, $Var(\hat{\beta}_{IV}) = E\left[(Z'X)^{-1} Z'\varepsilon\varepsilon' Z (Z'X)^{-1}\right] \overset{T\to\infty}{\to} \frac{\sigma^2}{T} \Sigma_{ZX}^{-1} \Sigma_{ZZ} \Sigma_{ZX}^{-1}$ and asymptotically normally distributed.

A particular IV estimator, which can be used also when $q>k$, is obtained by regressing in a first stage the explanatory variables X on the instruments Z, and in a second stage the dependent variable y the on fitted values of the first stage regression. This is the Two-Stage Least Squares (TSLS) estimator, which has the same properties of IV estimators. In particular, the better the explanatory capacity of Z for X, the greater the precision of $\hat{\beta}_{TSLS}$.

The necessary, but not sufficient, condition for identifiability of the model parameters β is that the number of instruments is at least equal to the number of endogenous regressors. This is often referred to as *order condition*. The sufficient condition for identifiability is instead called *rank condition* and, basically, it requires invertibility of a specific matrix that can be defined as follows. If y depends on k endogenous and p exogenous regressors, X and W respectively, and there are q instruments, Z, with $q \geq k$ (order condition), the TSLS estimator in the regression $y = (\hat{X}\ \ W)\eta + \xi = R\eta + \xi$ is $\hat{\eta}_{TSLS} = (R'R)^{-1}(R'y)$. Hence, for $\hat{\eta}_{TSLS}$ to be unique, the rank of the matrix $(R'R)$ must be equal to $k+p$, which is the rank condition.

If q is strictly greater than k, we can build an over-identification test whose null hypothesis is that the $(q-k)$ instruments in addition to the k strictly needed to satisfy the order condition are uncorrelated with the error term. The test is performed by calculating the residuals of the regression of y on R, $\hat{\xi} = y - R\hat{\eta}_{TSLS}$, regressing $\hat{\xi}$ on all the variables uncorrelated with the error

ε, $\hat{\xi} = Z\alpha + W\phi + v$ and computing $TR_{\xi}^2 \overset{a}{\sim} \chi^2(s)$, where s =*q*-*k* and the rejection region is one-sided.

Finally, we have considered a statistic to test whether the regressors are endogenous or not, the Hausman test. The rationale of the test is that, defining $H_0 : \text{cov}(X, \varepsilon) = 0$, $H_1 : \text{cov}(X, \varepsilon) \neq 0$, under H_0 the OLS estimator is consistent and efficient while the IV estimator is also consistent but inefficient. Instead, under the alternative hypothesis, the IV estimator remains consistent while OLS loses this property. The Hausman statistic is $H = \left(\hat{\beta}_{IV} - \hat{\beta}_{OLS}\right)' \left[V\left(\hat{\beta}_{IV}\right) - V\left(\hat{\beta}_{OLS}\right)\right]^{-1} \left(\hat{\beta}_{IV} - \hat{\beta}_{OLS}\right) \underset{H_0}{\sim} \chi^2(k)$. If H_0 is true, the statistic will take lower values because both estimators will be close to the true value of the coefficients, vice versa under H_1. Hence, the rejection of the test is one-sided, $[\chi_c^2, \infty)$, where χ_c^2 is the critical value taken from the $\chi^2(k)$ distribution for a chosen significance level, and *k* is the number of regressors.

An alternative version of the Hausman test, with the same null and alternative hypotheses, requires to run an OLS regression of each variable in *X* on all the instruments in *Z*: $X_i = Z\gamma_i + v_i$, *i*=*1*,…,*k*, to group the *k* residuals of these regressions in the vector $\hat{v}$, to estimate by OLS the model $y = X\beta + \hat{v}\delta + u$, and, finally, to test $\delta = 0$ with an *F*-statistic, which is indeed equivalent to the Hausman statistic.

Exercises

1) Consider the linear regression model under OLS assumptions, with stochastic regressors.
a) Show that if E (ε | X) = 0, then COV (ε, X) = 0
(b)) Is it true that COV (ε, X) = 0 $\Rightarrow$E (ε | X) = 0?

2) Consider the model:

$$\mathrm{Y} = \mathrm{X}\beta + \varepsilon \text{ with } \varepsilon \sim \mathrm{iid}(0, \sigma_\varepsilon^2 I_T)$$

$$\frac{X'X}{T} \rightarrow \Sigma_{XX}$$

$$\mathrm{cov}(\varepsilon, X) = \Sigma_{\varepsilon X}$$

a) Suppose you do not know $\Sigma_{\varepsilon X}$ but you have a variable Z such that:

$$\frac{Z'Z}{T} \rightarrow \Sigma_{ZZ}$$

$$\mathrm{cov}(\varepsilon, Z) = 0$$

and

$$\frac{Z'X}{T} \rightarrow \Sigma_{ZX}$$

Derive the IV estimator, show if it is consistent or not, and derive the formula for its asymptotic variance and for a consistent estimator of this variance.

b) Suppose now that $\Sigma_{\varepsilon x}$ is a known value, with $\Sigma_{\varepsilon x}$>0. Show whether the OLS estimator under- or over-estimates β when $T\rightarrow\infty$. Using the available information, propose an alternative estimator β^* that corrects OLS in such a way that β^* is consistent for β.

3) Consider the linear regression model under OLS assumptions, with stochastic regressors and no intercept. You have the following dataset:

obs	Y	x	z_1	z_2
1	2	1	0.284	2.2373
2	9	3	2.5561	2.2373
3	10	3	2.5561	2.9831
4	15	4	4.5442	3.7288

where x is an endogenous explanatory variable for Y, and z_1 and z_2 are valid instruments for x. Compute the TSLS estimator and the formula for its variance.

4) Using the workfile "example_simul2_chap6.wf1", consider the following model:

$$y = \beta_0 + \beta_1 x + \varepsilon$$

where y is private spending in Research and Development (R&D), X is total public spending, the variables are expressed in logs and the elasticity is 20%.

a) Estimate the model by OLS. Do you get reasonable results? If not, what is a probable cause of the problems?
b) Suppose that GDP growth is a valid instrument for public spending, use this variable to implement a TSLS estimator and comment on the results.
c) In this example, is TSLS better than OLS? Apply a test to support your argument.

5)

a) Using data in the workfile "example_inv_chap3.wf1", estimate by OLS an investment function, allowing for a possible change in the parameters due to the financial crisis, and comment on the results.
b) Estimate the same regression by TSLS, using lryd(-1), rr(-1), lrc(-1), lrc(-2), and crisis as instruments, and comment on the results
c) Compute the Hausman test and comment on the results.

6) Show that in

$$y = \beta_1 X_1 + \beta_2 X_2 + \varepsilon$$

with $\text{cov}(X_1, \varepsilon) = 0$, $\text{cov}(X_2, \varepsilon) \neq 0$, the OLS estimator for β_1 in general is not consistent.

7. Dynamic Models

7.1 Dynamic models: a classification

Dynamic models are a particularly important class of models in econometrics. The peculiarity of dynamic models is that lagged values of the dependent variable appear as regressors (possibly) together with other explanatory variables. All the underlying assumptions of the linear regression model are maintained.

The dependence of the dependent variable on its lags is justified by both microeconomic phenomena, such as the presence of adjustment costs or the persistence of consumers' habits, and macroeconomic ones, such as bargaining mechanisms that only allow periodic adjustments or no-arbitrage conditions that grant financial markets efficiency. For example, central banks change only periodically, and in a limited way, the reference interest rate. As we will see in detail below, this creates correlation between the contemporaneous value of the interest rate and its lags. Similarly, aggregate consumption, investment as well as wages and inflation are often very persistent and, typically, today's price of a security is equal to its yesterday's price plus a random error.

The presence of lagged values of the dependent variable makes the regressors necessarily stochastic. Hence, the relevant econometric theory to study dynamic models is the one developed in Chapter 6. We will first provide a classification of dynamic models and then consider in detail their specification, estimation, inference and diagnostic control.

A general specification for a dynamic model is called *autoregressive distributed lags* (*AD*). In the *AD(p, q)* model the dependent variable, y, depends on $y_{t-1}, \ldots, y_{t-p}$ and $x_t, x_{t-1}, \ldots, x_{t-q}$. In particular, the *AD(1.1)* model is:

$$y_t = \beta_0 x_t + \beta_1 x_{t-1} + \alpha_1 y_{t-1} + u_t \tag{7.1.1}$$

Let us now consider various types of models that arise by imposing certain restrictions on the parameters of the *AD(1,1)* model in (7.1.1) and that are commonly used to model economic variables.

$$\textit{Static regression:} \quad \alpha_1 = \beta_1 = 0 \Rightarrow y_t = \beta_0 x_t + u_t \qquad (7.1.2)$$

This is the model we focused on so far. As the name suggests, it is a good model when the relationship between the explanatory and dependent variables is static. Many economic models are formulated as static relations, consider for example the Keynesian theory of consumption or the whole structure of the IS-LM model. Even no-arbitrage conditions, such as purchasing power parity or uncovered parity of interest rates, are expressed in static terms. In fact, economic theory deals primarily with fundamental relations between variables, those that should be valid in the long-run equilibrium, while the adjustment mechanisms to the equilibrium are much less precisely defined from a statistical point of view. For example, if there are deviations from the purchasing power parity, the required adjustment can be more or less fast, or if there are changes in fiscal or monetary policy, the effects on economic growth and inflation may not be immediate but occur with different lags. Thus, by carefully analyzing the dynamic component of the models, econometrics can also help to get a better understanding of the adjustment processes of economic systems.

Note that if the restrictions (7.1.2) are not met, and in particular if $\alpha_1 \neq 0$, we are omitting from the set of explanatory variables some regressors that are correlated over time. In addition to the usual omitted variable bias, this will also result in correlation in the errors of the statistical model, and possibly other violations of the assumptions, such heteroscedasticity and non-Normality of the errors.

$$\textit{AR (1) model:} \quad \beta_0 = \beta_1 = 0 \Rightarrow y_t = \alpha_1 y_{t-1} + u_t \qquad (7.1.3)$$

In the case of the AR(1) model, the variability of *y* is explained only by its past. This is the simplest specification in the class of so-called *time series* models, which we only briefly mention, as the objective of the book is to provide a first introduction to econometrics.

For example, as already mentioned above, financial market efficiency requires that the price of an asset today is equal to its price of yesterday plus a random error. Asset prices should then follow the model (7.1.3) with the additional restriction $\alpha = 1$. We will see in Section 7.3 that $\alpha = 1$ is a fundamental hypothesis that radically changes the characteristics of the variable *y* and the resulting model, called *random walk*, must be handled with special care.

$$\textit{Model in differences: } \alpha_1 = 1,\ \beta_0 = -\beta_1 \Rightarrow \Delta y_t = \beta_0 \Delta x_t + u_t \tag{7.1.4}$$

$$\Delta y_t = y_t - y_{t-1},\ \ \Delta x_t = x_t - x_{t-1}$$

In this case, the difference between the value at time t and that at t-1 of y is explained by the corresponding change in x. This is an interesting formulation: many economic variables show marked temporal trends which, in turn, may alter the results of regressions in levels (see Section 7.4). In this case, models in differences can instead generate more reliable conclusions.
Moreover, remember that if the variables are expressed in logarithms, then their first difference is a good approximation of their growth rate when the latter is small enough. Therefore, model (7.1.4) can also be used to explain, for example, the relationship between consumption growth and disposable income growth or the link between the change in the price of a product and the corresponding change in the demanded quantity.

$$\textit{"Leading indicator" model: } \alpha_1 = \beta_0 = 0 \Rightarrow y_t = \beta_1 x_{t-1} + u_t \tag{7.1.5}$$

This specification assumes that the variable x is *leading* (it moves before) y. It is therefore particularly useful for predicting future values of the dependent variable: y_{t+1} depends on x_t, whose value is known in period t, i.e., at the time we want to make the prediction (see Chapter 5 for details on forecasting with the linear model).
For example, y may be the growth rate of GDP and x an index of consumers' and firms' expectations, or the spread between long-term and short-term interest rates (typically the spread shrinks before a recession, anticipating reductions in future short-term interest rates). Or y might be the consumer price index and x the price index of raw materials.

$$\textit{Distributed-lag model: } \alpha_1 = 0 \Rightarrow y_t = \beta_0 x_t + \beta_1 x_{t-1} + u_t \tag{7.1.6}$$

This specification has received much attention in the econometric literature. As previously mentioned, the presence in the economy of adjustment costs and other frictions implies that the effects of the variable x on the dependent variable y do not occur instantly.
In the specification in (7.1.6) a change today in x affects y both today and tomorrow but, in general, more lags – possibly an infinite number of them – may be included as regressors. Take as an example the model with geometric weights (*GDL*, Geometric Distributed Lags) that can be written as

$$y_t = \alpha + \beta\left(x_t + \omega x_{t-1} + \omega^2 x_{t-2} + \ldots\right) + \varepsilon_t \qquad 0 < \omega < 1$$
$$= \alpha + \beta \sum_{i=0}^{\infty} \omega^i x_{t-i} + \varepsilon_t \qquad (7.1.7)$$

The exponentially decreasing weight structure is represented in Figure 7.1

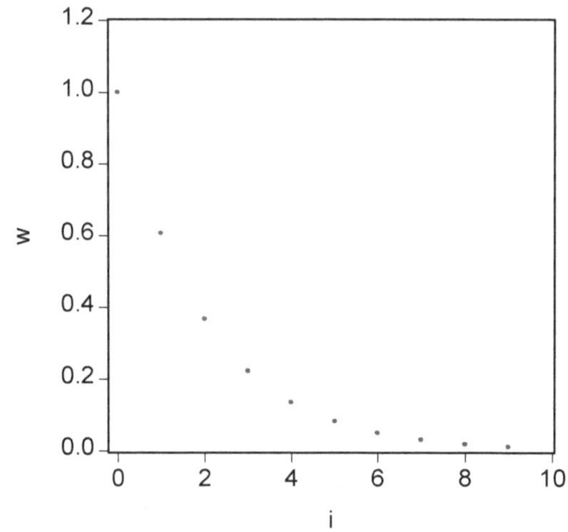

Figure 7.1: Structure of weights in GDL

The long-term response of *y* to a change in *x* is equal to:

$$\beta \sum_0^{\infty} \omega^i = \frac{\beta}{1-\omega}$$

So the closer to 1 the weight ω is, the greater the difference between the long-term and impact response, β.
The average delay in the response of *y* is calculated as:

$$Mean \;\; lag = \frac{\sum_0^{\infty} i\beta\omega^i}{\sum_0^{\infty} \beta\omega^i} = \frac{\omega/(1-\omega)^2}{1/1-\omega} = \frac{\omega}{1-\omega}$$

For example, if $\omega = 0.5$, $ml = 1$, i.e. half of the cumulative effect of a change of *x* on *y* occurs during the first period.
The *GDL* model in (7.1.7) can be rewritten as

$$y_t - \omega y_{t-1} = \alpha(1-\omega) + \beta x_t + u_t \quad u_t = \varepsilon_t - \omega\varepsilon_{t-1}$$

If we define $\beta = \beta_0 \; and \; \omega = \alpha_1$ the *GDL* in (7.1.8) then becomes a special case of that *AD (1.1)* in (7.1.1).

$$\textit{Partial adjustment model: } \beta_1 = 0 \;\Rightarrow\; y_t = \beta_0 x_t + \alpha_1 y_{t-1} + u_t \tag{7.1.8}$$

The name of this model is due to the fact that it can be derived as follows. Suppose there is a desired level of y_t given by

$$y_t^* = \alpha' + \beta' x_t + \varepsilon'_t \tag{7.1.9}$$

where y_t^* can represent, for example, the central bank's optimal interest rate, or the optimal production level for a firm. The y variable cannot be adjusted immediately to reach y_t^*, for example for technical reasons, transaction costs, etc. We then have:

$$y_t - y_{t-1} = \gamma(y_t^* - y_{t-1}) \qquad 0 < \gamma < 1 \tag{7.1.10}$$

The closer to 1 the parameter γ is, the faster the y adjustment to the desired value, y_t^*. Replacing y_t^* with the expression obtained from (7.1.10) in (7.1.9), we have:

$$y_t = \underbrace{\gamma\alpha'}_{c} + \underbrace{\gamma\beta'}_{\beta_0} x_t + \underbrace{(1-\gamma)}_{\alpha_1} y_{t-1} + \underbrace{\gamma\varepsilon'_t}_{u_t} \tag{7.1.11}$$

which is a partial adjustment specification similar to the one in (7.1.8).

$$\begin{aligned}&\textit{Error correction model: } \beta_0 + \beta_1 + \alpha_1 = 1 \;\Rightarrow \\ &\Delta y_t = \beta_0 \Delta x_t - (1-\alpha_1)(y_{t-1} - x_{t-1}) + u_t\end{aligned} \tag{7.1.12}$$

The Error Correction Model (ECM) is a fundamental model in modern econometrics and provides a good representation for a wide range of economic variables. This model will be analyzed in greater detail in section 7.5.

The intuition underlying the model is that the variables *y* and *x* move together, at least in the long run, i.e. deviations of *y* from *x* must be reabsorbed. In this particular case, the dependent variable *y* (and possibly also *x*) behaves accordingly, responding not only to changes in *x* but also to deviations from the equilibrium relationship, *y-x*.

$$\text{"Dead-Start" and VAR:}\quad \beta_0 = 0 \;\Rightarrow\; y_t = \beta_1 x_{t-1} + \alpha_1 y_{t-1} + u_t \qquad (7.1.13)$$

In this case the explanatory variable *x* has no contemporaneous effect on *y*. If we add an equation similar to (7.1.13) for *x*,

$$x_t = \beta_2 x_{t-1} + \alpha_2 y_{t-1} + v_t \qquad (7.1.14)$$

the joint model (7.1.13), (7.1.14) for *y* and *x* is called *Vector Autoregression (VAR)*. This model represents the multivariate version of the autoregressive model (7.1.3) and it is widely used in modern econometrics. The ECM model can be also obtained by imposing constraints on the VAR. Given the introductory nature of this book, which focuses mainly on univariate models, we will omit the study of *VAR* models.

$$\text{Model with autoregressive errors: } \alpha_1\beta_0 + \beta_1 = 0 \;\Rightarrow$$
$$y_t = \beta_0 x_t + \varepsilon_t\,,\quad \varepsilon_t = \alpha_1 \varepsilon_{t-1} + u_t \qquad (7.1.15)$$

An extremely popular empirical model, the static model with AR (1) errors, is a very particular case of the *AD (1,1)* specification, which is obtained by means of a nonlinear restriction on its parameters. The restriction $\alpha_1\beta_0 + \beta_1 = 0$ is usually called COMFAC.

If one of the tests for no correlation of errors analyzed in Chapter 4 rejects the null hypothesis in a static model, the best solution is to change the dynamics of the model, instead of using GLS, unless the COMFAC restriction is satisfied.

7.2 Dynamic models: specification, estimation, inference and diagnostic control

To discuss the specification of dynamic models we need to know first how to estimate their parameters. Hence, we begin this section by evaluating whether and under what conditions the OLS estimators are also applicable in a dynamic model.

Since the lags of the dependent variable are in general among the regressors of the dynamic model, the first question we ask is whether or not the regressors and the error can be considered as independent. From Section 6.1 we know that if this is the case then all properties of the OLS estimators will remain valid.
For simplicity, let us consider the AR(1) model in (7.1.3):

$$y_t = \alpha_1 y_{t-1} + u_t \tag{7.2.1}$$

Substituting for y_{t-1} its expression, we obtain:

$$y_t = \alpha^2_1 y_{t-2} + \alpha_1 u_{t-1} + u_t \tag{7.2.2}$$

Continuing the substitution yields:

$$y_t = u_t + \alpha_1 u_{t-1} + \alpha^2_1 u_{t-2} + \alpha^3_1 u_{t-3} + \ldots \tag{7.2.3}$$

and

$$y_{t-1} = u_{t-1} + \alpha_1 u_{t-2} + \alpha^2_1 u_{t-3} + \alpha^3_1 u_{t-4} + \ldots \tag{7.2.4}$$

It follows therefore that, by defining $x_t = (y_{t-1}, y_{t-2}, y_{t-3}, \ldots)'$ and $e_t = (u_t, u_{t-1}, u_{t-2}, \ldots)'$, x_t and e_t are not independent. For example, y_{t-1} depends on u_{t-1}, u_{t-2}, ... As a consequence, we cannot apply what we studied in Section 6.1 to the estimation of dynamic models. However, it can be shown that, if the errors of the dynamic model are uncorrelated, then the cross-moment conditions between regressors and errors seen in Section 6.2 are still met. Therefore, the OLS estimator for the parameters of the dynamic model remains consistent and asymptotically normally distributed.
Instead, when the errors are correlated and the lagged dependent variable appears among the regressors, typically the conditions in Section 6.2 are not met and the OLS estimation method does not produce consistent estimators. For example, consider the case where the errors of the AR(1) model in (7.2.1) are

$$u_t = \varepsilon_t - \omega \varepsilon_{t-1} \tag{7.2.5}$$

In this case, y_{t-1} is correlated with ε_{t-1} and hence with u_t. We must then find good instruments for y_{t-1} and use an IV estimation method to obtain consistent estimators of the parameters.

Once we have obtained consistent and asymptotically Normal estimators for the parameters of the dynamic model, for hypothesis testing we can still use the t-test or the F-test. These statistics have an asymptotic justification. Hence, in order to derive the distribution of the F-statistic to test q restrictions, it would be theoretically more correct to use the result

$$qF \underset{H_0}{\overset{a}{\sim}} \chi^2(q) \tag{7.2.6}$$

In practice, however, the F distribution typically remains a better approximation for the distribution of the F-test in finite samples.
Particularly interesting hypotheses on the parameters are those that allow the simplification of the AD model into one of those analyzed in Section 7.1. It is also important to remember that, as in the static case, before applying the tests the correct specification of the model should be checked.
Even nonlinear restrictions, such as the COMFAC $\alpha_0\beta_0 + \beta_1 = 0$ that transform the model into one with autoregressive errors (7.1.15), can be tested with the F-statistic.
Hypothesis testing can also be used for model specification. In particular, the t-test and the F-test can be applied sequentially to determine the significance of the lags. For example, starting with an AD (p, p) model, we can test whether $y_{t-p}$and x_{t-p} are not significant. If the null hypothesis is not rejected, we consider a new model with p-1 lags and test for the significance of $y_{t-p+1}$and x_{t-p+1}. We proceed sequentially and stop when the null hypothesis is rejected.
To determine p, the initial lag length, we must take into account that:

- if the number of lags is too high, higher than the true one, there are problems of collinearity of the regressors and reduced efficiency of estimators (case of irrelevant regressors in Chapter 3);
- if the number of lags is too low, lower than the true one, the estimators are inconsistent (case of omitted regressors in Chapter 3);
- the initial specification must be congruent, meaning that diagnostics controls should not reject the null hypotheses related to the correct specification of the model.

There are no generally valid rules, but typical initial values for p are 2-3 for annual data, 4-5 for quarterly data, and 12-13 for monthly data. It should be also considered that proper sequential testing, which controls for the overall size of the procedure, is very complex. The suggested iterative procedure for lag length determination should be considered more as an empirical technique than as a formal statistical procedure.

An alternative method for model selection is based on the so-called information criteria. Information criteria combine a measure of goodness of fit with a penalization for the number of parameters. For example,

$$\log\left(\hat{\sigma}^2\right)+kg(T), \tag{7.2.7}$$

where $\hat{\sigma}^2$ is the estimator of the variance of the error, k the number of parameters and $g(T)$ a function of the number of observations, T. The idea is to choose the model that minimizes the criterion, in order to achieve a good compromise between explanatory power and a parsimonious specification. Two criteria often used in practice are:

$$\textit{Akaike Information Criterion (AIC)}:\ kg(T)=k/T \tag{7.2.8}$$

$$\textit{Schwarz Information Criterion (SIC)}:\ kg(T)=k\log(T)/T \tag{7.2.9}$$

It can be shown that, under appropriate conditions, some of these criteria select the right model with probability that tends to 1 when T goes to infinity. This is the case for the Schwarz criterion but not for the AIC. However, in finite samples, there is no general ranking of the performance of these criteria.
Finally, also in the case of dynamic models it is very important to check that the assumptions underlying the model are satisfied. Tests for lack of correlation, homoscedasticity and linearity seen in Chapter 4, as well as those for parameter stability seen in Chapter 5, are also applicable in this context (with the exception of the Durbin-Watson test that applies only to static models). The procedures have an asymptotic justification, as in the case of testing hypotheses about the model parameters.

7.3 An empirical analysis with stationary simulated data

The data for this analysis are contained in the "example_simul1_chap7.wf1" workfile. The aim is to find a proper model to characterize the relationship between the variables plotted in Figure 7.2, which are clearly correlated.
We start by checking whether a static regression of y on an intercept and x is a good model. OLS estimation yields the results in Table 7.1. The regressors are both significant, but the model fit is unsatisfactory and the Durbin-Watson test leads us to reject the null hypothesis of no serial correlation of the errors (against the alternative hypothesis of serial correlation).

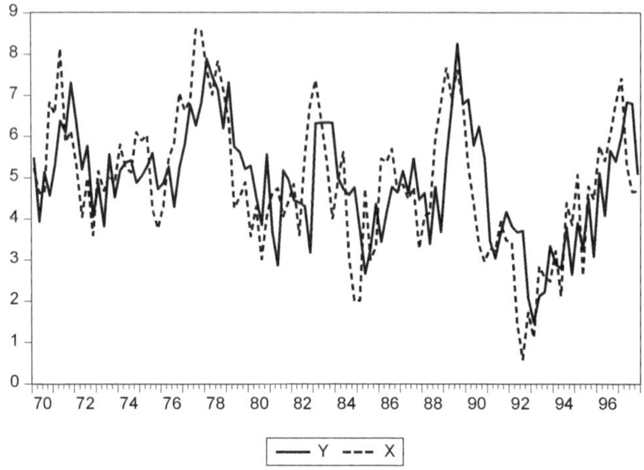

Figure 7.2: Plot of the dependent and explanatory variables

Dependent variable: Y
Least Squares Method:
Sample: 1997q4 1970q1
Included observations: 112

Variable	Coefficient	Std. Error	*t*-Statistic	Prob.
(C)	2.281570	0.305797	7.461052	0.0000
X	0.534458	0.059311	9.011124	0.0000

R-squared	0.424687	Mean dependent var	4.896378
Adjusted R-squared	0.419457	S.D. dependent var	1.340199
S.E. of regression	1.021142	Akaike information criterion	2.897417
Sum squared resid	114.7005	Schwarz criterion	2.945962
Log likelihood	-160.2553	*F*-statistic	81.20035
Durbin-Watson stat	1.089783	Prob (*F*-statistic)	0.000000

Table 7.1: Static regression of y on x

To tackle the problem of correlated errors we can introduce dynamics in the model. Initially, we will assume that y does not depend on x but only on its own lag, obtaining an AR(1) model. Estimation results are reported in Table 7.2. The coefficient of determination increases and the DW statistic takes a value closer to two, although we know that this test is not reliable when applied in dynamic models.

Two alternative ways to take into account the correlation of the errors in a static model are either the specification of a model in differences or the use of a GLS correction in the static model to introduce a correlation structure for the errors. To evaluate the first possibility, we create the first differences of the variables using the commands:

```
genr dy = y-y ( -1)
genr dx = x-x ( -1)
```

Next, we execute the linear regression using *dy* and *dx*, getting the output in Table 7.3. It turns out that R^2 decreases considerably, and this is not surprising since the dependent variable is different from before: it is much less persistent and therefore more difficult to explain. The DW test still suggests serial correlation in the errors, implying that we still miss a satisfactory specification.

Dependent variable: Y
Least Squares Method:
Sample: 1997q4 1970q1
Included observations: 112

Variable	Coefficient	Std. Error	*t*-Statistic	Prob.
(C)	1.364591	0.335053	4.072764	0.0001
Y (-1)	0.721452	0.066034	10.92541	0.0000

R-squared	0.520414	Mean dependent var	4.896378
Adjusted R-squared	0.516054	S.D. dependent var	1.340199
S.E. of regression	0.932325	Akaike information criterion	2.715426
Sum squared resid	95.61536	Schwarz criterion	2.763971
Log likelihood	-150.0639	*F*-statistic	119.3646
Durbin-Watson stat	2.349011	Prob (*F*-statistic)	0.000000

Table 7.2: AR(1) model for y

Dependent variable: DY
Least Squares Method:
Sample: 1970: 1 1997: 4
Included observations: 112

Variable	Coefficient	Std. Error	*t*-Statistic	Prob.
DX	0.170578	0.084928	2.008513	0.0470
(C)	0.000905	0.093260	0.009705	0.9923

R-squared	0.035376	Mean dependent var	0.000989
Adjusted R-squared	0.026607	S.D. dependent var	1.000369
S.E. of regression	0.986971	Akaike information criterion	2.829344
Sum squared resid	107.1523	Schwarz criterion	2.877889
Log likelihood	-156.4433	*F*-statistic	4.034124
Durbin-Watson stat	2.604277	Prob (*F*-statistic)	0.047039

Table 7.3: Model in differences for y

Let us therefore go back to the static equation, and try to impose an AR(1) model for errors:

$$y_t = \alpha + \beta x_t + \varepsilon_t, \qquad \varepsilon_t = \rho\varepsilon_{t-1} + u_t, \qquad |\rho| < 1, \; u_t \;\; iid \;\; (0, \sigma^2)$$

By subtracting ρy_{t-1} to both sides of the model we get:

$$y_t - \rho y_{t-1} = \alpha + \beta x_t - \rho(\alpha + \beta x_{t-1} + \varepsilon_{t-1}) + \varepsilon_t,$$

and, using the equation that defines the AR(1) process for ε_t, we can set $u_t = \varepsilon_t - \rho\varepsilon_{t-1}$. Finally, isolating y_t on the left-hand side, we get the AD(1,1) model:

$$y_t = \alpha - \rho\alpha + \beta x_t - \rho\beta x_{t-1} + \rho y_{t-1} + u_t, \qquad u_t \;\; iid, \text{ that is}$$
$$y_t = \tilde{\alpha} + \beta x_t - \tilde{\rho} x_{t-1} + \rho y_{t-1} + u_t, \qquad u_t \;\; iid$$

We then regress *y* on a constant, *x*, *x* (-1) and *y* (-1) and interpret the estimated constant term as $\alpha(1-\rho)$. According to the values reported in Table 7.4, the regression results improve, but the correlation in the errors seems to persist.

Dependent variable: Y
Least Squares Method:
Sample: 1970: 1 1997: 4
Included observations: 112

Variable	Coefficient	Std. Error	*t*-Statistic	Prob.
X	0.181186	0.070482	2.570677	0.0115
(C)	0.711372	0.297580	2.390519	0.0186
X (-1)	0.256639	0.082841	3.097958	0.0025
Y (-1)	0.417351	0.072005	5.796147	0.0000
R-squared	0.681466	Mean dependent var		4.896378
Adjusted R-squared	0.672618	S.D. dependent var		1.340199
S.E. of regression	0.766826	Akaike information criterion		2.341946
Sum squared resid	63.50631	Schwarz criterion		2.439035
Log likelihood	-127.1490	*F*-statistic		77.01782
Durbin-Watson stat	2.774948	Prob (*F*-statistic)		0.000000

Table 7.4: AD(1,1) model for y

We can then verify whether additional lags of y and x are significant. Let us consider an AD(2,2) specification, whose estimation results are shown in Table 7.5.

Dependent variable: Y

Variable	Coefficient	Std. Error	*t*-Statistic	Prob.
X	0.069148	0.044371	1.558407	0.1221
(C)	0.829942	0.198387	4.183444	0.0001
X (-1)	0.030698	0.055229	0.555834	0.5795
X (-2)	0.660466	0.054459	12.12769	0.0000
Y (-1)	-0.002224	0.060457	-0.036787	0.9707
Y (-2)	0.072588	0.051836	1.400350	0.1643
R-squared	0.880570	Mean dependent var		4.896378
Adjusted R-squared	0.874937	S.D. dependent var		1.340199
S.E. of regression	0.473952	Akaike information criterion		1.396662
Sum squared resid	23.81083	Schwarz criterion		1.542296
Log likelihood	-72.21305	*F*-statistic		156.3099
Durbin-Watson stat	2.104679	Prob (*F*-statistic)		0.000000

Table 7.5: AD (2,2) model for y

The coefficient of x (-2) is highly significant, R^2 significantly improves and, importantly, the DW and the LM tests for the null hypothesis of no serial autocorrelation against the alternative hypothesis of autocorrelation of order m = 1 no longer reject the null hypothesis.
In the AD (2.2) specification, however, many regressors are not significant. If we exclude them from the regression, we get the "Leading indicators" model of Table 7.6, whose residuals still pass all the diagnostic checks.

Dependent variable: Y

Variable	Coefficient	Std. Error	*t*-Statistic	Prob.
(C)	1.145950	0.144156	7.949364	0.0000
X (-2)	0.765898	0.027937	27.41492	0.0000

R-squared	0.872327	Mean dependent var	4.896378
Adjusted R-squared	0.871167	S.D. dependent var	1.340199
S.E. of regression	0.481042	Akaike information criterion	1.391973
Sum squared resid	25.45420	Schwarz criterion	1.440518
Log likelihood	-75.95049	*F*-statistic	751.5781
Durbin-Watson stat	2.070914	Prob (*F*-statistic)	0.000000

Table 7.6:"Leading indicator" model for y

7.4 An empirical analysis of the determinants of the FED decisions

The objectives of the US Central Bank, the Federal Reserve (FED), are to stabilize inflation and limit economic fluctuations around the level of potential output. The tool is the interest rate and the relation between the instrument and the objectives can be characterized by the so-called Taylor rule, named after the economist who first proposed it:

$$r_t^* = \bar{r} + \beta(\pi_{t+12}^e - \pi_t^*) + \gamma(y_t - y_t^*) \qquad (7.4.1)$$

In the equation (7.4.1) r_t^* represents the monetary policy target, $\bar{r}$ the equilibrium interest rate, π_{t+12}^e the expected inflation today for the period *t+12*, y_t real output, and π_t^*, y_t^* the desired values of inflation and output.

The parameter β indicates whether monetary policy is accommodating ($\beta <$ 1) or stabilizing ($\beta >$ 1), while γ measures how important output stabilization is for the Central Bank.
To prevent sudden changes in the markets, the interest rate will progressively adjust against the target, following the partial adjustment mechanism:

$$r_t = (1-\rho_1-\rho_2)r_t^* + \rho_1 r_{t-1} + \rho_2 r_{t-2} + u_t \qquad (7.4.2)$$

Combining equations (7.4.1) and (7.4.2) we get:

$$r_t = \alpha + (1-\rho_1-\rho_2)\beta(\pi_{t+12}^e - \pi_t^*) + (1-\rho_1-\rho_2)\gamma(y_t - y_t^*) + \qquad (7.4.3)$$
$$+ \rho_1 r_{t-1} + \rho_2 r_{t-2} + u_t$$

where $\alpha = (1-\rho_1-\rho_2)\bar{r}$. We can rewrite (7.4.3) as

$$r_t = \alpha + (1-\rho_1-\rho_2)\beta(\pi_{t+12} - \pi_t^*) + (1-\rho_1-\rho_2)\gamma(y_t - y_t^*) + \qquad (7.4.4)$$
$$+ \rho_1 r_{t-1} + \rho_2 r_{t-2} + \varepsilon_t$$

where $\varepsilon_t = (1-\rho_1-\rho_2)\beta(\pi_{t+12}^e - \pi_{t+12}) + u_t$. This allows us to express the dependent variable in terms of observable regressors, while π_{t+12}^e in (7.4.3) is not observable. To estimate this equation, we have at our disposal a monthly sample from 1990 to 2007 that contains data about the actual rate on Federal Funds, inflation calculated on the Consumer Price Index (CPI), a measure of output obtained as the logarithm of the index of industrial production, the FED's inflation target $\pi_t^* = 2\%$, and the output gap computed as deviation of output from its trend (where the trend is calculated using the so-called Hodrick Prescott filter, available in EViews). The data are contained in the "example_fed_chap7.wf1" workfile. Note that the sample stops in June 2007 excluding the problematic period of the financial crisis, when the FED put in place policies of quantitative easing, which could bias our results.
It is not possible to estimate the equation (7.4.4) by OLS because the regressor $\pi_{t+12} - \pi_t^*$ is correlated with the error ε_t by construction, generating endogeneity and making OLS estimation inconsistent. To address this issue, we must find good instruments and apply the instrumental variables estimation method. First, we need to choose a valid set of instruments, that is, a set of variables correlated with the endogenous variable (in this case, π_{t+12}) and not

correlated with the errors. Let us try with some lags of inflation, Federal Funds rate, and let us also include the growth rate of GDP. After selecting the most relevant lags we obtain the specification reported in Table 7.7, from which the instruments are significant in explaining inflation at time $t + 12$.

Dependent variable: BIRD (12)
Least Squares Method:
Sample (adjusted): 1991M02 2007M06

Variable	Coefficient	Std. Error	*t*-Statistic	Prob.
(C)	0.800034	0.107193	7.463524	0.0000
BIRD (-1)	0.525286	0.035684	14.72032	0.0000
FEDFUNDS (-1)	0.443905	0.126973	3.496055	0.0006
FEDFUNDS (-2)	-0.401036	0.127687	-3.140774	0.0020
USq1 (-1)	0.037846	0.012540	3.017917	0.0029

R-squared	0.814188	Mean dependent var	2.634874
Adjusted R-squared	0.810317	S.D. dependent var	0.778906
S.E. of regression	0.339233	Akaike information criterion	0.700797
Sum squared resid	22.09523	Schwarz criterion	0.784127
Log likelihood	-64.02846	Hannan-Quinn criterion	0.734529
F-statistic	210.3262	Durbin-Watson stat	0.262566
Prob (*F*-statistic)	0.000000		

Table 7.7: Regression of endogenous variable on the instruments

We now use these instruments to estimate our equation (7.4.4) by two-stage least squares (TSLS). We use robust standard errors to take into account possible heteroscedasticity and correlation of the errors in (7.4.4). In particular, from the EViews window in Figure 7.3, we click "Options" and in the resulting window (reported in Figure 7.4), we select "Coefficient covariance matrix" and specify the method of White.

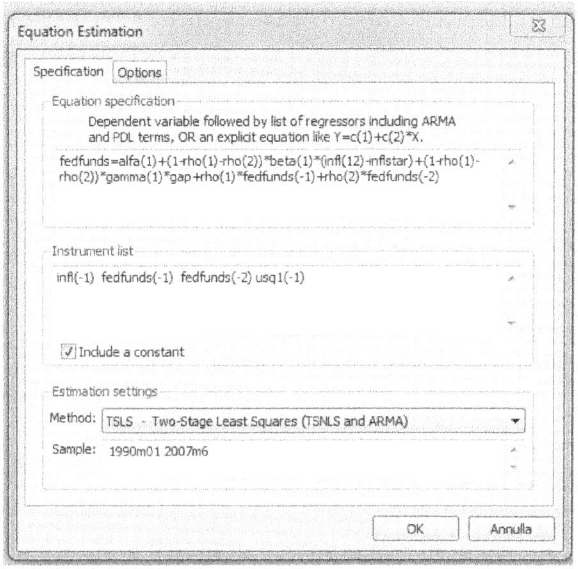

Figure 7.3: IV Estimation with EViews

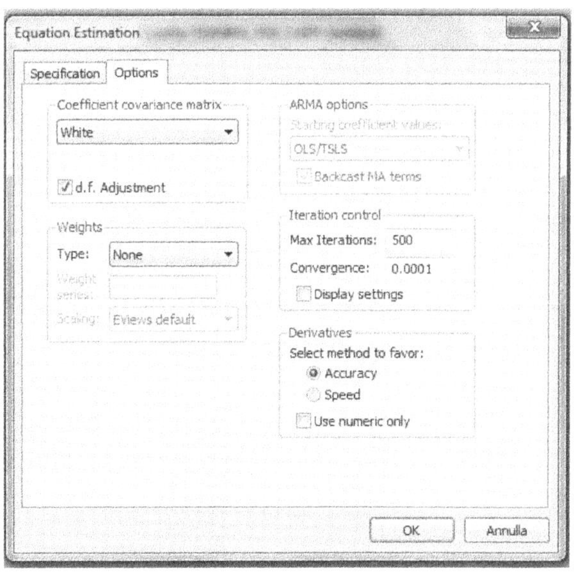

Figure 7.4: Choice of White robust estimator for the coefficient covariance matrix

The results presented in Table 7.8 are reasonable, implying a policy of inflation stabilization ($\beta > 1$) combined with a substantial interest in the output target, $\gamma = 3.99$. The result for the coefficient of the difference between the expected inflation rate and the target is close to that proposed by Taylor (β =1.5, which is within the 90% confidence interval). However, the original Taylor hypothesis that the value of the parameter γ is equal to 0.5 is strongly rejected by the Wald test.

Dependent variable: FEDFUNDS
Method: Two-Stage Least Squares
Sample (adjusted): 1991M02 2007M06
Included observations: 197 after adjustments
Estimation settings: tol = 0.00010, derivs = analytic
FEDFUNDS = alpha (1) + (1-(1)-RHO RHO (2)) * (1) * BETA (INFL (12)-INFLSTAR)
+ (1-(1)-RHO RHO (2)) * (1) * GAP + RHO (1) * FEDFUNDS (-1)
+ RHO (2) * FEDFUNDS (-2)
Instrument specification: c BIRD (-1) (-1) FEDFUNDS FEDFUNDS (-2) USq1 (-1)

	Coefficient	Std. Error	*t*-Statistic	Prob.
Alpha (1)	-0.018289	0.144079	-0.126938	0.8991
RHO (1)	0.794983	0.269288	2.952162	0.0035
RHO (2)	0.148989	0.243176	0.612680	0.5408
BETA (1)	1.777015	1.001456	1.774432	0.0776
RANGE (1)	3.990747	0.971516	4.107753	0.0001

R-squared	0.979420	Mean dependent var	4.123096
Adjusted R-squared	0.978991	S.D. dependent var	1.652074
S.E. of regression	0.239457	Sum squared resid	11.00922
Durbin-Watson stat	0.990007	J-statistic	and 7.28-19
Instrument rank	5		

Table 7.8: robust estimation equation IV (7.4.4)

Finally, looking at the model residuals in Figure 7.5, we see how these show some signs of autocorrelation and the presence of some outliers.

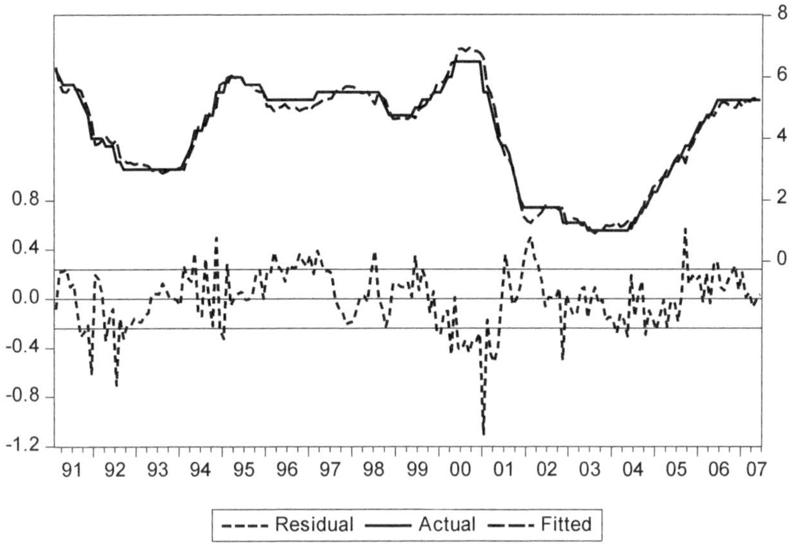

Figure 7.5: Residuals, fitted and actual values for the regression in Table 7.8

Actually, this is not surprising given that the errors ε in (7.4.4) include the prediction error $\pi_{t+12} - \pi^e_{t+12}$ and are thus correlated by construction. It can be shown that, if the model in (7.4.4) is correct, then the errors should follow the MA(11) model in (7.4.5), with $_t\ v\ iid \sim (0,\ \sigma^2)$:

$$\varepsilon_t = \theta_1 v_{t-1} + \theta_2 v_{t-2} + \ldots + \theta_{11} v_{t-11} + v_t \tag{7.4.5}$$

Dependent variable: RES
Least Squares Method:
Sample (adjusted): 1991M02 2007M06
Included observations: 197 after adjustments
Convergence achieved after 41 iterations
MA Backcast: 1990M03 1991M01

Variable	Coefficient	Std. Error	t-Statistic	Prob.
(C)	0.001544	0.040148	0.038462	0.9694
MA (1)	0.225985	0.070837	3.190205	0.0017
MA (2)	0.222366	0.071922	3.091738	0.0023
MA (3)	0.399610	0.072481	5.513335	0.0000
MA (4)	0.261722	0.076620	3.415831	0.0008
MA (5)	0.259902	0.074272	3.499307	0.0006
MA (6)	0.396420	0.070842	5.595826	0.0000
MA (7)	0.389001	0.074627	5.212609	0.0000
MA (8)	0.128978	0.077280	1.668969	0.0968
MA (9)	0.141861	0.071992	1.970519	0.0503
MA (10)	-0.099898	0.070615	-1.414678	0.1588
MA (11)	-0.238870	0.069946	-3.415083	0.0008

R-squared	0.413219	Mean dependent var	E-2.03 13
Adjusted R-squared	0.378329	S.D. dependent var	0.237001
S.E. of regression	0.186866	Akaike information criterion	-0.457870
Sum squared resid	6.460002	Schwarz criterion	-0.257877
Log likelihood	57.10016	Hannan-Quinn criterion	-0.376911
F-statistic	11.84358	Durbin-Watson stat	1.905922
Prob (F-statistic)	0.000000		

Table 7.9: MA(11) model for the residuals of the regression in Table 7.8

To check if this is the case, we perform regression (7.4.5), using the residuals in place of ε_t, obtaining Table 7.9. Actually, the MA(11) model seems to give good results, with most lags of estimated residuals statistically significant. However, the errors of this model remain mildly correlated, suggesting that perhaps additional lags should be inserted in the specification of the empirical Taylor rule.

7.5 Unit roots and stochastic trends

Consider the AR (1) model

$$y_t = \alpha_1 y_{t-1} + u_t, \tag{7.5.1}$$

and assume that all the assumptions underlying the linear model are met, possibly excluding the one on the Normality of errors. In the previous section we saw that we can write the model as

$$y_t = u_t + \alpha_1 u_{t-1} + \alpha^2{}_1 u_{t-2} + \alpha^3{}_1 u_{t-3} + \ldots \tag{7.5.2}$$

that is

$$y_t = \sum_{i=0}^{\infty} \alpha_1^i u_{t-i} \tag{7.5.3}$$

Therefore, the errors u_t, often associated with economic shocks, have an effect on the dependent variable which declines exponentially when $| a | < 1$.
If $\alpha = 1$, i.e. there is a *unit root*, then

$$y_t = \sum_{i=0}^{\infty} u_{t-i} \tag{7.5.4}$$

and the shocks have persistent effects, with the dependent variable determined by the sum of present and past shocks. Under the assumption $y_0 = 0$, or $\sum_{i=t}^{\infty} u_{t-i} = 0$, we can rewrite y_t as the sum of shocks from time 1 to time t: $y_t = \sum_{i=0}^{t-1} u_{t-i} = u_1 + u_2 + \ldots + u_t$. Note also that in this case the variance of y increases continuously over time, being

$$\text{var}(y_t) = t\sigma_u^2, \tag{7.5.5}$$

while if $| \alpha | < 1$ the variance stabilizes:

$$\text{var}(y_t) = \sigma_u^2 / (1 - \alpha_1^2) \tag{7.5.6}$$

Moreover

$$\operatorname{cov}(y_t, y_{t-k}) = \frac{\sigma_u^2 \alpha_1^k}{1-\alpha_1^2} \tag{7.5.7}$$

so that y_t and y_{t-k} become less and less correlated if $| a | < 1$. Instead, when $\alpha = 1$, it is:

$$\operatorname{cov}(y_t, y_{t-k}) = (t-k)\sigma_u^2$$

The AR (1) model with the hypothesis of unit root, $\alpha = 1$, and non-correlated and homoscedastic errors is defined *random walk*:

$$y_t = y_{t-1} + u_t \tag{7.5.8}$$

The term

$$ST_t = \sum_{i=0}^{\infty} u_{t-i} \text{ or } ST_t = \sum_{j=0}^{t} u_j \tag{7.5.9}$$

is defined a *stochastic trend*, and is sometimes opposed to a *deterministic linear trend*:

$$DT_t = t \tag{7.5.10}$$

where $t = 1.2, ..., T$.
If the variable y_t depends on a deterministic rather than stochastic trend,

$$y_t = \alpha\, t + u_t \tag{7.5.11}$$

then, given that the trend is a deterministic variable, both its variance and the covariance with y_{t-k} are constant and respectively equal to

$$\operatorname{var}(y_t) = \sigma_u^2, \ \operatorname{cov}(y_t, y_{t-k}) = 0. \tag{7.5.12}$$

These properties are therefore very different from the case where y_t depends on the stochastic trend.
The variable y_t may also depend both on a deterministic trend and stochastic one. For example, for the so-called *random walk with drift* model:

$$y_t = c + y_{t-1} + u_t \tag{7.5.13}$$

we have

$$y_t = ct + \sum_{i=0}^{t} \alpha_1^i u_{t-i} = cDT_t + ST_t . \tag{7.5.14}$$

The stochastic trend dominates the deterministic one and the properties of y_t are similar to those seen in the case of the random walk.
What we have seen so far for the AR(1) model also applies to more general dynamic models. For example, consider the AR(2) model:

$$y_t = \alpha_1 y_{t-1} + \alpha_2 y_{t-2} + u_t \tag{7.5.15}$$

and rewrite it as:

$$(y_t - y_{t-1}) = -(1 - \alpha_1 - \alpha_2) y_{t-1} - \alpha_2 (y_{t-1} - y_{t-2}) + u_t , \tag{7.5.16}$$

or, by defining,

$$\Delta y_t = y_t - y_{t-1}, \tag{7.5.17}$$

as:

$$\Delta y_t = -(1 - \alpha_1 - \alpha_2) y_{t-1} - \alpha_2 \Delta y_{t-1} + u_t \tag{7.5.18}$$

If

$$\alpha_1 + \alpha_2 = 1 , \tag{7.5.19}$$

which is the condition to have a unit root in y_t (the counterpart of the condition $\alpha = 1$ in the AR (1) model), then imposing the parameter restriction and summing both sides of (7.5.18) we get:

$$y_t = -\alpha_2 y_{t-1} + \sum_{i=0}^{t} u_i . \tag{7.5.20}$$

Therefore, y_t depends on its own past and on a stochastic trend.
When a variable depends on a stochastic trend, it is called *integrated of order 1*, *I(1)*. Otherwise, the variable is called *integrated of order 0*, *I(0)*, or *weakly stationary* (possibly around a linear deterministic trend).
Note that taking the first differences of a *I(1)* variable we get an *I(0)* variable. For example, for the random walk in (7.5.8), it is:

$$\Delta y_t = u_t \tag{7.5.21}$$

while for the AR(2) model with unit root in (7.5.15), it is:

$$\Delta y_t = -\alpha_2 \Delta y_{t-1} + u_t \tag{7.5.22}$$

It follows that Δy_t is stationary in both cases (unless it is $\alpha_2 = -1$ in (7.5.22), in which case we say that *y* is integrated of order two, *I(2)*).
If we take first differences of a stationary variable we introduce a particular form of correlation in the errors. For example, in the case of the AR(1) model in (7.5.1) with $| a | < 1$, we have:

$$\Delta y_t = \alpha_1 \Delta y_{t-1} + v_t, \qquad v_t = u_t - u_{t-1}. \tag{7.5.23}$$

From an economic perspective, it can be important to decide whether a variable is integrated or not because different theories often have different implications in this respect. For example, there is a heated debate in macroeconomics on the relative role of supply shocks (such as technological innovations) and demand shocks (e.g., fiscal and monetary policies) in determining the level of production. Supply shocks are persistent, while demand shocks typically have effects that decay over time. So, if supply shocks are more important than demand ones, we would expect that production is an integrated variable. Similarly, the intertemporal consumption theory predicts that consumption is a random walk, while financial markets efficiency requires that stock prices are a random walk.
Whether or not the variable is integrated also has important implications for econometric theory since, as we shall see in the next sections, the presence of unit roots can fundamentally change the properties of the estimators and test statistics.

7.6 Implications for estimation and inference

We have seen in Section 7.2 that, in general, the OLS estimator of the parameters of dynamic model is consistent and asymptotically normally distributed. In the presence of a unit root, the consistency property remains valid. Actually, the OLS estimator of the parameters associated with the unit root (e.g. α_1 in the AR(1) model in (7.5.1) or $\alpha_1 + \alpha_2$ in the A (2) model in (7.5.15)) is called super-consistent, because the speed of convergence to the true value of the parameter is greater than in the standard case, T vs $\sqrt{T}$.

However, the asymptotic distribution of the OLS estimator for the parameters associated with the unit root is quite different from the Normal one and there is no closed form expression for it. In particular, the distribution is centered on the correct value (one), but it is asymmetric with more probability in the left tail than in the right one. An example is shown in Figure 7.6.
This feature complicates also testing the hypothesis of the presence of a unit root. For example, if we test in the AR(1) model the hypothesis:

$$H_0\colon \alpha = 1 \text{ against } H_1\colon \alpha < 1, \tag{7.6.1}$$

that is the integration of order 1 against the weak stationarity of the variable, we can still use the t-statistic:

$$t_{DF} = (\hat{\alpha}_{OLS} - 1) / \hat{s}.e.(\hat{\alpha}_{OLS}) \tag{7.6.2}$$

but its asymptotic distribution is not Student-t because the OLS estimator is not asymptotically normally distributed under the null hypothesis of a unit root.
The asymptotic distribution of the t-test for a unit root is called the Dickey-Fuller distribution, after the name of the scholars who derived it, and the test t_{DF} is also known as the *Dickey-Fuller test.* The asymmetry in the OLS estimator distribution in the numerator of test statistic determines an asymmetric shape also for the distribution of the test, as plotted in Figure 7.7.
Note that, given the asymmetry of the Dickey-Fuller distribution, the critical value for the test is lower than it would be with a Student t-distribution. Therefore, if we used the Student-t distribution to define the rejection region of the test, we would reject the null hypothesis of unit root much more often than the chosen significance level. Appropriate critical values for testing the unit root hypothesis are available in most econometric software.
It is important to note that the distribution of the test changes depending on the deterministic component of the model. For example, the distribution is different depending on whether or not a constant appears among the regressors. Hence, we have to select the appropriate deterministic component and tables with critical values for the model at hand. Econometric software such as EViews typically report directly the p-value of the Dickey-Fuller test for a unit root, which facilitates the decision to reject or not the null hypothesis.

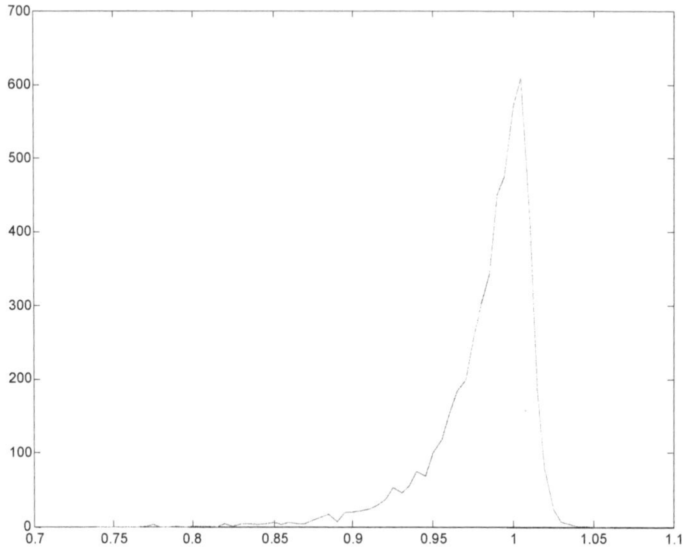

Figure 7.6: Empirical distribution in 5000 samples of the OLS estimator of a in the AR(1) model when $\alpha = 1$, T = 100

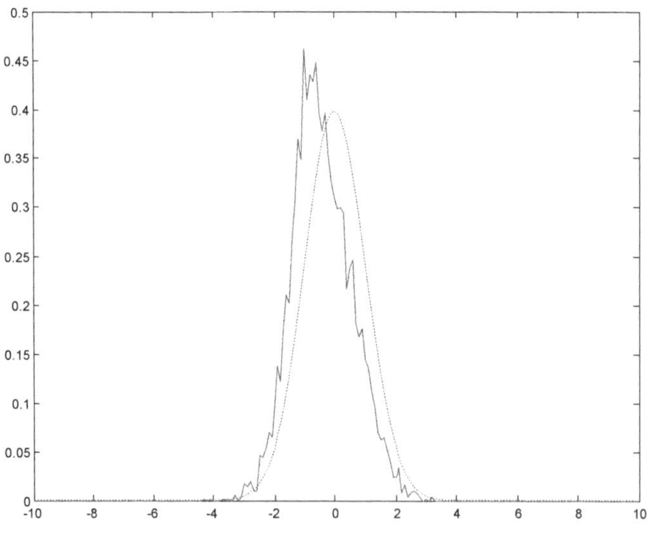

Figure 7.7: Empirical distribution in 5000 samples of the t-statistic for $\alpha = 1$ in the AR(1) model when $\alpha = 1$ (solid line) and Student-t density with 99 degrees of freedom (dotted line)

The Dickey-Fuller test is specific for the AR(1) model. Let us now see how we can test the unit root hypothesis in a more general autoregressive model. In the previous section we saw that, given the AR(2) model

$$y_t = \alpha_1 y_{t-1} + \alpha_2 y_{t-2} + u_t \tag{7.6.3}$$

we can rewrite it as

$$\Delta y_t = -(1-\alpha_1-\alpha_2)y_{t-1} - \alpha_2 \Delta y_{t-1} + u_t = \gamma \quad y_{t-1} + \beta_1 \Delta y_{t-1} + u_t \tag{7.6.4}$$

and the unit root hypothesis is equivalent to the restriction

$$\alpha_1 + \alpha_2 = 1 \text{ or } \gamma = 0. \tag{7.6.5}$$

It can be shown that this result holds for a generic AR(*p*) model. Therefore,

$$y_t = \alpha_1 y_{t-1} + \alpha_2 y_{t-2} + \ldots + \alpha_p y_{t-p} + u_t \tag{7.6.6}$$

is equivalent to

$$\Delta y_t = \gamma \quad y_{t-1} + \beta_1 \Delta y_{t-1} + \ldots + \beta_{p-1} \Delta y_{t-p+1} + u_t \tag{7.6.7}$$

with

$$\gamma = -(1-\alpha_1-\alpha_2-\ldots-\alpha_p), \tag{7.6.8}$$

and the presence of a unit root requires

$$\gamma = 0, \text{ or } \alpha_1 + \alpha_2 + \ldots + \alpha_p = 1. \tag{7.6.9}$$

Dickey and Fuller have also derived the *t*-test for $\gamma = 0$ in the model (7.6.7), labeled *t*-ADF for *Augmented Dickey-Fuller test*, and shown that it has the same asymptotic distribution of *t*-DF in (7.6.2). Then, we can use the same critical values for the two tests. As in the case of the DF test, the distribution of the ADF test also depends on the deterministic component included in the model. For the actual significance level of the test to correspond to the nominal one it is necessary that the number of lags used in the model (7.6.7) is the true one. It can be shown that having fewer lags than needed increases the actual significance level of the test, i.e. the null hypothesis is rejected too often.

Instead, inserting irrelevant lags reduces the power of the test, i.e. the ability to reject the null hypothesis when it is false.
The power of the unit root test is also very low when the relevant parameter value is next to 1, but not exactly 1. For example, if $\alpha = 0.98$, the probability of rejecting the null hypothesis $\alpha = 1$ is very low. This represents an important problem when the test is used to provide evidence against or in favor of a certain economic theory that predicts that the variable of interest is integrated. However, from a practical perspective, if the DF or ADF test does not reject the null hypothesis, the behavior of the variable is very similar to the case of integration and the distribution of the estimators and tests is very closely approximated by the non-standard distribution for the case of unit root.
Parameter changes can also significantly reduce the power of the unit root tests, creating spurious persistence in the model residuals. Consider for example the case where

$$\begin{aligned} y_t &= \alpha_1 t + u_t, \qquad t = 1,\ldots,t^* \\ y_t &= \alpha_2 t + u_t, \qquad t = t^* + 1,\ldots,T \end{aligned} \tag{7.6.10}$$

but the adopted regression model is

$$y_t = \alpha t + u_t, \qquad t = 1,\ldots,T \tag{7.6.11}$$

This situation is represented graphically in Figure 7.8. The model residuals are very persistent, as if they were generated by a stochastic trend, but here the persistence is spurious and due to the change in slope of the deterministic trend. There are tests that permit to test the joint hypotheses of unit root and stability of parameters, but they are too complex for an introductory discussion.
Finally, another interesting but complex test is the Kwiatkowski, Phillips, Schmidt, and Shin test (*KPSS test*). In this case, the null hypothesis is stationary while the alternative hypothesis is integration. The idea behind the test is that, as we saw in (7.5.23), taking first differences of a stationary variable creates a particular correlation in the errors, $v_t = u_t - u_{t-1}$, namely a Moving Average (MA) with a negative root. Then, we can test whether the residuals of a model in differences have this feature (i.e., we can test whether we impose the presence of a unit root by mistake).

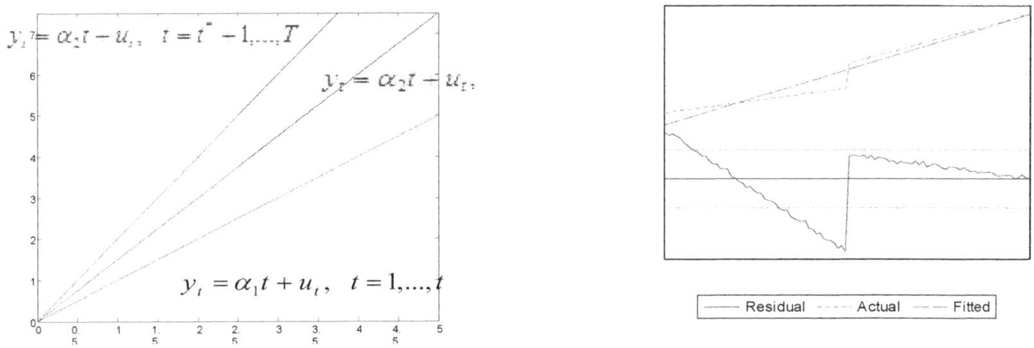

Figure 7.8: Linear segmented trends (coefficient changing in period t) and residuals of a regression assuming a single linear trend (no change in parameters)*

7.7 Cointegration: basics

Suppose we have two variables, *y* and *x*, with *y* integrated of order 1, *I(1)*, and *x* stationary. Let us examine, first, the model

$$y_t = \beta x_t + u_t . \tag{7.7.1}$$

The coefficient of determination in this model is

$$R^2 = \frac{ESS}{TSS} = \frac{\hat{\beta}^2 \sum_{t=1}^{T} x_t^2}{\sum_{t=1}^{T} y_t^2} . \tag{7.7.2}$$

Since *y* is integrated, we have

$$\frac{\sum_{t=1}^{T} x_t^2}{\sum_{t=1}^{T} y_t^2} \to 0 \text{ for } T \to \infty , \tag{7.7.3}$$

from which

$$R^2 \to 0 \text{ for } T \to \infty \tag{7.7.4}$$

and then the explanatory power of *x* is null. In other words, we cannot explain an integrated variable with a stationary one.
Let us evaluate now the model

$$x_t = \alpha y_t + e_t \tag{7.7.6}$$

The OLS estimator of the parameter α is

$$\hat{\alpha} = \frac{\sum_{t=1}^{T} x_t y_t}{\sum_{t=1}^{T} y_t^2} \tag{7.7.7}$$

and, since y is *I(1)*, the denominator grows faster than the numerator and we have

$$\hat{\alpha} \to 0 \text{ for } T \to \infty . \tag{7.7.8}$$

Therefore, we also cannot explain a stationary variable with an integrated one. Stationary variables must be explained by stationary variables (the case that we implicitly considered in the first six chapters of the book), and integrated variables by integrated variables. This is what is called a *balanced regression.*
Let us now carefully consider the case in which both the dependent variable and the regressor are integrated of order 1. Specifically, let us suppose that:

$$\begin{matrix} y_t = y_{t-1} + e_t \\ x_t = x_{t-1} + u_t \end{matrix} \qquad \begin{matrix} e_t \\ u_t \end{matrix} \overset{i.i.d.}{\sim} \left(\begin{pmatrix} 0 \\ 0 \end{pmatrix}, \begin{pmatrix} \sigma_e^2 & 0 \\ 0 & \sigma_u^2 \end{pmatrix} \right), \tag{7.7.9}$$

so that *y* and *x* are completely independent from each other.
Let us examine the regression

$$y_t = \beta x_t + u_t \tag{7.7.10}$$

It can be shown that in this regression, when $T \to \infty$, then:

$$R^2 \to 1 \tag{7.7.11}$$

$$\hat{\beta}_{OLS} \to \text{to a random variable,} \tag{7.7.12}$$

$$DW \to 0\,. \tag{7.7.13}$$

Then, even if y and x are not correlated, the fact that they are both integrated of order 1 causes the coefficient of determination to asymptotically converge to 1 (instead of 0) and the OLS estimator to converge to a random variable (instead of 0). This phenomenon is called *spurious regression*. The most important indicator of a spurious regression is the presence of very high correlation in the regression residuals: the fact that the Durbin-Watson test for the no correlation of errors converges to 0 indicates that errors in (7.7.10) are integrated of order 1:

$$u_t = u_{t-1} + v_t\,, \tag{7.7.14}$$

If the errors of the regression of one integrated variable on another integrated variable are stationary, for example,

$$u_t = \alpha u_{t-1} + v_t \text{ with } |\,\alpha\,| < 1, \tag{7.7.15}$$

then the variables are said to be *cointegrated*. *Cointegration* implies the existence of a linear combination of the integrated variables that is stationary. In our example, y and x are $I(1)$ but $y - \beta x = u$ is stationary.

From an economic point of view, cointegration is a property that should apply very often and in very different fields. For example, consumption, analyzed by itself, could be a random walk, but it should be cointegrated with disposable income. Government spending and taxes may be integrated of order 1, but their difference, the deficit, should be stationary so that the national debt is sustainable, not growing without limits (in relation to GDP). Production inputs and the output can be integrated when analyzed individually, but should be cointegrated for the existence of a production function. Interest rates at different maturities can be integrated but their differences, the spreads, should be stationary, and hence the rates should be cointegrated. And so on.

From an econometric point of view, it is possible to test whether two variables are cointegrated, making sure that the errors of the regression of one variable on the other are stationary, for example by ensuring that condition (7.7.15) is satisfied. This is the intuition behind the *cointegration test of Engle and Granger*.

The Engle and Granger test requires two stages. In the first one, we estimate by OLS the regression

$$y_t = \beta x_t + u_t \tag{7.7.16}$$

where x can also be a vector of regressors instead of a single variable, and we calculate the residuals

$$\hat{u}_t = y_t - \hat{\beta}_{OLS} x_t \tag{7.7.17}$$

The result of super-consistency of the OLS estimator given in Section 7.6 also applies to the estimator of the parameters of the cointegration relation, in this case $\hat{\beta}_{OLS}$ converges to the true value β at the rate T instead of $\sqrt{T}$.
In the second stage, we estimate an *AR(q)* model for the residuals:

$$\hat{u}_{tt} = c + \alpha_1 \hat{u}_{t-1} + \ldots + \alpha_q \hat{u}_{t-q} + v_t \tag{7.7.18}$$

and implement a *t*-test for $\alpha_1 + \alpha_2 + \ldots + \alpha_q = 1$, which coincides with the ADF unit root test (or DF when $q = 1$). The null hypothesis is no cointegration between y and x (as the errors are integrated when y and x are not cointegrated), while cointegration is the alternative.
We saw in Section 7.6 that the *t*-test distribution is not standard due to the presence of a unit root under the null hypothesis. In this case, the *t*-test distribution is also different from the Dickey-Fuller one, because in the AR(q) model in (7.7.18) we use the residuals of the first-stage regression as dependent variable, since the errors u_t in (7.7.16) are not observable. If we used the critical values of the DF or ADF tests, the test would be biased towards the rejection of the null hypothesis of unit root, because the OLS estimation method applied in the first stage generates residuals with minimum variance and hence "maximally stationary". Appropriate critical values for the cointegration test of Engle and Granger (i.e. for the *t*-test for $\alpha_1 + \alpha_2 + \ldots + \alpha_q = 1$ in (7.7.18)) are reported in Table 7.10 for different model characteristics, while *p*-values are typically directly available from econometric software.

Number of variables	Test-statistic	Significance level		
		0.01	0.05	0.10
N=2	τ_μ	-3.90	-3.34	-3.04
	τ_τ	-4.32	-3.78	-3.50
N=3	τ_μ	-4.29	-3.74	-3.45
	τ_τ	-4.66	-4.12	-3.84
N=4	τ_μ	-4.64	-4.10	-3.81
	τ_τ	-4.97	-4.43	-4.15
N=5	τ_μ	-4.96	-4.42	-4.13
	τ_τ	-5.25	-4.72	-4.43
N=6	τ_μ	-5.25	-4.71	-4.42
	τ_τ	-5.52	-4.98	-4.70
Source: Davidson and MacKinnon (1993)[1]				

Table 7.10: Critical values of the Engle and Granger test

Note that cointegration means that the variables *y* and *x* taken individually can have an arbitrary evolution, driven by a stochastic trend, but analyzed together are linked by a relation that binds their evolution, making sure that one combination of the variables is no longer driven by a stochastic trend. Then, it will be

$$\begin{aligned} y_t &= \beta ST_t + e_{yt} \\ x_t &= ST_t + e_{xt} \end{aligned} \tag{7.7.19}$$

with e_{yt} and e_{xt} stationary, errors, so that

$$y_t - \beta x_t = e_{yt} - \beta e_{xt} = u_t \tag{7.7.20}$$

[1] R. Davidson, J.G. MacKinnon, *Estimation and Inference in Econometrics*, Oxford University Press, 1993.

does not depend on ST. Therefore, in order for two or more *I(1)* variables to be cointegrated, it is necessary that they are driven by a single stochastic trend. In the case of a spurious regression, the variables are instead driven each by a different stochastic trend.
In the short term, there may be quite persistent deviations from the cointegration relation $y_t - \beta x_t$, due to the error $e_{yt} - \beta e_{xt} = u_t$. Since $E(e_{yt} - \beta e_{xt}) = E(u_t) = 0$, in the long run, we instead expect $y_t = \beta x_t$. In this sense, the cointegration relation can be interpreted as a *long-term equilibrium* relationship between variables.
One last important result proved by Engle and Granger, known as the *Engle and Granger representation theorem*, is that two variables are cointegrated if and only if they can be represented as an Error Correction Model (ECM), as seen in Section 7.1. In particular, it must be:

$$\Delta y_t = \delta(y_{t-1} - \beta x_{t-1}) + \textit{lagged values of } \Delta y_t \textit{ and } \Delta x_t + u_{yt} \qquad (7.7.21)$$

and/or

$$\Delta x_t = \gamma(y_{t-1} - \beta x_{t-1}) + \textit{lagged values of } \Delta y_t \textit{ and } \Delta x_t + u_{yt}\,. \qquad (7.7.22)$$

Therefore, there can be deviations from the long-term equilibrium, i.e., from the cointegration relation $y_t - \beta x_t$, but *y*, or *x*, or both should move in such a way that the deviations will decrease over time and, on average, the condition $y_t = \beta x_t$ is satisfied. As noted in Section 7.1, ECM models play a major role in modern applied econometrics due to their connection with the presence of integration and cointegration between variables.
If the variables were not cointegrated, it would be $\delta = 0$ in (7.7.21) and $\gamma = 0$ in (7.7.22), because we would have a stationary dependent variable, Δy_t or Δx_t, and an integrated a regressor, $y_t - \beta x_t$. Therefore, if the variables are integrated, but not cointegrated, the appropriate model is specified in differences. Before cointegration became popular, models in differences were the most widely used tool for the analysis of economic variables. However, if the variables are indeed cointegrated, these models are not correctly specified, due to the omission of the stationary term $y_t - \beta x_t$, which, as seen in Chapter 3, creates bias in the estimators.
From an inferential point of view, the distribution of the OLS estimator in a regression with *I(1)* cointegrated variables, as $\hat{\beta}_{OLS}$ in (7.7.16), is non-standard.

Therefore, we cannot use the *t*- and *F*-statistics together with critical values from the *t*- and *F*- distributions to test hypotheses about parameters of the cointegration relation. However, given that all the variables that appear in the ECM are stationary, all results seen in Section 7-2 for estimation, inference and diagnostic control are directly applicable in this context.
From the applied point of view, the results in this section suggest:

- starting an empirical analysis by verifying that the dependent and the explanatory variables have the same order of integration;
- if the variables are *I(1)*, testing whether they are cointegrated and estimating the cointegration relation, for example $y_t - \hat{\beta}_{OLS} x_t$.;
- inserting the estimated cointegration relation(s) into an ECM for the variable of interest, such as $\Delta y_t = \delta(y_t - \hat{\beta}_{OLS} x_t) +$ *lagged values of* Δy_t *and* $\Delta x_t + u_{yt}$;
- checking that the resulting ECM model provides a coherent representation of the variable of interest, for example ensuring that the assumptions underlying the linear model are satisfied and the parameters are stable;
- testing the hypotheses of interest on the parameters of the ECM model using inferential procedures.

Finally, it is useful to mention the existence of more sophisticated techniques for the analysis of cointegration between more than two variables and the joint study of the evolution of different dependent variables with ECM models. Their complexity, though, makes them unsuitable for an introductory level analysis.

7.8 An empirical analysis with integrated simulated data

The workfile "example_simul2_chap7.wf1" contains the series *H*, *X*, *Y*, *W*, *Z* and *Q*. Let us start comparing, in Figure 7.9, the dynamics of an AR(1) stationary series, *W*, and those of a series with deterministic trend, *Q*.

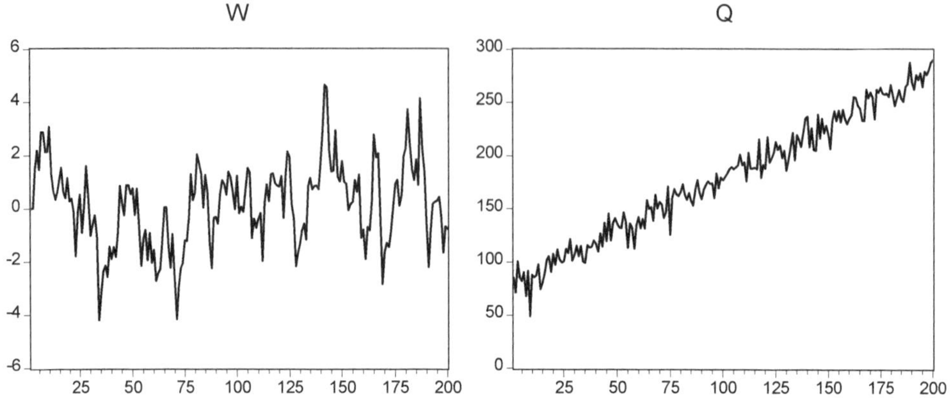

Figure 7.9: Stationary variables in levels (W) or around a deterministic trend (Q)

The first series, *W*, has no trending behavior, i.e., it neither grows nor declines over the long run. The variability neither rises nor shrinks over time, and positive values tend to be followed by positive values for a few observations, while negative values tend to be followed by negative values in the same way, until an exogenous shock of opposite sign does not bring the series close to the average (0, in this case).

The second variable *Q* grows over time along a linear trend, showing an explosive behavior in levels. Once cleaned from the effects of the trend, the series has rather stable variation, and detrended observations appear to be stationary, as they move around the trend randomly and with little persistence.

The other four series available in the workfile are plotted in Figure 7.10. For each series, deviations from the mean value or a linear trend are quite persistent. We are most likely observing processes with unit roots.

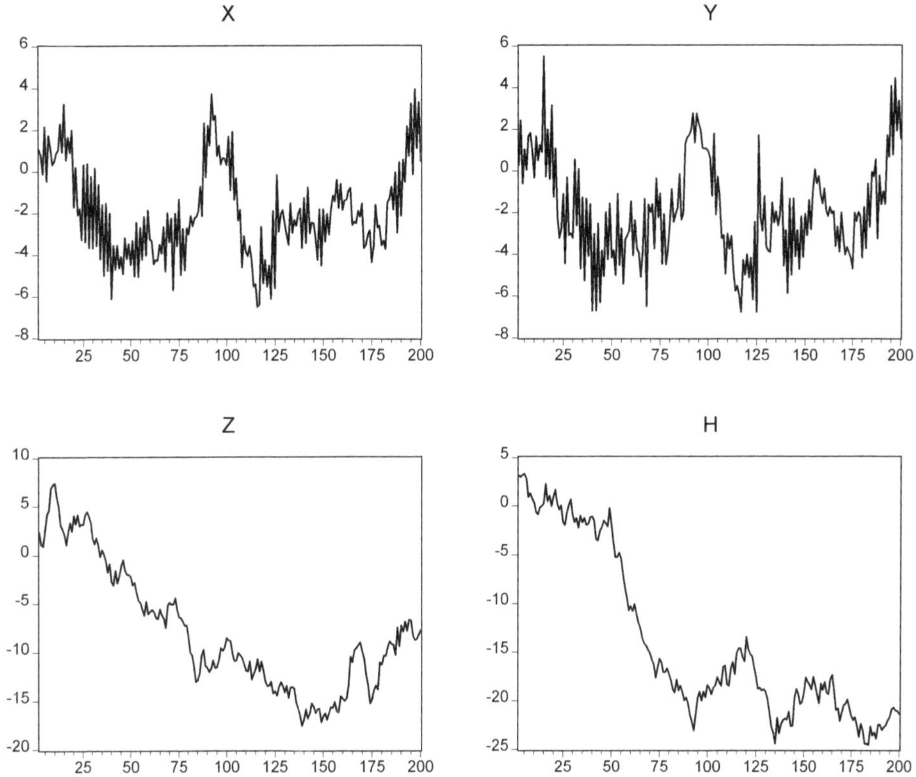

Figure 7.10: Plots of the variables X, Y, Z, H

To formally test for the unit root hypothesis, EViews allows to perform automatically different procedures, including the Dickey-Fuller (DF) test and its augmented version (ADF) described in Section 7.6. The null hypothesis is that the variable is integrated of order 1, against the alternative hypothesis of stationarity, and the underlying model is either an AR(1), in the case of DF, or an AR(p), in the case of the ADF. Taking as an example the X series, by clicking on "View" and selecting "Unit Root Test" we get the window reported in Figure 7.11. In the menu, we select the option "Augmented Dickey-Fuller", with the possibility to choose the number of lags. To implement the "Dickey-Fuller" test we must set the lag length to 0. This produces Table 7.11.
The test result leads to reject the null hypothesis of a unit root for the X variable. However, this finding could be due to the use of a misspecified model. In particular, the lag length could be too short. Hence, we implement the Augmented Dickey-Fuller test, using two lags, which produces the output in Table 7.12.

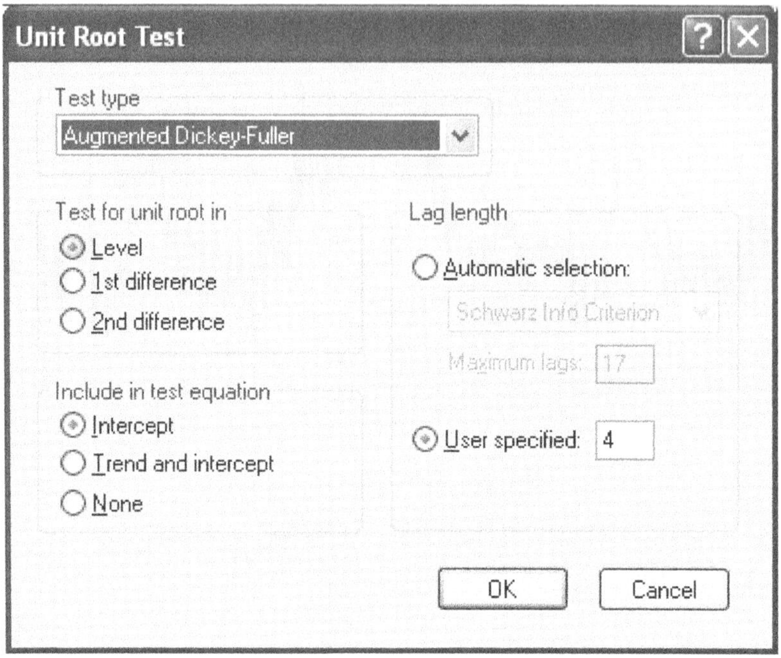

Figure 7.11: EViews window for unit root test

The ADF test now does not reject the null hypothesis of unit root, the *p*-value is about 0.22, in line with the observation made in Section 7.6 that increasing the number of lags typically reduces the probability of rejecting the null hypothesis.

Null Hypothesis: x has a unit root
Exogenous: Constant
Lag Length: 0 (Fixed)

		t-Statistic	Prob.*
Augmented Dickey-Fuller test statistic		-6.934683	0.0000
Critical values Test:	1% level	-3.463235	
	5% level	-2.875898	
	10% level	-2.574501	

Table 7.11: Dickey-Fuller Test for the variable X

Another test available in the unit root menu is due to Kwiatkowski-Phillips-Schmidt-Shin (KPSS test) and it has stationarity of the variable under the null hypothesis and unit root under the alternative, contrary to DF and ADF. The output of this test is reported in Table 7.13. The KPSS test does not reject the null hypothesis of stationarity of the variable *X*, because the test statistic is well below the relevant critical values. Hence, the result is in contrast with that of the ADF test but in line with that of the DF test.

The Dickey-Fuller test for *Y* gives a *p*-value of 0.0000, while the ADF test with two lags gives 0.0712 as *p*-value. Hence, as for the variable *X*, the DF test rejects the null hypothesis of a unit root while the ADF test does not, at the 5% significance level. In addition, also in this case the KPSS test does not reject the null hypothesis of stationarity: the test statistic has a value of 0.131 that is lower than the 5% critical value, 0.463.

For the variable *Z*, the *p*-values of the DF and ADF statistics are 0.4958 and 0.5063, respectively. The Kwiatkowski-Phillips-Schmidt-Shin test is 1.316, now well above the 5% critical value. Hence, all the procedures provide support for the presence of a unit root in the model generating the *Z* variable. A similar finding emerges for the variable *H*.

Null Hypothesis: X has a unit root
"Exogenous: Constant
Lag Length: 2 (Fixed)

		t-Statistic	Prob.*
Augmented Dickey-Fuller test statistic		-2.154189	0.2240
Critical values Test:	1% level	-3.463576	
	5% level	-2.876047	
	10% level	-2.574581	

* MacKinnon (1996) one-sided *p*-values.[2]

Table 7.12: Augmented Dickey-Fuller Test for the variable X

Finally, for the variable W, the null hypothesis of non-stationarity is strongly rejected by the DF and ADF tests, while the null hypothesis of stationarity is not rejected in the KPSS test. Hence, in this case all the procedures are in favor of stationarity of *W*.

[2] J.G. MacKinnon, «Numerical Distribution Functions For unit Root and Cointegration Tests». *Journal of Applied Econometrics*, 11(6), 2001, 601–618.

Null Hypothesis: X is stationary
"Exogenous: Constant
Bandwidth: 11 (Newey-West using Bartlett kernel)

		LM-Stat.
Kwiatkowski-Phillips-Schmidt-Shin test statistic		0.124136
Asymptotic critical values *:	1% level	0.739000
	5% level	0.463000
	10% level	0.347000

* Kwiatkowski-Phillips-Schmidt-Shin (1992, Table 1)

Residual variance (no correction)	4.981427
HAC corrected variance (Bartlett kernel)	36.42982

Table 7.13: KPSS Test for the null hypothesis of stationarity of the variable X

In summary, for *W* all the procedures are in favor of stationarity while for *Z* and *H* they all suggest a unit root. Instead, for *X* and *Y* only the ADF test suggests a unit root, with DF and KPSS are in favor of stationarity. As in this case we know the data generating process, we can say that not only *Z* and *H*, but also *X* and *Y* have a unit root, while *W* is indeed stationary.
We might wonder whether the integrated variables we have detected have a single unit root, and so they are *I(1)*, or a double unit root, and so they are *I(2)*. To assess the presence of a second unit root, we can run the DF and ADF tests on the first differences of the variables, see Figure 7.12. It turns out that the unit root hypothesis for the first differences is rejected for all variables, which can therefore be treated as *I(1)*.

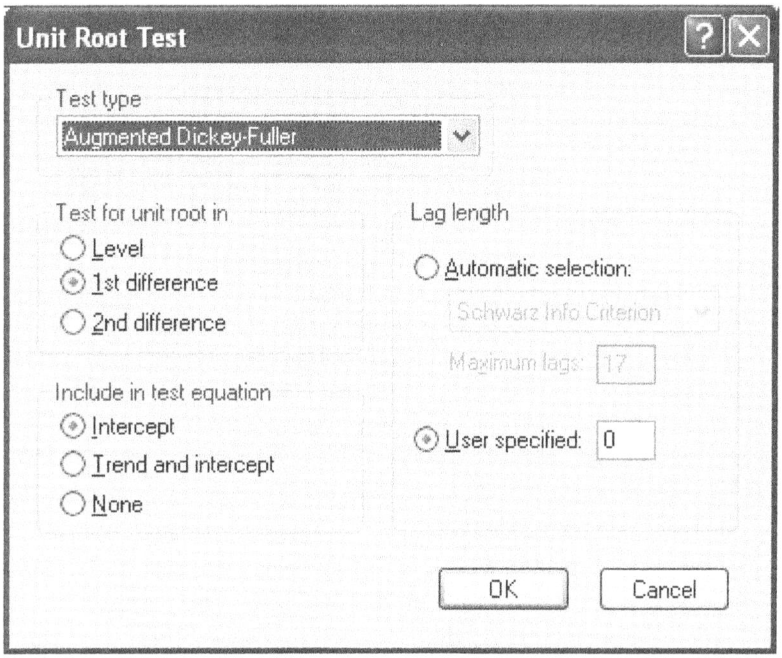

Figure 7.12: Choice of unit root tests

Let's try now to regress the integrated variable *Z* on the stationary variable *W*. Table 7.14, which collects the results, shows a R^2 close to 0. Similarly, if we regress the stationary variable *W* on the integrated variable *Z* we get a R^2 close to 0, see Table 7.15. Both results are in line with the theoretical considerations in Section 7.7.

Dependent variable: Z

Variable	Coefficient	Std. Error	*t*-Statistic	Prob.
(C)	-7.610810	0.453591	-16.77900	0.0000
W	-0.621494	0.292128	-2.127470	0.0346

R-squared	0.022348	Mean dependent var	-7.710320
Adjusted R-squared	0.017411	S.D. dependent var	6.436834
S.E. of regression	6.380553	Akaike information criterion	6.554336
Sum squared resid	8060.869	Schwarz criterion	6.587320
Log likelihood	-653.4336	*F*-statistic	4.526129
Durbin-Watson stat	0.032320	Prob (*F*-statistic)	0.034618

Table 7.14: Regression of the integrated variable on the stationary variable W

Dependent variable: W

Variable	Coefficient	Std. Error	*t*-Statistic	Prob.
(C)	-0.117143	0.169592	-0.690734	0.4905
Z	-0.035959	0.016902	-2.127470	0.0346

R-squared	0.022348	Mean dependent var	0.160113
Adjusted R-squared	0.017411	S.D. dependent var	1.548312
S.E. of regression	1.534774	Akaike information criterion	3.704593
Sum squared resid	466.3951	Schwarz criterion	3.737576
Log likelihood	-368.4593	*F*-statistic	4.526129
Durbin-Watson stat	0.487872	Prob (*F*-statistic)	0.034618

Table 7.15: Regression of the stationary variable W on the integrated variable Z

Now let us consider a model where an integrated variable, *Z*, is explained by another integrated variable, *H*, as in Table 7.16. The variable *H* is highly significant and the coefficient of determination is quite high, but the value of the DW statistic close to 0 suggests that this is a spurious regression.
As an example of a cointegration model, let use regress *Y* on *X*, obtaining Table 7.17. As we have seen, the *Y* and *X* variables are integrated, but the regression residuals appear stationary, supporting the hypothesis of cointegration between Y and X.
To formally test the hypothesis that the regression errors are stationary, we can use the Engle and Granger procedure. Having just performed the regression of *Y* on *X*, we save the residuals, labeling them, for example, "res":

```
series res = resid
```

Next we estimate an AR model for res, for example an AR(1) model with intercept:

```
ls d (res) c res ( -1)
```

which can be used to perform the DF test. The DF statistic is equal to -15.26806, to be compared with the critical values in Table 7.10 (not the usual DF critical values). For example, at the 1% level the proper critical value is -3.9 and since -15.26806 < -3.9 , we strongly reject the null hypothesis of a unit root in the errors, providing support for cointegration between *Y* and *X*, with the cointegration relation being approximately *Y-X*. An ADF test leads to the same conclusion.
When applying the Engle and Granger test to check for cointegration between *Z* and *H*, the null hypothesis of a unit root in the regression errors (no cointegration) cannot be rejected, in line with the low value for the DW test noted above.

Dependent variable: Z

Variable	Coefficient	Std. Error	*t*-Statistic	Prob.
(C)	1.282310	0.417006	3.075039	0.0024
H	0.650571	0.025500	25.51229	0.0000

R-squared	0.767655	Mean dependent var	-7.757806
Adjusted R-squared	0.766476	S.D. dependent var	6.417853
S.E. of regression	3.101387	Akaike information criterion	5.111575
Sum squared resid	1894.864	Schwarz criterion	5.144674
Log likelihood	-506.6017	*F*-statistic	650.8770
Durbin-Watson stat	0.151612	Prob (*F*-statistic)	0.000000

Table 7.16: Example of a spurious regression

Dependent variable: Y
Least Squares Method:
Sample: 2200
Included observations: 199

Variable	Coefficient	Std. Error	*t*-Statistic	Prob.
(C)	0.014865	0.091977	0.161612	0.8718
X	0.990040	0.031141	31.79217	0.0000

R-squared	0.836886	Mean dependent var	-1.897948
Adjusted R-squared	0.836058	S.D. dependent var	2.423807
S.E. of regression	0.981394	Akaike information criterion	2.810314
Sum squared resid	189.7375	Schwarz criterion	2.843413
Log likelihood	-277.6263	*F*-statistic	1010.742
Durbin-Watson stat	2.166193	Prob (*F*-statistic)	0.000000

Table 7.17: Example of a cointegration regression

Now, by applying the Engle and Granger representation theorem, we know that either *X* or *Y*, or both, must admit an ECM specification. The ECM (Error Correction Model) uses only stationary variables: first differences of the *I(1)* variables, which are *I(0)* by definition, and the stationary linear combinations of the integrated variables (cointegration relations).
To estimate the coefficients of the cointegration relation, $y - \beta x$, we can use the OLS estimates in the regression of *Y* on *X*, since the estimator is super-consistent. For example, from Table 7.17, we know that the estimated

coefficient of X is close to 1, and then we can use $(y-x)$ as the cointegration relation. The ECM for Y is then, using EViews notation:

$$d(y) \;\; c \;\; (y(-1)-x(-1)) \;\; d(x(-1)) \;\; d(y(-1))$$

and the output is shown in Table 7.18. The values are the expected ones, in particular the adjustment coefficient is negative and highly significant. Hence, if the difference between Y and X is positive, Y decreases in order to bring the system back to the equilibrium where $Y=X$. It is also worth mentioning that in the ECM, once conditioning upon the estimated cointegration relationships, the OLS estimators of the remaining parameters are asymptotically normally distributed, which makes all the previous results on inference and hypothesis testing applicable.

Dependent variable: D (Y)

Variable	Coefficient	Std. Error	*t*-Statistic	Prob.
(C)	0.045380	0.098148	0.462364	0.6443
Y (-1)-X (-1)	-0.887791	0.149002	-5.958229	0.0000
D (X (-1))	-0.622788	0.109540	-5.685505	0.0000
D (Y (-1))	-0.204204	0.100713	-2.027587	0.0440

R-squared	0.653710	Mean dependent var	0.006299
Adjusted R-squared	0.648355	S.D. dependent var	2.326140
S.E. of regression	1.379394	Akaike information criterion	3.501160
Sum squared resid	369.1290	Schwarz criterion	3.567590
Log likelihood	-342.6149	*F*-statistic	122.0747
Durbin-Watson stat	1.854676	Prob (*F*-statistic)	0.000000

Table 7.18: ECM for Y

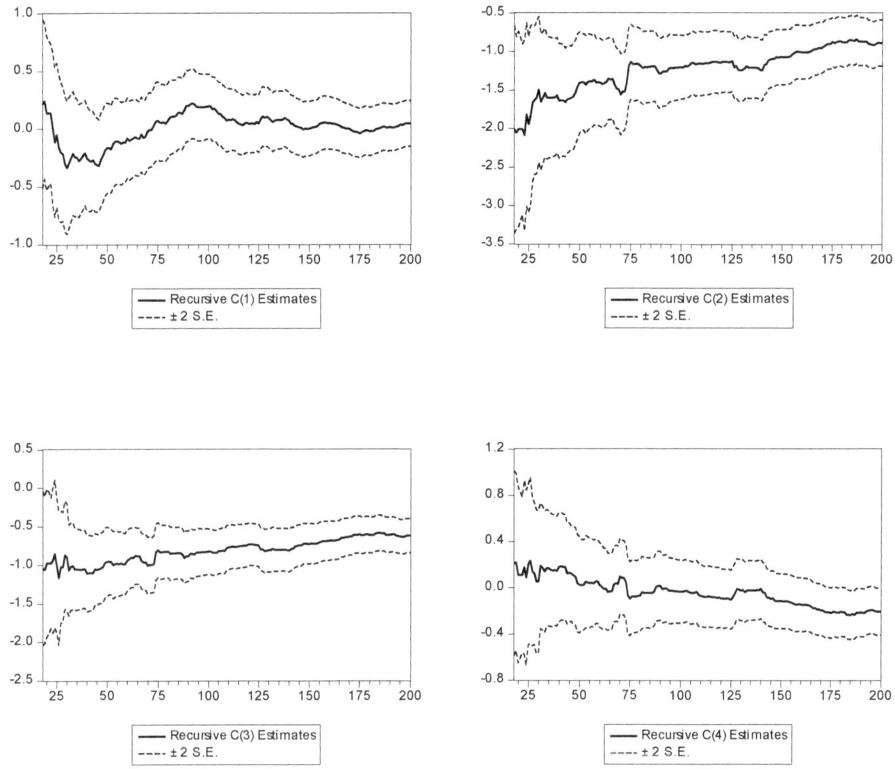

Figure 7.13: Recursive estimators of the ECM parameters

Finally, the LM test for the null hypothesis of no autocorrelation of the errors and the White test for the null hypothesis of homoscedasticity of the errors cannot reject the respective null hypotheses at the 5% significance level. The recursive estimators of the parameters, shown in Figure 7.13, support their stability.

7.9 An empirical analysis of the determinants of aggregate consumption

We will now try to formulate a dynamic model for aggregate consumption in Italy that addresses many of the problems of the static specifications of the previous chapters. To get a long enough sample, let us focus on the relation between disposable income and consumption. Data are available in the "example_cons_chap7.wf1" workfile.

We start by verifying whether consumption and income are integrated of order 1 or stationary. Tables 7.19 and 7.20 show that the null hypothesis of unit root is not rejected, while the hypothesis of stationarity can be rejected at the 5%

level for consumption and at the 10% for disposable income. Hence, we continue our analysis assuming that both the dependent and the explanatory variable are *I(1)*.

Null Hypothesis: LRC has a unit root
"Exogenous: Constant
Lag Length:1(Automatic - based on SIC, maxlag = 11)

		t-Statistic	Prob.*
Augmented Dickey-Fuller test statistic		-1.527008	0.5154
Critical values Test:	1% level	-3.507394	
	5% level	-2.895109	
	10% level	-2.584738	

* MacKinnon (1996) one-sided *p*-values.

Null Hypothesis: LRC is stationary
"Exogenous: Constant
Bandwidth: 7 (Newey-West automatic) using Bartlett kernel

		LM-Stat.
Kwiatkowski-Phillips-Schmidt-Shin test statistic		0.511960
Asymptotic critical values *:	1% level	0.739000
	5% level	0.463000
	10% level	0.347000

* Kwiatkowski-Phillips-Schmidt-Shin (1992, Table 1)[3]

Table 7.19: Unit root and stationarity test for consumption

[3] D. Kwiatkowski, P.C.B. Phillips, P. Schmidt, Y. Shin, «Testing the Null Hypothesis of Stationarity Against the Alternative of a Unit Root», *Journal of Econometrics*, 54(1-3), 1992, 159-178.

Null Hypothesis: LRYD has a unit root
"Exogenous: Constant
Lag Length: 0 (Automatic - based on SIC, maxlag = 11)

		t-Statistic	Prob.*
Augmented Dickey-Fuller test statistic		-1.721020	0.4172
Critical values Test:	1% level	-3.506484	
	5% level	-2.894716	
	10% level	-2.584529	

Null Hypothesis: LRYD is stationary
"Exogenous: Constant
Bandwidth: 7 (Newey-West automatic) using Bartlett kernel

		LM-Stat.
Kwiatkowski-Phillips-Schmidt-Shin test statistic		0.387874
Asymptotic critical values *:	1% level	0.739000
	5% level	0.463000
	10% level	0.347000

Table 7.20: Unit root and stationarity test for disposable income

To verify the presence of cointegration, we linearly regress consumption in levels on a constant and disposable income, and we include also the dummy CRISIS to allow for changes both in the intercept and in the slope. Having performed the regression, we save the residuals (labeling them ECM), and test the null hypothesis of unit root for ECM using the proper critical values for this Engle and Granger test. According to Table 7.21, we strongly reject the null hypothesis of unit root for ECM, and thus we can conclude that income and consumption are cointegrated and their stationary linear combination is given by ECM.

Null Hypothesis: ECM has a unit root
"Exogenous: None
Lag Length: 0 (Fixed)

		t-Statistic	Prob.*
Augmented Dickey-Fuller test statistic		-5.238341	0.0000
Critical values Test:	1% level	-2.591505	
	5% level	-1.944530	
	10% level	-1.614341	

* MacKinnon (1996) one-sided p-values.

Table 7.21: Engle and Granger cointegration test

We can now formulate an Error Correction Model for consumption. We start from a specification where the growth rate of consumption is regressed on the growth rate of income, lags of both growth rates, a lagged value of ECM, and the dummy associated with the crisis. After verifying that this model satisfies all checks for the residuals and eliminating non-significant regressors, we get to the formulation in Table 7.22. All regressors are highly significant. Moreover, the coefficient of the error correction term has the expected negative sign, implying that if consumption exceeds its equilibrium value, it will be reduced in the next period.

Dependent variable: D (LRC)
Least Squares Method:
Sample (adjusted): 1990q3 2012q1
Included observations: 87 after adjustments

Variable	Coefficient	Std. Error	*t*-Statistic	Prob.
D (LRYD)	0.848102	0.040899	20.73640	0.0000
ECM (-1)	-0.382221	0.092153	-4.147668	0.0001
D (LRC (-1))	-0.240758	0.097537	-2.468388	0.0157
(D) (LRYD (-1))	0.276856	0.094974	2.915078	0.0046
CRISIS	0.003499	0.001854	1.887292	0.0627

R-squared	0.865068	Mean dependent var	8.25 E-05
Adjusted R-squared	0.858486	S.D. dependent var	0.021136
S.E. of regression	0.007951	Akaike information criterion	-6.775280
Sum squared resid	0.005184	Schwarz criterion	-6.633562
Log likelihood	299.7247	Hannan-Quinn criterion	-6.718215
Durbin-Watson stat	2.027971		

Table 7.22: Error correction model for aggregate consumption

Based on Figure 7.14, the fit of the model is good, although it is difficult to perfectly capture the entire variability of the consumption growth rate. The residuals do not show any obvious problems, apart from some abnormal observations at the beginning and at the end of the sample. Indeed, tests for homoscedasticity and no correlation of the errors do not reject the respective null hypotheses, see Table 7.23. Tests of Normality and linearity also do not reject their null hypotheses at the 5% level of significance.

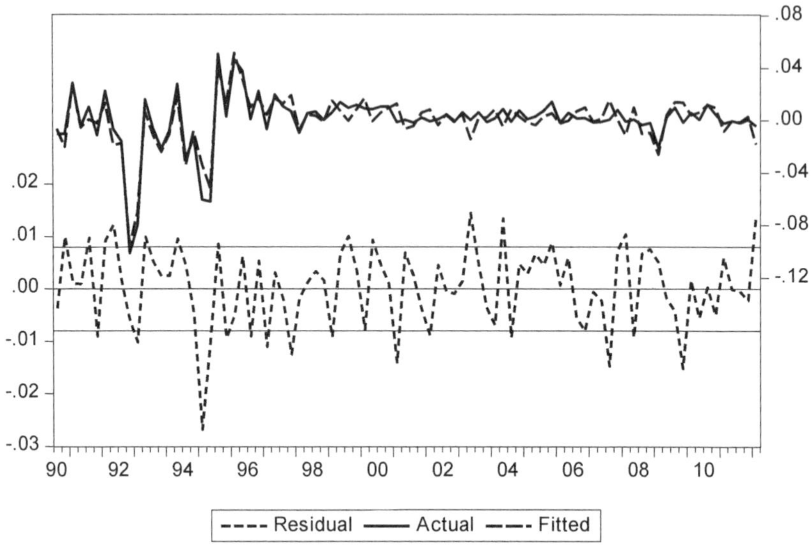

Figure 7.14: ECM fitted values and residuals

Breusch-Godfrey Serial Correlation LM Test:

F-statistic	2.353045	Prob. F (2.80)	0.1016
Obs * R-squared	3.585802	Prob. Chi-square (2)	0.1665

Heteroscedasticity Tests: White

F-statistic	1.307086	Prob. F (15.71)	0.221200
Obs * R-squared	18.82592	Prob. Chi-Square (15)	0.221700
Scaled explained SS	24.55824	Prob. Chi-Square (15)	0.056200

Ramsey RESET Test

	Value	Df	Probability
t-statistic	2.517404	81.00000	0.013800
F-statistic	6.337322	(1, 81)	0.013800
Likelihood ratio	6.553600	1.000000	0.010500

Table 7.23: Test for correct specification of the ECM

Figure 7.15 shows the recursive estimators of the parameters, providing no indications of instability. Hence, overall, the ECM specification in Table 7.22 seems to represent a good econometric model for the evolution of aggregate consumption.

Finally, it is interesting to use this ECM model to reconsider the effects of the introduction of the euro on consumption. In the static model in Chapter 5, these effects were statistically significant, but that model was not correctly specified. By introducing two additional regressors in the dynamic model of Table 7.22, i.e. EURODU * DLRYD and EURODU * ECM (-1), these remain individually significant at the 1% level, and jointly significant at the 5% level, while the dummy for the crisis loses its significance. We can therefore conclude that the introduction of the euro has brought changes in aggregate consumption, a claim that is now better supported by statistical evidence than in Chapter 5, as it is based on a properly specified econometric model.

7.10 An empirical analysis of short-term interest rates

The "example_int_chap7.wf1" workfile contains the monthly (average of daily) series of the overnight interest rate and one-month interest rate for Italy, for the period January 1990 to February 2006.
For the overnight rate we do not reject the null hypothesis of non-stationarity according to the ADF test (with lag length determined by the Schwarz Information Criterion), the *p*-value is 0.47. The assumption of stationarity is rejected by the KPSS test, with a value of 8.02 against a critical value of 0.73 (at the 1% level). The corresponding values for the one-month rate are 0.22 (*p*-value for the ADF tests), and 6.79 (value of the KPSS test). Hence, we will treat both variables as *I(1)*.
To check whether the two series are cointegrated, we apply the Engle-Granger test, estimating by OLS the linear relationship between the two interest rates, saving the resulting residuals, and testing for a unit root in them with the ADF test. The EViews commands are:

```
ls one_month c overnight
series res = resid
ls d (res) c res ( -1) d (res ( -1))
```

We obtain a value of the ADF test of -19.4 , which should be compared with a critical value of -3.9. Hence, we can strongly reject the null hypothesis of integration and proceed considering monthly and overnight rates as cointegrated.

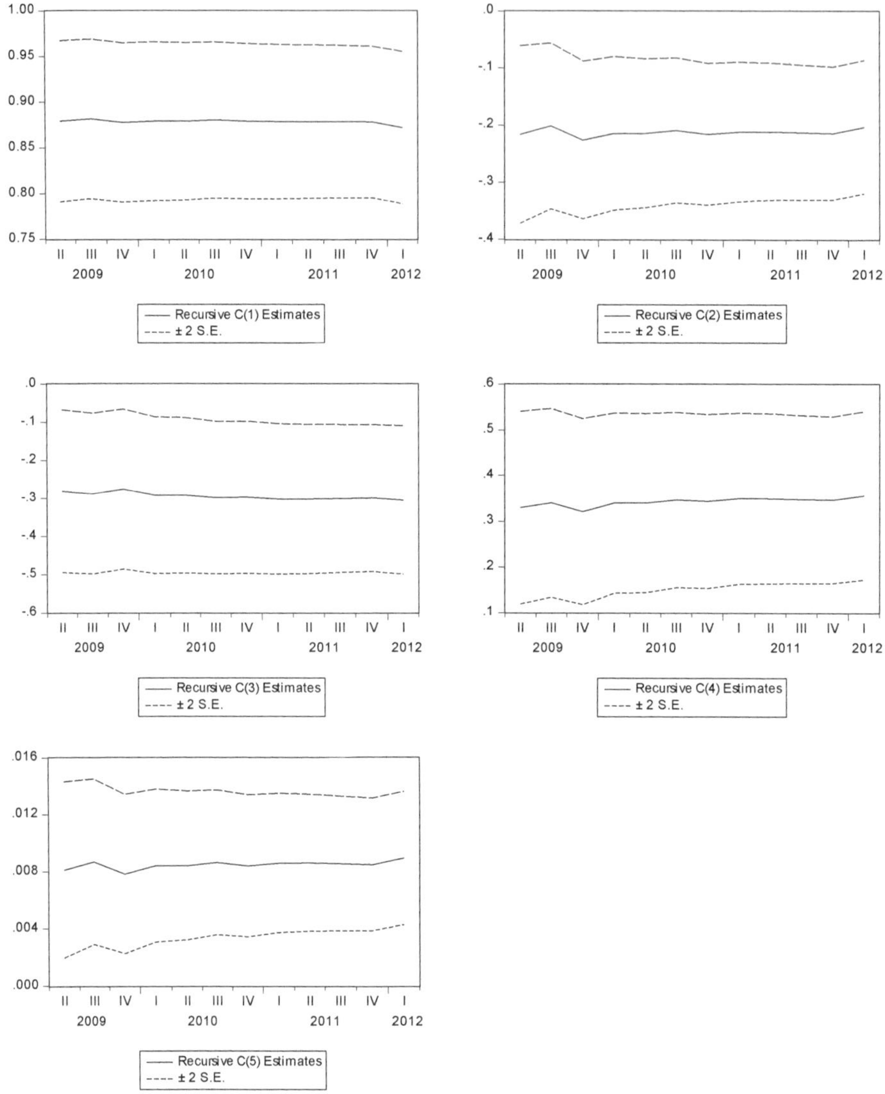

Figure 7.15: Recursive estimation of the ECM parameters

The coefficient for the overnight interest rate in long-term regression is equal to 0.98. We know that it is not possible to use the *t*-test to verify the hypothesis $\beta = 1$, because the distribution of the OLS estimator of the parameter is non-standard. However, we can check whether the spread, built as "spread = one_month–overnight", is stationary. The ADF test, combined with the Schwarz Information Criterion for lag length selection, yields a *p*-value of 0.000, and the KPSS test with the automatic selection criteria returns a value of

2.11 against a threshold value of 0.73 at the 1% significance level. Therefore, we can conclude that the null hypothesis of stationarity of the spread can be accepted.

At this point, it is interesting to evaluate whether both series adjust following a deviation from the long equilibrium, or just one of them does so. We start from two ECM models: one explaining the variation in the one-month rate and the other explaining the variation in the overnight rate. Each has up to three lags and dummy variables corresponding to excessively high residuals (referred to as DO_data and DOM_data for the overnight rate and for the one-month rate regressions, respectively). Eliminating the non-significant regressors, we get the estimation output in Tables 7.24 and 7.25.

Dependent variable: D (ONE_MONTH)

Variable	Coefficient	Std. Error	*t*-Statistic	Prob.
(C)	0.001647	0.000861	1.913034	0.0558
D (OVERNIGHT)	0.097668	0.005531	17.65791	0.0000
SPREAD (-1)	-0.036554	0.003336	-10.95656	0.0000
DOM02	0.645517	0.055768	11.57500	0.0000

R-squared	0.087930	Mean dependent var	-0.002287
Adjusted R-squared	0.087388	S.D. dependent var	0.058370
S.E. of regression	0.055762	Akaike information criterion	-2.934674
Sum squared resid	15.69909	Schwarz criterion	-2.929507
Log likelihood	7418.454	Hannan-Quinn criterion	-2.932864
F-statistic	162.2524	Durbin-Watson stat	1.918093
Prob (*F*-statistic)	0.000000		

Table 7.24: ECM for the one-month rate

Dependent variable: D (OVERNIGHT)

Variable	Coefficient	Std. Error	*t*-Statistic	Prob.
(C)	-0.018741	0.002032	-9.221885	0.0000
D (ONE_MONTH)	0.565360	0.032111	17.60656	0.0000
SPREAD (-1)	0.179116	0.007691	23.28942	0.0000
DON92	-0.847763	0.132721	-6.387575	0.0000
DON93_1	-1.445402	0.132680	-10.89391	0.0000
DON93_2	-1.402422	0.132770	-10.56280	0.0000
DON96	-1.532348	0.132726	-11.54523	0.0000

R-squared	0.188134	Mean dependent var	-0.002239
Adjusted R-squared	0.187168	S.D. dependent var	0.147120
S.E. of regression	0.132639	Akaike information criterion	-1.200980
Sum squared resid	88.77537	Schwarz criterion	-1.191937
Log likelihood	3041.277	Hannan-Quinn criterion	-1.197813
F-statistic	194.8847	Durbin-Watson stat	2.055331
Prob (*F*-statistic)	0.000000		

Table 7.25: ECM for the overnight rate

It turns out that only the one-month rate adjusts to the disequilibrium, and the coefficient has the expected negative sign.

The main concepts of this chapter

This chapter covered models with stochastic regressors in which lagged values of the dependent variable appear as regressors, while all the other assumptions of the linear model are respected.
A general specification for a dynamic model is called autoregressive distributed lags (ARDL). In the ARD(p, q) model, the dependent variable, y, depends on y_{t-1}, ..., y_{t-p} and x_t, x_{t-1}, ..., x_{t-q}.
Among the most significant special cases of ARDL models, we have seen the AR model: $y_t = \alpha_1 y_{t-1} + u_t$; the model in differences: $\Delta y_t = \beta_0 \Delta x_t + u_t$; the "leading indicator" model: $y_t = \beta_1 x_{t-1} + u_t$; the distributed lag model: $y_t = \beta_0 x_t + \beta_1 x_{t-1} + u_t$; the Geometric Distributed Lag Model(GDL) $y_t = \alpha + \beta \sum_{i=0}^{\infty} \omega^i x_{t-i} + \varepsilon_t$, $0 < \omega < 1$, which is a special case of the ARDL(1.1) model since it can be re-written as

$y_t = \alpha(1-\omega)+\omega y_{t-1} + \beta x_t + u_t$, $u_t = \varepsilon_t - \omega\varepsilon_{t-1}$. We have then studied the partial adjustment model: $y_t = \beta_0 x_t + \alpha_1 y_{t-1} + u_t$; the Error Correction Model (ECM): $\Delta y_t = \beta_0 \Delta x_t - (1-\alpha_1)(y_{t-1} - x_{t-1}) + u_t$; the "Dead Start" model: $y_t = \beta_1 x_{t-1} + \alpha_1 y_{t-1} + u_t$. Adding an equation similar to the one of the "Dead Start" model for x, $x_t = \beta_2 x_{t-1} + \alpha_2 y_{t-1} + v_t$, and considering the two equations together, we get a Vector Autoregression model (VAR). Finally, we have seen the model with autoregressive errors: imposing the COMFAC restrictions $\alpha_1\beta_0 + \beta_1 = 0$ into the ARDL(1.1) , we get $y_t = \beta_0 x_t + \varepsilon_t$, $\varepsilon_t = \alpha_1 \varepsilon_{t-1} + u_t$.
If the COMFAC restriction is satisfied, the model with autoregressive errors can be estimated by GLS. If the restriction is not met, it is appropriate to change the dynamics of the model.
In a dynamic model, the OLS estimators remain consistent and asymptotically Normal if the errors are not autocorrelated. On the contrary, if the errors are correlated, the OLS estimation method does not produce consistent estimators. In this case, if good instruments for the regressors are available, we can use IV estimation to obtain consistent estimators of the parameters. Once consistent and asymptotically Normal estimators for the dynamic model parameters are available, we can still use the *t*- or *F*- statistics for hypothesis testing.
Having a valid theory for hypothesis testing can, in turn, improve the model specification. Starting with a congruent initial specification (i.e., diagnostics tests should not reject the null hypothesis related to the correct specification of the model) *t*- and *F*-tests may be sequentially applied to determine the statistical significance of the included regressors and their lags.
As an alternative for model selection, we can use information criteria, which combine a measure of goodness of fit of the model with a function of the number of its parameters. The idea is to choose the model that minimizes the criterion, in order to achieve a good compromise between fit and parsimony of specification. Among the most used criteria, we mention the Akaike Information Criterion (*AIC*) and the Schwarz Information Criterion (*SIC*). Using $\log(\hat{\sigma}^2) + kg(T)$ as a criterion, AIC assumes $kg(T) = k/T$, while SIC $kg(T) = k\log(T)/T$. SIC selects the correct model with probability tending to 1 when *T* diverges to infinity.
Concerning diagnostic tests, i.e., tests for error no correlation, homoscedasticity, Normality, linearity and stability, with the exclusion of DW, they are also still applicable in the dynamic model context, though they have an asymptotic justification.

We then saw that in the AR (1) model: $y_t = \alpha_1 y_{t-1} + u_t = \sum_{i=-\infty}^{t} \alpha_1^i u_{t-i}$, if $|a| < 1$, the effects of the u_t errors on the dependent variable decline exponentially over time, and the unconditional variance of the variable is stable. Instead, if $\alpha = 1$, that is if the process has a unit root, we have $y_t = \sum_{i-\infty}^{t} u_{t-i}$, $\mathrm{var}(y_t) = t\sigma_u^2$ and $\mathrm{cov}(y_t, y_{t-k}) = \sigma_u^2$. So, in the case of a unit root, past shocks have persistent effects, and the variance of the variable grows over time. We have also defined the stochastic trend $ST_t = \sum_{i=-\infty}^{t} u_{t-i}$ or $ST_t = \sum_{i=0}^{t} u_{t-i}$, and compared its properties with those of the linear deterministic trend, $DT_t = t$. We have then seen that the unit root condition can be generalized for AR(q) models and, when it is verified, the series always depends on a stochastic trend. When a unit root is present, the series is said to be integrated of order 1: *I(1)*. Taking first differences of an *I(1)* variable yields a (weakly) stationary series, also said to be integrated of order 0, *I(0)*.

The presence of a unit root has important implications. If the series is integrated, the OLS estimator of the parameters associated with the unit root it is super-consistent, because the speed of convergence to the true values is greater than in the standard case, T vs $\sqrt{T}$. The asymptotic distribution of the estimator is centered on the correct value (one), but it is asymmetric and not available in closed form. To test the null hypothesis H_0: $\alpha = 1$ against H_1: $|\alpha| =< 1$ in an AR(1) model (integration of order 1 against weak stationarity) we can use the DF test statistics $t_{DF} = (\hat{\alpha}_{OLS} - 1)/\hat{s}.e.(\hat{\alpha}_{OLS})$, whose asymptotic distribution is the Dickey-Fuller distribution, for which tables of critical values are available. Using the *t*-student distribution to define the rejection region, would induce us to reject the null hypothesis of unit root much more often than the chosen significance level.

For the generic AR(*p*) model, which can always be rewritten as:

$$\Delta y_t = \gamma \quad y_{t-1} + \beta_1 \Delta y_{t-1} + \ldots + \beta_{p-1} \Delta y_{t-p+1} + u_t$$

the presence of a unit root is equivalent to the restriction $\gamma = 0$. The test statistic for the null hypothesis $\gamma = 0$ if $p > 1$ is called Augmented Dickey-Fuller (ADF), and it is asymptotically distributed as the DF statistic under H_0. The power of the unit root test (ability to reject H_0 when it is false) is

significantly reduced by the presence of changes in parameters and it is in any case low when α is close to 1, or the sample is short, or we use too many lags.
We have seen that it is also possible to test the null hypothesis of stationarity against the alternative hypothesis of integration, for example by means of the Kwiatkowski, Phillips, Schmidt, and Shin (*KPSS*) test, based on the study of the residuals of the model in differences. If the variable in levels is stationary, the residuals of a model for its difference should be of the type $v_t - v_{t-1}$, presenting a unit MA root.
We have then studied the features of regressions based on *I(0)/I(1)* variables. Regressing an integrated variable on a stationary one (or vice versa), we get an unbalanced regression whose explanatory power is asymptotically 0. In a balanced regression, a stationary (integrated) variable is regressed on another stationary (integrated) variable. When both the dependent variable and the regressors are integrated of order 1, but independent, we have a spurious regression, which has undesirable properties: $R^2 \to 1$ instead of 0, $\hat{\beta}_{OLS} \to$ to a random variable instead of 0 and $DW \to 0$, that is, the errors are integrated of order 1. On the other hand, when the variables are integrated of order 1 but they can be linearly combined to produce an *I(0)* variable, the variables are said to be cointegrated. In this case, the OLS estimator of the regression parameters is super-consistent, but again its asymptotic distribution is non-standard.
To test the null hypothesis of no cointegration between two or more series, we studied the two-stage Engle and Granger procedure. First, we run the regression $y_t = \beta x_t + u_t$, and compute the residuals $\hat{u}_t = y_t - \hat{\beta}_{OLS} x_t$. Then, in the second stage, we consider an AR*(q)* model for the residuals and perform the ADF test (the *t*-test for the significance of $\hat{u}_{t-1}$ to explain $\Delta\hat{u}_t$ together with $\Delta\hat{u}_{t-2},\dots,\Delta\hat{u}_{t-q}$), which has a specifically tabulated asymptotic distribution (different from that of the ADF test as it is conducted on a generated variable). If the first-stage regression errors are integrated, then *y* and *x* are not cointegrated.
Engle and Granger also showed that, for the variables to be cointegrated, they must be driven by common stochastic trends. In the short term there may be (even persistent) deviations from the cointegration relation, but in the long run they should disappear. The cointegration relation is then interpreted as a long-term equilibrium relation between variables. For this reason, cointegration relations play a key role in economics.
Finally, the Engle and Granger representation theorem states that two variables are cointegrated if and only if they can be represented with an ECM: $\Delta y_t = -\delta(y_{t-1} - \beta x_{t-1}) +$ (lagged values of Δy_t and Δx_t) $+ u_{yt}$. If the

variables are integrated, but not cointegrated, the appropriate model is specified in differences ($\delta = 0$). If the variables are cointegrated, however, the omission of the stationary term, $y_{t-1} - \beta x_{t-1}$, from the model in differences leads to inconsistent estimators, because of the exclusion of relevant variables. All variables that appear in the ECM are stationary, and hence standard techniques for inference and diagnostic controls are directly applicable in this context.

Exercises

1)
Given an AR(3) model:
(a) define the conditions for the presence of a unit root;
(b) indicate how you would test this condition;
(c) what would happen if you assumed that the test distribution were Student-*t*?
(d) how could you determine the number of lags to use in the AR model?

2)
Given the model

$$\begin{aligned} \Delta y_t &= (x_{t-1} - y_{t-1}) + e_t \\ \Delta x_t &= u_t \end{aligned} \qquad \begin{matrix} e_t \\ u_t \end{matrix} \overset{i.i.d.}{\sim} \left(\begin{pmatrix} 0 \\ 0 \end{pmatrix}, \begin{pmatrix} \sigma_e^2 & 0 \\ 0 & \sigma_u^2 \end{pmatrix} \right),$$

(a) discuss if x is $I(1)$ or not;
(b) discuss if y and x are cointegrated;
(c) show that y and x are guided by the same trend.

3)
Consider the linear regression:

$$y_t = bx_t + e_t$$

What happens to the OLS Estimator and R^2 when $T \to \infty$ and the following conditions are true:
(a) y is $I(1)$ and x is $I(0)$;
(b) y is $I(0)$ and x is $I(1)$;
(c) y and x are both $I(1)$ and not cointegrated;
(d) y and x are $I(1)$ and cointegrated.

4)
Given the model:

$$\begin{aligned} y_t &= \sum_{i=1}^{t-1} u_i + e_t \\ x_t &= \sum_{i=1}^{t-1} u_i + u_t \end{aligned} \qquad \begin{pmatrix} e_t \\ u_t \end{pmatrix} \overset{i.i.d.}{\sim} \left(\begin{pmatrix} 0 \\ 0 \end{pmatrix}, \begin{pmatrix} \sigma_e^2 & 0 \\ 0 & \sigma_u^2 \end{pmatrix} \right),$$

(a) explain if y and x are I(1);
(b) explain if y and x are cointegrated;
(c) obtain the cointegration relation;
(d) write the ECM for y.

5)
The EViews "workfile example_expectations_chap7" contains monthly data for Italian inflation, "INFL", and Italian consumers' inflation expectations "INFL_EXP" over the period 2001-2010. The variables have been constructed from the survey data collected by the European Commission, and are plotted in the following figure.

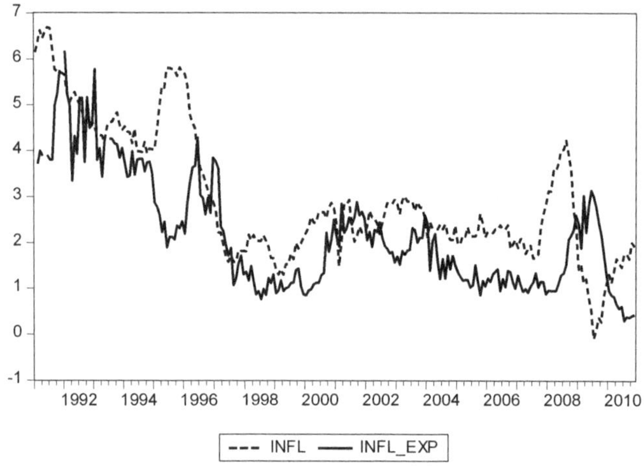

Taking into account the possible instability induced by the introduction of the euro and the global financial crisis, formulate an appropriate dynamic model for the relation between inflation and expected inflation. Comment on the results.

6)
Using the data in the EViews workfile example_prod_chap3, formulate an appropriate dynamic model for labor productivity and comment on the results.

8. Models for Panel Data

Sometimes data with both a longitudinal (cross-sectional) and a temporal dimension are available, known as panel data. Examples of panel data include time series of income or consumption of several households, of investment or profits of a number of companies, of educational attainments and wages of many workers, of sales and prices of various products, of GDP growth in a set of countries, etc.

The general problem is to explain the dependent variable y_{it} with a set of explanatory variables x_{it}, where t= 1,2, ..., T and i= 1,2, ..., N. In this chapter we will consider three basic econometric models for panel data:

- the Seemingly Unrelated Regression (SUR) model;
- the Fixed Effects (FE) model;
- the Random Effects (RE) model.

The goal is to use all the available data to obtain more efficient parameter estimators than those derived from unit-by-unit estimation, taking into account possible differences across units.

8.1 The Seemingly Unrelated Regression (SUR) model

The *Seemingly Unrelated Regression (SUR) model* for each unit is:

$$\underset{T\times 1}{y_i} = \underset{T\times k}{x_i}\ \underset{k\times 1}{\beta_i} + \underset{T\times 1}{e_i}, \qquad i=1, ..., N, \tag{8.1.1}$$

with all the standard assumptions of the linear regression model satisfied, plus the condition:

$$E(e_i e_j^{'}) = \sigma_{ij} I_T \quad (8.1.2)$$

Therefore, each unit depends on *k* unit-specific regressors, x_i, with unit-specific parameters, β_i. In addition, for each unit the errors are 0-mean, $E(e_i) = 0$, unrelated and homoscedastic $E(e_i e'_i) = \sigma_{ii} I$. The units are only connected through the (time invariant) correlation of the errors, $E(e_i e'_j) = \sigma_{ij} I$.

These hypotheses can be weakened to allow for a different number of regressors in each unit, k_i, heteroscedastic errors and/or serially correlated errors, $E(e_i e'_i) = \Omega_i$, or time-varying correlations between different unit errors, $E(e_i e'_j) = \Sigma_{ij}$. These extensions can be easily accommodated in the basic framework we use.

Let us assume that we are interested in estimating the parameters $\beta_i, i = 1, \ldots, N$. The first possibility is to use equation-by-equation OLS estimation. We know that $\hat{\beta}_{OLS}$ is the B.L.U.E. estimator – but it is B.L.U.E. only among those estimators which are a linear function of y_i. We might be able to do better using information on all the units, y_1, y_2, …, y_N. To evaluate whether this it is the case, we write the *N* models in compact form as:

$$\begin{bmatrix} y_1 & (T\times 1) \\ y_2 & (T\times 1) \\ & . \\ & . \\ & . \\ y_N & (T\times 1) \end{bmatrix} = \begin{bmatrix} x_1 & 0 & 0 & 0 & 0 & 0 \\ 0 & x_2 & & & & \\ . & & . & & & \\ . & & & . & & \\ . & & & & . & \\ 0 & 0 & . & . & . & x_N \\ & & & & & (T\times K) \end{bmatrix} \begin{bmatrix} \beta_1 & (K\times 1) \\ \beta_2 & (K\times 1) \\ & . \\ & . \\ & . \\ \beta_N & (K\times 1) \end{bmatrix} + \begin{bmatrix} e_1 & (T\times 1) \\ e_2 & (T\times 1) \\ & . \\ & . \\ & . \\ e_N & (T\times 1) \end{bmatrix} \quad (8.1.3)$$

that is,

$$\underset{NT\times 1}{Y} = \underset{NT\times NK}{X} \underset{NK\times 1}{\beta} + \underset{NT\times 1}{e} \tag{8.1.4}$$

with

$$\underset{NT\times NT}{Var(e)} = E(ee') = \Phi = E\left[\begin{pmatrix} e_1 \\ e_2 \\ . \\ . \\ . \\ e_N \end{pmatrix} \begin{pmatrix} e_1 & e_2 & . & . & . & e_N \end{pmatrix}\right]$$

$$= E\begin{bmatrix} e_1e'_1 & e_1e'_2 & . & . & . & e_1e'_N \\ e_2e'_1 & e_2e'_2 & . & . & . & e_2e'_N \\ . & & e_3e'_3 & & & \\ . & & & . & & . \\ . & & & & . & \\ e_Ne'_1 & e_Ne'_2 & . & . & . & e_Ne'_N \end{bmatrix} = \begin{bmatrix} \sigma_{11}I & \sigma_{12}I & . & . & . & \sigma_{1N}I \\ \sigma_{21}I & \sigma_{22}I & . & . & . & \sigma_{2N}I \\ . & & & & & \\ . & & & & & \\ . & & & & & \\ \sigma_{N1}I & \sigma_{N2}I & . & . & . & \sigma_{NN}I \end{bmatrix} \tag{8.1.5}$$

where

$$\sigma_{ij} = \sigma_{ji} \tag{8.1.6}$$

We can then apply GLS estimation to the model in (8.1.4), which yields the *SUR estimator*:

$$\begin{aligned} \hat{\beta}_{SUR} &= (X'\Phi^{-1}X)^{-1}X'\Phi^{-1}Y \\ V(\hat{\beta}_{SUR}) &= (X'\Phi^{-1}X)^{-1} \end{aligned} \tag{8.1.7}$$

The SUR estimator is B.L.U.E. in this framework, and all the properties of the GLS estimator studied in Chapter 4 apply.

It can be easily shown that $\hat{\beta}_{SUR} = \hat{\beta}_{OLS}$ in two cases:

- $\sigma_{ij} = 0$ for $i \neq j$ (no correlation between units, $\Phi = I$);

- $x_i = x$ for each i (same regressors in each unit).

In practice, as in the case of GLS estimation, the matrix Φ is unknown and must be estimated. Hence, we can estimate each equation by OLS and use the residuals to compute the estimators:

$$\hat{\sigma}_{ij} = \frac{1}{T}\hat{e}_i'\hat{e}_j = \frac{1}{T}\sum_{t=1}^{T}\hat{e}_{it}\hat{e}_{jt}, \qquad i, j = 1, \ldots, N \tag{8.1.8}$$

The estimator $\hat{\sigma}_{ij}$ is biased (because we divide by T) but consistent. Replacing each σ_{ij} in the Φ matrix with its estimator $\hat{\sigma}_{ij}$, we get $\hat{\Phi}$, which we then use in place of Φ in the formula in (8.1.7) to find the feasible estimator $\hat{\beta}_{SUR}$.
As in the case of the feasible GLS estimator (FGLS), we can iterate this procedure, constructing $\hat{\hat{\sigma}}_{ij}$ using the residuals $y_i - x_i\hat{\beta}_{SUR}$, and then using $\hat{\hat{\sigma}}_{ij}$ to obtain another estimator for Φ, $\hat{\hat{\Phi}}$, and for the model parameters, $\hat{\hat{\beta}}$. The latter is in turn used to obtain another set of residuals $y_i - x_i\hat{\hat{\beta}}$, and so on. The process is repeated until the parameter estimates vary very little from one iteration to the next. This iterated procedure is not asymptotically relevant, but in finite samples it can increase the efficiency of the estimator.
As in the case of OLS estimation in the presence of heteroscedasticity, considered in Chapter 4, the $\hat{\beta}_{OLS}$ estimator of the parameters of (8.1.4) is consistent but inefficient. However, if the temporal dimension of the sample is small, the matrix Φ is imprecisely estimated and $\hat{\beta}_{OLS}$ might be more efficient than $\hat{\beta}_{SUR}$.
Asymptotically, we have:

$$\hat{\beta}_{SUR} \sim N\left[\beta, (X'\Phi^{-1}X)^{-1}\right] \tag{8.1.9}$$

Thus, we can construct confidence intervals, hypothesis testing, etc. using the theory we have studied for the simple linear regression model. The results are only approximate in finite samples.

Finally, let us consider a test for the null hypothesis $\sigma_{ij} = 0$ for $i \neq j$. This is a test that indicates whether or not we should use SUR instead of OLS. Let us assume for the sake of simplicity that $N = 3$. Then:

$$\begin{aligned} &H_0 : \sigma_{12} = \sigma_{13} = \sigma_{23} = 0 \\ &H_1 : at \quad least \quad one \quad \sigma_{ij} \neq 0 \end{aligned} \tag{8.1.10}$$

Let us define the square of the estimated correlation coefficient between e_i and e_j as

$$\hat{r}_{ij}^{\ 2} = \frac{\hat{\sigma}_{ij}^{\ 2}}{\hat{\sigma}_{ii}\hat{\sigma}_{jj}} \tag{8.1.11}$$

The test is

$$\lambda = T(\hat{r}_{12}^{\ 2} + \hat{r}_{13}^{\ 2} + \hat{r}_{23}^{\ 2}) \tag{8.1.12}$$

Under H_0, when the temporal dimension diverges, we have:

$$\lambda \overset{a}{\sim} \chi^2(3) \tag{8.1.13}$$

If H_0 is true, the correlations between the errors will be very low and λ will be a small number, while otherwise it will be large. The test is therefore one-sided and we reject H_0 when λ is greater than the critical value for a certain significance level. Note the similarity with the LM test for uncorrelated errors studied in Chapter 4.

In the case of N units, the test statistic becomes

$$\lambda = T\sum_{i=2}^{N}\sum_{J=1}^{i-1}\hat{r}_{ij}^{\ 2} \tag{8.1.14}$$

with

$$\lambda \underset{H_0}{\overset{a}{\sim}} \chi^2(N(N-1)/2)\,. \tag{8.1.15}$$

8.2 The Fixed Effects model

The *Fixed Effects (FE) model* is specified as:

$$y_{it} = \beta_{1i} + \sum_{j=2}^{K} \beta_j x_{jit} + e_{it} \tag{8.2.1}$$

with 0-mean, homoscedastic and serially uncorrelated errors in each unit, $E(e_i) = 0$ and $E(e_i e'_i) = \sigma_{ii} I$, and uncorrelated errors across units for each *t*, $Cov(e_{it} e_{jt}) = 0$. These assumptions on the errors can be easily relaxed to allow for more general correlation schemes. The terms β_{1i} capture potentially omitted but unobservable variables for each unit.
The two main characteristics of the FE model are:

- the intercept is different for each unit (but deterministic – "fixed" – and constant over time);
- the coefficients β_j are equal between units (in contrast with the SUR model).

The FE model is suitable for analyzing relationships where the explanatory variables are expected to have the same effects across units on the dependent variable, with differences across the units captured by the unit-specific intercepts. For example, we may think that the marginal propensity to consume is very similar across different families, while their autonomous consumption can be heterogeneous due to differences in their wealth (assumed constant over time). As another example, the effects of investment in research and development on the profits of companies in the same industry may be equal, but profits may differ for different (unobservable, but stable over time) marketing policies. As yet another example, public investment in infrastructure may have similar effects on the growth rate of regions or countries with a similar initial level of infrastructure, but their growth rates may differ because of differences in the endowment of natural resources or in institutional structures. And so on.
Let us rewrite the FE model in vector notation as:

$$\underset{T\times 1}{y_i} = \underset{T\times 1}{J} \underset{1\times 1}{\beta_{1i}} + \underset{T\times(K-1)}{x_{si}} \underset{(K-1)\times 1}{\beta_s} + \underset{T\times 1}{e_i} \tag{8.2.2}$$

where

$$y_i = \begin{bmatrix} y_{i1} \\ y_{i2} \\ . \\ . \\ y_{iT} \end{bmatrix}; \qquad J = \begin{bmatrix} 1 \\ 1 \\ . \\ . \\ 1 \end{bmatrix};$$

$$x_{si} = \begin{bmatrix} x_{2i1} & x_{3i1} & . & . & x_{Ki1} \\ x_{2i2} & x_{3i2} & . & . & x_{Ki2} \\ . & . & & & \\ . & . & & & \\ x_{2iT} & x_{3iT} & . & . & x_{KiT} \end{bmatrix}; \qquad i = 1,2,\ldots,N. \tag{8.2.3}$$

We can now collect all the equations for $i = 1,2,\ldots,N$ in the model:

$$\underset{NT\times 1}{\begin{bmatrix} y_1 \\ (T\times 1) \\ y_2 \\ . \\ . \\ .. \\ y_N \\ (T\times 1) \end{bmatrix}} = \underset{NT\times(N+(K-1))}{\begin{bmatrix} \underset{(T\times 1)}{J} & 0 & 0 & . & . & 0 & \underset{(T\times(k-1))}{x_{s1}} \\ 0 & J & 0 & 0 & 0 & 0 & x_{s2} \\ 0 & 0 & J & & & & . \\ . & & & . & & & . \\ . & & & & . & & . \\ 0 & 0 & 0 & . & . & \underset{(T\times 1)}{J} & \underset{(T\times(k-1))}{x_{sN}} \end{bmatrix}} \underset{(N+(K-1))}{\begin{bmatrix} \beta_{11} \\ \beta_{12} \\ . \\ . \\ \beta_{1N} \\ \beta_s \end{bmatrix}} \tag{8.2.4}$$

where β_{1i} are the unit-specific parameters and β_s those common across units. In compact notation:

$$Y = X\beta + e \tag{8.2.5}$$

with

$$Var(e) = \sigma_e^{\,2} I_{NT\times NT} \tag{8.2.6}$$

We can then estimate the parameters $\beta = [\beta_{11}\beta_{12}\ldots\beta_{1N}\beta_s]'$ of the model in (8.2.4) by OLS, obtaining the *Fixed Effects estimator*:

$$\hat{\beta}_{FE} = (X'X)^{-1} X'y. \tag{8.2.7}$$

We can therefore apply all the estimation and inferential theory we have studied for the simple linear regression model to the FE model too.

When the number of units, N, is very large we can face numerical problems in the computation of the OLS estimator $\hat{\beta}_{FE}$ due to matrix inversion, as $(X'X)^{-1}$ appears in the formula of $\hat{\beta}_{FE}$ in (8.2.7). We can then use an alternative method for the calculation of $\hat{\beta}_{FE}$ (which yields exactly the same estimator as in 8.2.7). Let us start from the model in (8.2.1), that is,

$$y_{it} = \beta_{1i} + \sum_{j=2}^{K} \beta_j x_{jit} + e_{it} \tag{8.2.8}$$

Taking the averages over time of both sides of (8.2.8) yields:

$$\frac{\sum_{t=1}^{T} y_{it}}{T} = \beta_{1i} + \sum_{j=2}^{K} \beta_j \frac{\sum_{t=1}^{T} x_{jit}}{T} + \frac{\sum_{t=1}^{T} e_{it}}{T} \tag{8.2.9}$$

that is,

$$\bar{y}_i = \beta_{1i} + \sum_{j=2}^{K} \beta_j \bar{x}_{ji} + \bar{e}_i \tag{8.2.10}$$

Subtracting this average from the starting model in (8.2.8), we obtain:

$$y_{it} - \bar{y}_i = \sum_{j=2}^{K} \beta_j (x_{jit} - \bar{x}_{ji}) + (e_{it} - \bar{e}_i), \tag{8.2.11}$$

or, in vector notation,

$$\underset{(T\times 1)}{\tilde{y}_i} = \underset{(T\times(K-1))}{\tilde{x}_{si}} \underset{((K-1)\times 1)}{\beta_s} + \underset{(T\times 1)}{\tilde{e}}, \tag{8.2.12}$$

with

$$\begin{bmatrix} \tilde{y}_{1,} \\ \tilde{y}_2 \\ \cdot \\ \cdot \\ \cdot \\ \tilde{y}_N \end{bmatrix} = \begin{bmatrix} \tilde{x}_{s1} \\ \tilde{x}_{s2} \\ \cdot \\ \cdot \\ \cdot \\ \tilde{x}_{sN} \end{bmatrix} \underset{((K-1)\times 1)}{\beta_s} + \begin{bmatrix} \tilde{e}_1 \\ \tilde{e}_2 \\ \cdot \\ \cdot \\ \cdot \\ \tilde{e}_N \end{bmatrix} \qquad (8.2.13)$$

$$(NT \times 1) \quad (NT \times (K-1))$$

Grouping, we obtain:

$$\tilde{Y} = \tilde{X}_s \beta_s + \tilde{e}\,. \qquad (8.2.14)$$

Estimating the model (8.2.14) by OLS we get $\hat{\beta}_{s,OLS}$. From equation (8.2.10) we can then derive $\hat{\beta}_{1i}$ as:

$$\hat{\beta}_{1i} = \bar{y}_i - \sum_{j=2}^{K} \hat{\beta}_{j,OLS} \bar{x}_{ji}\,. \qquad (8.2.15)$$

Finally, we can estimate the variance of the error term of the FE model as:

$$\hat{\sigma}_e^{\,2} = \frac{\hat{e}'\hat{e}}{NT - (N + K - 1)}, \qquad (8.2.16)$$

where NT is the total number of observations and $N + K$-1 is the total number of parameters in the conditional means.

8.3 The Random Effects (RE) model

The specification of the *Random Effects (RE) model* is similar to that of the FE model:

$$y_{it} = \beta_{1i} + \sum_{j=2}^{K} \beta_j x_{jit} + e_{it} \qquad (8.3.1)$$

with 0-mean, homoscedastic and serially uncorrelated errors in each unit, $E(e_i) = 0$ and $E(e_i e'_i) = \sigma_{ii} I$, and uncorrelated errors across units for each t, $Cov(e_{it} e_{jt}) = 0$. In the RE model, however, the term that differentiates the units, β_{1i}, is stochastic, with:

$$\beta_{1i} = \overline{\beta}_1 + \mu_i \tag{8.3.2}$$

$$\begin{array}{ll} E(\mu_i) = 0 & Var(\mu_i) = \sigma_\mu^{\ 2} \\ E(\mu_i \mu_j) = 0 & E(\mu_i e_{jt}) = 0 \end{array} \tag{8.3.3}$$

Then, the model (8.3.1) can be rewritten as:

$$y_{it} = \overline{\beta}_1 + \sum_{j=2}^{K} \beta_j x_{jit} + \mu_i + e_{it} \tag{8.3.4}$$

and the N units can be considered as a random sample from a larger population, characterized by the parameters $\beta = (\overline{\beta}_1, \beta_2, ..., \beta_K)$ that we want to estimate.

Using a more compact notation, we express the RE model in (8.3.4.) as:

$$\underset{(T \times 1)}{y_i} = \underset{(T \times 1)}{J} \underset{(1 \times 1)}{\overline{\beta}_1} + \underset{(T \times (K-1))}{x_{si}} \underset{((K-1) \times 1)}{\beta_s} + \underset{(1 \times 1)}{\mu_i} \tag{8.3.5}$$

where

$$\underset{(T \times 1)}{J'} = [1 \quad 1 \quad ... \quad 1] \tag{8.3.6}$$

Let us explore in detail the structure of the variance-covariance matrix, V, of the error term:

$$\underset{(T \times 1)}{\mu_i J + e_i} \tag{8.3.7}$$

We have:

$$V = E\left[(\mu_i J + e_i)(\mu_i J + e_i)'\right] =$$
$$= E\left[\mu_i^2\right]JJ' + J\underbrace{E\left[\mu_i e_i'\right]}_{0} + \underbrace{E\left[\mu_i e_i\right]}_{0}J' + E\left[e_i e'_i\right] \tag{8.3.8}$$
$$= \underset{(T\times T)}{\sigma_\mu^2 JJ'} + \underset{(T\times T)}{\sigma_e^2 I}$$

with:

$$JJ' = \begin{bmatrix} 1 \\ 1 \\ . \\ . \\ 1 \end{bmatrix} \begin{bmatrix} 1 & 1 & . & . & 1 \end{bmatrix} = \begin{bmatrix} 1 & 1 & . & . & 1 \\ 1 & 1 & . & . & 1 \\ . & . & . & . & . \\ . & . & . & . & . \\ 1 & 1 & . & . & 1 \end{bmatrix} \tag{8.3.9}$$

It follows that:

$$V = \begin{bmatrix} \sigma_\mu^2 + \sigma_e^2 & \sigma_\mu^2 & \sigma_\mu^2 & . \; . & \sigma_\mu^2 \\ \sigma_\mu^2 & \sigma_\mu^2 + \sigma_e^2 & \sigma_\mu^2 & . \; . & \sigma_\mu^2 \\ . & . & \sigma_\mu^2 + \sigma_e^2 & . \; . & . \\ . & . & . & . \; . & . \\ . & . & . & . \; . & . \\ \sigma_\mu^2 & \sigma_\mu^2 & \sigma_\mu^2 & . \; . & \sigma_\mu^2 + \sigma_e^2 \end{bmatrix}. \tag{8.3.10}$$

Note two important features of the matrix V:

- the correlation is not declining over time, but it is constant;
- the matrix V is the same for all units.

By grouping the constant term J and the unit-specific explanatory variables x_{si} in X_i, we rewrite the model (8.3.5) as:

$$\underset{(T\times 1)}{y_i} = \underset{(T\times k)}{X_i} \underset{(k\times 1)}{\beta} + \mu_i J + e_i \tag{8.3.11}$$

The model including all units is then:

$$\underbrace{\begin{bmatrix} y_{1_{T\times 1}} \\ y_2 \\ . \\ . \\ y_N \end{bmatrix}}_{\substack{Y \\ NT\times 1}} = \underbrace{\begin{bmatrix} x_{1_{T\times K}} \\ x_2 \\ . \\ . \\ x_N \end{bmatrix}}_{\substack{X \\ NT\times 1}} \beta_{K\times 1} + \underbrace{\begin{bmatrix} \mu_1 J_{T\times 1} \\ \mu_2 J \\ . \\ . \\ \mu_N J \end{bmatrix}}_{\substack{\mu \\ NT\times 1}} + \underbrace{\begin{bmatrix} e_{1_{T\times 1}} \\ e_2 \\ . \\ . \\ e_N \end{bmatrix}}_{\substack{e \\ NT\times 1}} \tag{8.3.12}$$

$$Y = X\beta + \mu + e$$

Since

$$E\left[(\mu_i J + e_i)(\mu_j J + e_j)'\right] = 0, \quad i \neq j \tag{8.3.13}$$

the variance-covariance matrix of the error term $\varepsilon = \mu + e$ is block diagonal:

$$\underbrace{\Gamma}_{NT\times NT} = E(\varepsilon\varepsilon') = \begin{bmatrix} \underbrace{V}_{T\times T} & 0 & \dots & 0 \\ 0 & \underbrace{V}_{T\times T} & \dots & 0 \\ \dots & & \dots & 0 \\ 0 & 0 & \dots & \underbrace{V}_{T\times T} \end{bmatrix} \tag{8.3.14}$$

Assuming that $\sigma_\mu^{\ 2}$ and $\sigma_e^{\ 2}$ are known, so that V is known, we can estimate the parameters of interest $\beta = (\overline{\beta}_1, \beta_2, ..., \beta_K)$ by GLS, and obtain the *random effects estimator*:

$$\hat{\beta}_{RE} = \left(X'\Gamma^{-1}X\right)^{-1}(X'\Gamma^{-1}y) \tag{8.3.15}$$

In a similar fashion to the fixed effects model, there may be numerical problems in computing $\hat{\beta}_{RE}$ when the number of units in the sample, N, is high. In the fixed effects model, it was possible to estimate the parameters β in the transformed model

$$y_{it} - \bar{y}_i = \sum_{j=2}^{K} (x_{jit} - \bar{x}_{ji})\beta_j + (e_{it} - \bar{e}_i) \tag{8.3.16}$$

In the model with stochastic effects the transformed model to be used is:

$$y_{it} - \alpha\bar{y}_i = (1-\alpha)\bar{\beta}_1 + \sum_{j=2}^{K} (x_{jit} - \alpha\bar{x}_{ji})\beta_j + \varepsilon_{it} - \alpha\bar{\varepsilon}_i \tag{8.3.17}$$

Where

$$\alpha = 1 - \frac{\sigma_e}{\sigma_1} \tag{8.3.18}$$

With

$$\sigma_1^{\,2} = T\sigma_\mu^{\,2} + \sigma_e^{\,2} \tag{8.3.19}$$

It can be shown that this transformation deletes the correlation in the errors and then the model parameters in (8.3.17) can be estimated by OLS.
In practice, we need to estimate $\sigma_e^{\,2}$ and $\sigma_\mu^{\,2}$ to use GLS in (8.3.12), or to transform the model as in (8.3.17) so that errors are uncorrelated and OLS can be applied. For $\sigma_e^{\,2}$, we can use the estimator obtained using $\hat{\beta}_{FE}$, that is,

$$\hat{\sigma}_e^{\,2} = \frac{\hat{e}'\hat{e}}{NT - (N + K - 1)} \tag{8.3.20}$$

where $\hat{e}$ are the Fixed Effects estimation residuals. For estimation of $\sigma_\mu^{\,2}$, let us consider again the starting model

$$y_{it} = \bar{\beta}_1 + \sum_{j=2}^{K} \beta_j x_{jit} + \mu_i + e_{it} \tag{8.3.21}$$

and take the average over time of each term:

$$\bar{y}_i = \bar{\beta}_1 + \sum_{j=2}^{K} \beta_j \bar{x}_{ji} + \mu_i + \bar{e}_i \qquad i = 1,2\ldots, N \tag{8.3.22}$$

The variance of the error term is:

$$Var(\mu_i + \bar{e}_i) = \sigma_\mu^2 + \frac{\sigma_e^2}{T} = \frac{\sigma_1^2}{T} \quad (8.3.23)$$

Therefore, we can estimate the model (8.3.22) by OLS, and use the residuals to obtain an estimator of $\frac{\sigma_1^2}{T}$. In particular,

$$\frac{\hat{\sigma}_1^2}{T} = \frac{\sum_{i=1}^{N} (\mu_i + \bar{e}_i)^2}{N - K} \quad (8.3.24)$$

Using $\hat{\sigma}_1^2$ and $\hat{\sigma}_\mu^2$, we obtain $\hat{\alpha}$ and we use it to construct the variables in equation (8.3.17), whose parameters can be then estimated by OLS. The resulting estimators are called *feasible random effects estimators*, $\hat{\beta}_{FRE}$.
Finally, note that starting from $\hat{\sigma}_1^2$ we can construct:

$$\hat{\sigma}_\mu^2 = \frac{\hat{\sigma}_1^2 - \hat{\sigma}_e^2}{T} \quad (8.3.25)$$

which is an unbiased estimator of σ_μ^2. However, it might be $\hat{\sigma}_\mu^2 < 0$. When this happens empirically, generally the RE model is not correctly specified.

8.4 Some additional considerations on fixed and random effects

A problem that we have not faced yet is how to choose between fixed and random effects. When $T \to \infty$ and N is fixed, and all the other model assumptions are valid, $\hat{\beta}_{FE}$ and $\hat{\beta}_{FE}$ become identical. Then, if the temporal dimension of the sample is much larger than the longitudinal one, it makes sense to use $\hat{\beta}_{FE}$ that is easier to calculate. If T is small and N is large, the differences between $\hat{\beta}_{FE}$ and $\hat{\beta}_{RE}$ are generally significant, and the choice between the two methods becomes relevant.

For this choice, it is useful to remember that the model with random effects suggests that the units under analysis are extracted from a larger population. For example, if we use data for all regions of Italy, we use up the population and so fixed effects seem more appropriate. On the other hand, the Italian regions can be thought of as a subsample of European regions, and therefore also the random effects would be justifiable.
If we choose to adopt the RE model, we should also consider whether x_i and μ_i in (8.3.1) and (8.3.2) are correlated. If they are, the $\hat{\beta}_{RE}$ estimator is no longer consistent. The Fixed Effects estimator, $\hat{\beta}_{FE}$, remains consistent both when x_i and μ_i are correlated and when they are not, but in the latter case it is not efficient (as when using OLS instead of GLS in the presence of heteroscedastic errors, Chapter 4). We can use these last two considerations to test the hypothesis H_0: random effects uncorrelated with the regressors, against H_1: random effects correlated with the regressors. In particular, we can use a statistic similar to the Hausman's test we have studied in Chapter 6:

$$H = (\hat{\beta}_{FE} - \hat{\beta}_{RE})'[V(\hat{\beta}_{FE}) - V(\hat{\beta}_{RE})]^{-1}(\hat{\beta}_{FE} - \hat{\beta}_{RE}) \underset{H_0}{\overset{a}{\sim}} \chi^2_{(k)} \qquad (8.4.1)$$

Under H_0 , the statistic is expected to take small values, because both $\hat{\beta}_{FE}$ that $\hat{\beta}_{RE}$ are consistent. Then, the rejection region is one-sided.
It should be then emphasized that with both FE and RE the parameters that measure the effects of the explanatory variables, β_s, are treated as invariant across units. When this assumption is not met, that is the effects vary across units, both $\hat{\beta}_{FE}$ and $\hat{\beta}_{RE}$ are biased and inconsistent, as in the case of the OLS estimator applied in the presence of changing parameters of the linear model (Chapter 5). The SUR model is more general in this respect, but, because of this, estimation of its parameters requires panel data with a considerable temporal dimension.
Another strong hypothesis underlying all the three models for panel data that we have seen is that there are no significant interactions across the unit, i.e., x_j and y_j do not affect y_i. This assumption is reasonable for panels with a very wide longitudinal dimension, such as hundreds of families or firms, because in

this case the correlation across units is probably very small (possibly after conditioning on some common variables). In the case of a limited number of units, such as the G7 countries or major financial markets, it is very likely that there are direct effects of one unit on the others, which may not be captured only via cross-correlation of the errors. The omission of significant regressors implies, as seen in Chapter 3 for the linear model, bias and inconsistency of the parameter estimators.
To conclude, a lagged dependent variable is not included among the regressors in the models we have analyzed. In both models with fixed effects and with stochastic effects the dynamics may generate many problems in those samples with a limited time dimension, but there are appropriate ways to consider these cases (known as Dynamic Panel Data, or DPD).

8.5 An empirical analysis with simulated data

The data for this example are contained in the workfile "example_simul_1_chap8.wf1" and they consist of a dependent and an explanatory variable for five units. Each variable consists of 212 observations. Performing simple regressions, e.g., $y1$ on a constant and $x12$, and $y2$ on a constant and $x22$, we get highly correlated residuals, which should not happen if the two errors were uncorrelated.
We can be more rigorous in the analysis of the cross-correlation of residuals. Let us open the $r1$, ..., $r5$ series as a group and, by clicking on "View" from the Group window choose "Correlations". We get the symmetric matrix of sample correlations shown in Table 8.1.

	R1	R2	R3	R4	R5
R1	1.000000	0.881906	-0.925257	-0.962444	-0.860205
R2	0.881906	1.000000	-0.809935	-0.890185	-0.673982
R3	-0.925257	-0.809935	1.000000	0.938988	0.912979
R4	-0.962444	-0.890185	0.938988	1.000000	0.873963
R5	-0.860205	-0.673982	0.912979	0.873963	1.000000

Table 8.1: Correlations across the OLS residuals of five units of the panel

All values are very high in absolute value, with the lowest (between $r2$ and $r5$) being -0.67. To test whether these values are significantly different from 0, we construct the LM test for the null hypothesis "all 0 correlations" against the alternative hypothesis "at least one non-0 correlation", at the 5% confidence level. The test statistic is

$$\lambda = T\sum_{i=2}^{N}\sum_{J=1}^{i-1}\hat{r}_{ij}^{\ 2}$$

Where

$$\hat{r}_{ij}^{\ 2} = \frac{\hat{\sigma}_{ij}^{\ 2}}{\hat{\sigma}_{ii}\hat{\sigma}_{jj}}, \quad i,j = 1,\ldots, N$$

Since we have five units, we can build a short EViews program that computes this test. After clicking on "New" in the main window, choose program and type the following commands in the new window (you find the explanations of each program string within brackets in italics):

```
scalar N = 5 ' (indicate the number of units)
scalar s = 0 ' (we construct the scalar for the sum of R2)
for! i = 2 to N
        for! 1 to! j = i-1 ' (we calculate the square of correlations for each pair of residuals)
        scalar r_ i_! j!
r _! _! i j = (@cor (r! i, r! j)) ^ 2
        next
next
for! i = 2 to N
        for! 1 to! j = i-1
        s = s + r_ i_! j!
        scalar lambda = @obs (resid) * s ' (we calculate the test statistic λ)
        scalar pvalue = @chisq (lambda (N * (N-1)/2))
        next
next
for! i = 2 to N
        for! 1 to! j = i-1
        delete r_ i_! j!
        next
next
delete s
        delete n ' (delete all intermediate scalars)
        = pvalue ' (it shows the p-value of our LM test)
```

You can save this program in the desired folder and then, keeping only the workfile window and the main window of the software open, click on "Run". In the workfile, only the value of the test statistic and its *p*-value, which also appears in the lower left part of the EViews' window, will be stored. Note that the program can be used in every workfile to run the same test if you have already performed the single regressions and you have saved the residuals as "ri", $i = 1 ...5$; you only have to change the first line, putting "N =" according to the number of regressions. If you have not performed the individual regressions, you must add appropriate command lines.
In this case, the *p*-value is very close to 0 (the computer returns 0 and lambda = 1628.55), and therefore we strongly reject the null hypothesis of no correlation among the errors. We can then assume that the five models (y_i regressed on a constant and x_{i2}, i = 1, ..., .5) have correlated errors, and separate OLS regressions are less efficient than an analysis conducted using panel data models. We have discussed that if:

$$E(e_i e_j') = \sigma_{ij} I_T \qquad i = 1, ..., .5$$

and the coefficients of single units are potentially different, then we can more efficiently estimate them using the SUR model. EViews helps us in the estimation of the following type of SUR model:

$$\begin{bmatrix} y_{1,} & (T\times 1) \\ y_2 & (T\times 1) \\ & . \\ & . \\ & . \\ y_N & (T\times 1) \end{bmatrix} = \begin{bmatrix} x_1 & 0 & 0 & 0 & 0 & 0 \\ 0 & x_2 & & & & \\ . & & . & & & \\ . & & & . & & \\ . & & & & . & \\ 0 & 0 & . & . & . & x_N \\ & & & & & (T\times k) \end{bmatrix} \begin{bmatrix} \beta_1 & (K\times 1) \\ \beta_2 & (K\times 1) \\ & . \\ & . \\ & . \\ \beta_N & (K\times 1) \end{bmatrix} + \begin{bmatrix} e_1 & (T\times 1) \\ e_2 & (T\times 1) \\ & . \\ & . \\ & . \\ e_N & (T\times 1) \end{bmatrix}$$

N = 5

Specifically, it computes:

$$\hat{\beta}_{SUR} = (X'\Phi^{-1}X)^{-1}X'\Phi^{-1}Y,$$
$$V(\hat{\beta}_{SUR}) = (X'\Phi^{-1}X)^{-1},$$

where

$$\underset{NT\times NT}{\Phi = Var(e)} = E(ee') = E\left[\begin{pmatrix} e_1 \\ e_2 \\ . \\ . \\ . \\ e_N \end{pmatrix}\begin{pmatrix} e_1 & e_2 & . & . & . & e_N \end{pmatrix}\right] = \begin{bmatrix} \sigma_{11}I & \sigma_{12}I & . & . & . & \sigma_{1N}I \\ \sigma_{21}I & \sigma_{22}I & . & . & . & \sigma_{2N}I \\ . & & & & & \\ . & & & & & \\ . & & & & & \\ \sigma_{N1}I & \sigma_{N2}I & . & . & . & \sigma_{NN}I \end{bmatrix}$$

To obtain the estimator, we right-click on the white screen in the workfile, select "New Object" in the new window, as shown in Figure 8.1, and then choose "System".

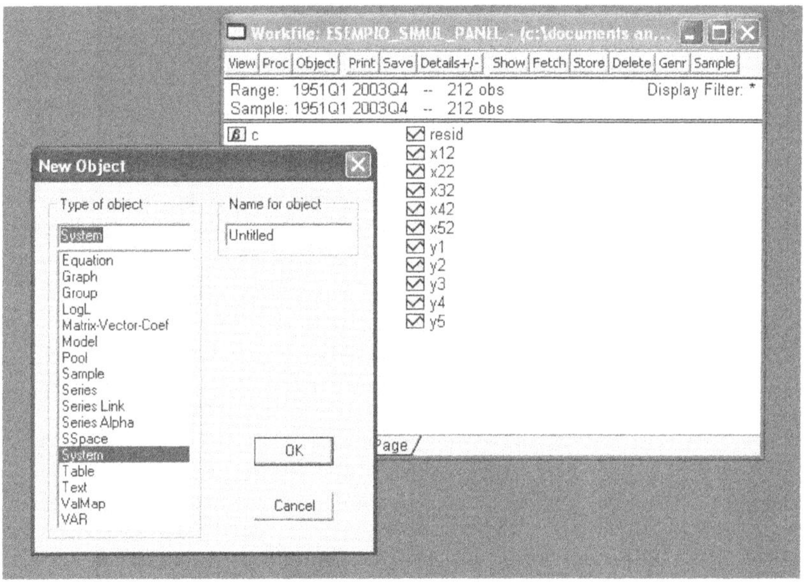

Figure 8.1: Creating a new object in EViews.

In the new window, we write all the equations of the SUR system:

$y1 = c(1) + c(6) * x12$
$y2 = c(2) + c(7) * x22$
$y3 = c(3) + c(8) * x32$
$y4 = c(4) + c(9) * x42$
$y5 = c(5) + c(10) * x52$

After renaming the system by clicking on "Name" in the system window, for example "sys01", in the command line we type the following command:
sys01.sur (i, showopts, m = 1)
This allows us to obtain (feasible) GLS estimates with only one iteration ("m = 1" in the arguments of the command).
Table 8.2 shows the main results of the SUR regression and of the single regressions. The estimated coefficients are statistically significant and can be compared with the OLS estimates of the "stacked" model, which assumes uncorrelated errors across units and unconstrained coefficients. By clicking again on "Estimate", we choose "Ordinary Least Square". Clicking "Ok" we get, in another window, the results of the estimation of five individual models, shown in Table 8.3.

System: SYS01
Estimation Method: Seemingly Unrelated Regression
Sample: 1951q1 2003q4
Included observations: 212
Total system (balanced) observations 1060
Estimation settings: tol = 0.00010, derivs = analytic (linear)

	Coefficient	Std. Error	*t*-Statistic	Prob.
(C) (1)	2.375907	0.022280	106.6409	0.0000
(C) (6)	1.248695	0.007949	157.0872	0.0000
(C) (2)	-0.560843	0.009699	-57.82467	0.0000
(C) (7)	1.199450	0.004547	263.8110	0.0000
(C) (3)	0.658250	0.016323	40.32713	0.0000
C (8)	-1.251055	0.005216	-239.8507	0.0000
(C) (4)	1.232824	0.030996	39.77311	0.0000
(C) (9)	1.005657	0.006187	162.5321	0.0000
(C) (5)	-0.456741	0.007461	-61.21699	0.0000
C (10)	1.511771	0.006223	242.9274	0.0000

Residual covariance method	Determinant	2.21 E-11

Equation: Y1 = C (1) + C (6) * X 12
Observations: 212

R-squared	0.875592	Mean dependent var	4.228783
Adjusted R-squared	0.874999	S.D. dependent var	0.783820
S.E. of regression	0.277123	Sum squared resid	16.12740
Durbin-Watson stat	1.870988		

Equation: Y2 = C (2) + C (7) * X 22
Observations: 212

R-squared	0.979629	Mean dependent var	-0.612640
Adjusted R-squared	0.979532	S.D. dependent var	0.992000
S.E. of regression	0.141921	Sum squared resid	4.229708
Durbin-Watson stat	1.880107		

Equation: Y3 = C (3) + C (8) * X 32
Observations: 212

R-squared	0.955764	Mean dependent var	1.878311
Adjusted R-squared	0.955554	S.D. dependent var	1.076335
S.E. of regression	0.226916	Sum squared resid	10.81313
Durbin-Watson stat	1.993359		

Equation: Y4 = C (4) + C (9) * 42
Observations: 212

R-squared	0.836223	Mean dependent var	0.135806
Adjusted R-squared	0.835443	S.D. dependent var	1.096819
S.E. of regression	0.444931	Sum squared resid	41.57239
Durbin-Watson stat	1.913910		

Equation: Y5 = C (5) + C (10) * X 52
Observations: 212

R-squared	0.972856	Mean dependent var	-0.018381
Adjusted R-squared	0.972727	S.D. dependent var	0.642932
S.E. of regression	0.106177	Sum squared resid	2.367466
Durbin-Watson stat	2.053756		

Table 8.2: SUR estimation results

Overall, the coefficients are different, but the estimated standard error, in particular, is systematically lower in the SUR model, which is more efficient as a GLS estimator compared with the OLS estimator under the assumption of heteroscedasticity.

System: SYS01
Estimation Method: Least Squares

	Coefficient	Std. Error	*t*-Statistic	Prob.
(C) (1)	2.421160	0.050599	47.84970	0.0000
(C) (6)	1.218199	0.031607	38.54186	0.0000
(C) (2)	-0.560627	0.009757	-57.45993	0.0000
(C) (7)	1.204454	0.011980	100.5362	0.0000
(C) (3)	0.655139	0.023928	27.37966	0.0000
C (8)	-1.254246	0.018619	-67.36428	0.0000
(C) (4)	1.185170	0.044023	26.92184	0.0000
(C) (9)	0.961971	0.029192	32.95308	0.0000
(C) (5)	-0.462000	0.008884	-52.00490	0.0000
C (10)	1.529907	0.017589	86.98125	0.0000
Residual covariance method	Determinant	3.27 E-11		

Table 8.3: OLS estimation results

We can now apply the iterative GLS method: the estimated GLS residuals allow us to build $\hat{\hat{\sigma}}_{ij}$ and then $\hat{\hat{\Phi}}$ and a new GLS estimator, and so on until the differences in the estimators obtained in successive steps are negligible. We can choose one of the two options of "Iterate weights and coefs", which are equivalent in linear models like the one we are analyzing, in the iteration options window, as shown in Figure 8.2.

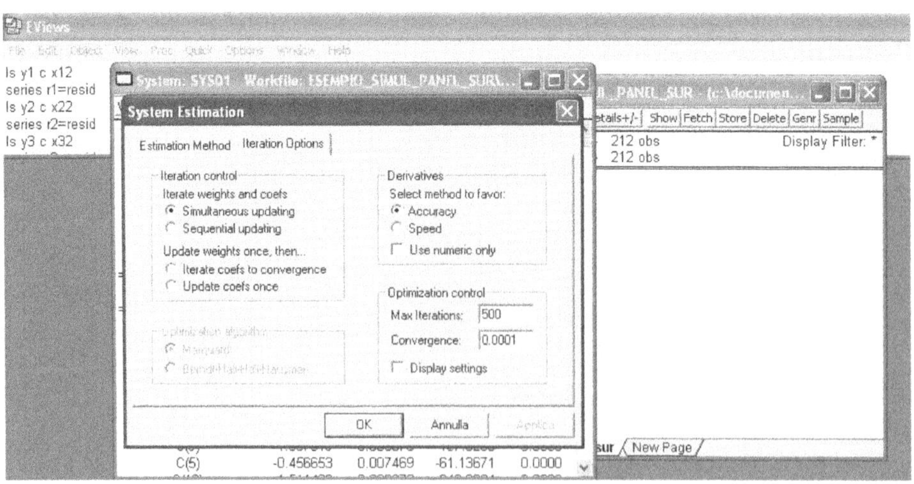

Figure 8.2: A window for selecting options for system estimation

Once we confirm, we obtain Table 8.4.

System: SYS01
Estimation Method: Seemingly Unrelated Regression

	Coefficient	Std. Error	*t*-Statistic	Prob.
(C) (1)	2.375907	0.022280	106.6409	0.0000
(C) (6)	1.248695	0.007949	157.0872	0.0000
(C) (2)	-0.560843	0.009699	-57.82467	0.0000
(C) (7)	1.199450	0.004547	263.8110	0.0000
(C) (3)	0.658250	0.016323	40.32713	0.0000
C (8)	-1.251055	0.005216	-239.8507	0.0000
(C) (4)	1.232824	0.030996	39.77311	0.0000
(C) (9)	1.005657	0.006187	162.5321	0.0000
(C) (5)	-0.456741	0.007461	-61.21699	0.0000
C (10)	1.511771	0.006223	242.9274	0.0000

Residual covariance Determinant method 2.21 E-11

Equation: Y1 = C (1) + C (6) * X 12
Observations: 212

R-squared	0.875592	Mean dependent var	4.228783
Adjusted R-squared	0.874999	S.D. dependent var	0.783820
S.E. of regression	0.277123	Sum squared resid	16.12740
Durbin-Watson stat	1.870988		

Equation: Y2 = C (2) + C (7) * X 22
Observations: 212

R-squared	0.979629	Mean dependent var	-0.612640
Adjusted R-squared	0.979532	S.D. dependent var	0.992000
S.E. of regression	0.141921	Sum squared resid	4.229708
Durbin-Watson stat	1.880107		

Equation: Y3 = C (3) + C (8) * X 32
Observations: 212

R-squared	0.955764	Mean dependent var	1.878311
Adjusted R-squared	0.955554	S.D. dependent var	1.076335
S.E. of regression	0.226916	Sum squared resid	10.81313
Durbin-Watson stat	1.993359		

Equation: Y4 = C (4) + C (9) * 42
Observations: 212

R-squared	0.836223	Mean dependent var	0.135806
Adjusted R-squared	0.835443	S.D. dependent var	1.096819
S.E. of regression	0.444931	Sum squared resid	41.57239

Durbin-Watson stat	1.913910

Equation: Y5 = C (5) + C (10) * X 52
Observations: 212

R-squared	0.972856	Mean dependent var	-0.018381
Adjusted R-squared	0.972727	S.D. dependent var	0.642932
S.E. of regression	0.106177	Sum squared resid	2.367466
Durbin-Watson stat	2.053756		

Table 8.4: Iterated SUR estimation results

In the case of Table 8.4, convergence is reached in a few iterations, and the estimation results are very similar to one-step estimation. We note, however, that the standard errors are often smaller with this procedure, which is thus more efficient than one-step estimation, as expected.
After having estimated the model, we can build confidence intervals and test hypotheses on the parameters. An interesting hypothesis concerns the coefficient of *x* in individual units. If the coefficients were all the same, it would be possible to estimate more efficiently the parameter with a "Fixed Effects" model, adjusted to take into account heteroscedasticity. To perform the test for equality of the coefficients, in the menu "View", "Coefficient Diagnostics", "Wald test", we write the null hypothesis as:
c(6) = c(7) = c(8) = c(9) =c (10)
and we get the output in Table 8.5, which leads us to reject the null hypothesis.

Wald Test:
System: SYS01

Test Statistic	Value	df	Probability
Chi-square	172055.4	4	0.0000

Null Hypothesis: C (6) = C (7) = C (8) = C (9) = C (10)
Null Hypothesis Summary:

Normalized Restriction (= 0)	Value	Std. Err.
(C) (6)-C (10)	-0.263076	0.009999
(C) (7)-C (10)	-0.312321	0.007698
C (8)-C (10)	-2.762826	0.008290
(C) (9) (10)	-0.506114	0.008826

Restrictions are linear in coefficients.

Table 8.5: Test for equality of the coefficients of x in the SUR model

Let us anyway try to estimate the fixed effects model *with correlated errors among equations*. In EViews, we simply rewrite the system as:

$y1 = c\,(1) + c\,(6) * x\,12$
$y2 = c\,(2) + c\,(6) * x\,22$
$y3 = c\,(3) + c\,(6) * x\,32$
$y4 = c\,(4) + c\,(6) * x\,42$
$y5 = c\,(5) + c\,(6) * x\,52$

and then estimate it again as SUR. As already shown by the Wald test, this approach does not fit our data well.

Another possibility for estimating fixed effects models with EViews requires the creation of a "pool" object, as we will see in the next example.

8.6 An empirical analysis with simulated data on the use of fixed and random effects methods

The data are contained in the "example_simul2_chap8" workfile. We consider again five units for which the dependent variable y linearly depends on $x2$ and $x3$. By performing the OLS regressions, we get the five the panels of Table 8.6. Let us rename the residuals with the command "series r1 = resid" for the first equation, and so on for the other ones, in order to test for cross-correlation of errors later on.

Y1	Coefficient	Std. Error	*t*-Statistic	Prob.
(C)	3.955607	0.062937	62.85001	0.0000
X 12	2.292129	0.153874	14.89617	0.0000
X 13	0.710899	0.050101	14.18920	0.0000
Y2				
(C)	4.103546	0.117750	34.84956	0.000000
X 22	1.944595	0.077792	24.99730	0.000000
X 23	0.788792	0.063023	12.51594	0.000000
Y3				
(C)	3.969804	0.086871	45.69755	0.000000
X 32	1.938267	0.047922	40.44592	0.000000
X 33	0.734343	0.065698	11.17750	0.000000
Y4				
(C)	3.836573	0.071767	53.45862	0.000000
X42	2.054392	0.057478	35.74218	0.000000
X43	0.607377	0.044885	13.53177	0.000000

Y5				
(C)	3.998591	0.046351	86.26717	0.000000
X 52	1.900509	0.049811	38.15436	0.000000
X53	0.757519	0.055375	13.67969	0.000000

Table 8.6: Unit-by-unit OLS regressions

The results show that the coefficients of $x2$ and $x3$ ($c(2)$ and $c(3)$, respectively) are very similar in the five models. For example, an F-test does not reject the null hypothesis $c(2) = 2$, $c(3) = 0.75$ in any unit at the 5% level. Moreover, running the test for no cross-correlation in the errors, using the program we have described in the previous section, we get a p-value of 0.51, so that we cannot reject this null hypothesis. Hence, the Fixed Effects (FE) model assumptions are not rejected and we can estimate the parameters using this methodology.
To estimate the FE model, we click on "Object", "New Object", and choose "Pool" in the window shown in Figure 8.3.

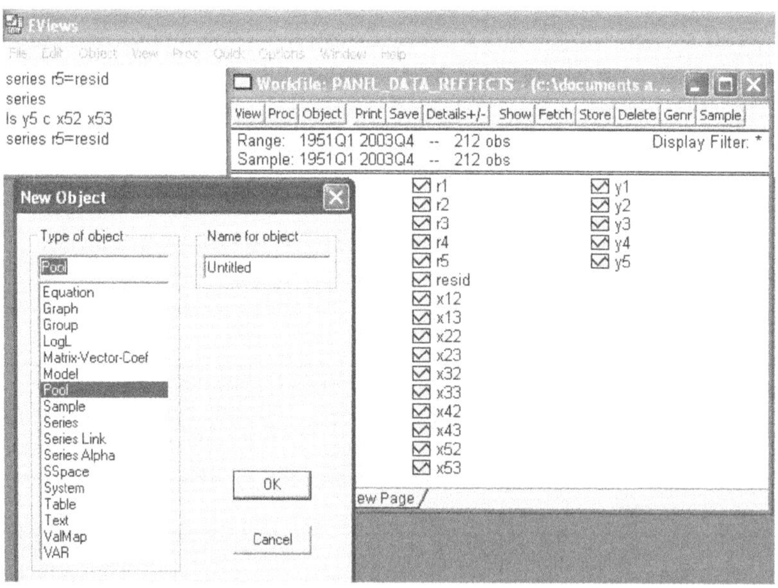

Figure 8.3: A window to create a panel object in EViews

We get a window where we list the "Identifiers", the elements of the set I indexed by i, where i is associated with unit i of the panel data. We write down the numbers from 1 to 5 in columns and then click on "Estimate"; the window shown in Figure 8.4 will appear.

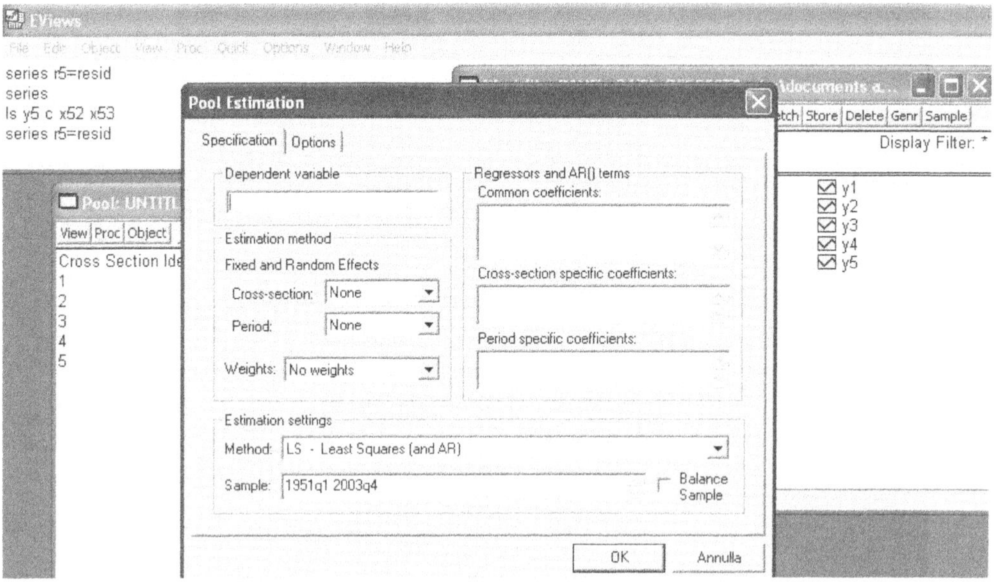

Figure 8.4: Definition and estimation of a panel model in EViews

In the "dependent variable" field, we write "y?", where the symbol "?" tells EViews to use the Identifiers. Among the "common coefficients" we put "c x?2 x?3" , namely, a constant and the unit-specific regressors. In the "Fixed and Random Effects" estimation option, we choose "Cross-Section: Fixed". Confirming, we get Table 8.7.

Table 8.7 shows that the regression is significant, and we do not reject the null hypothesis "c(2) = 2, c(3) = 0.75" at a 5% confidence level. The values of the intercepts for each unit, for example β_{1i}, are given by C + the corresponding fixed effect reported in the lower part of the Table.

Dependent variable: Y?
Method: Pooled Least Squares

Variable	Coefficient	Std. Error	*t*-Statistic	Prob.
(C)	3.977963	0.031331	126.9678	0.0000
X?2	1.951621	0.026511	73.61631	0.0000
X?3	0.718013	0.025160	28.53841	0.0000
Fixed Effects (Cross)				
1-C	-0.006960			
2-C	0.013842			
3-C	-0.028583			
4-C	0.027659			
5-C	-0.005958			

Effects Specification: fixed cross-section (dummy variables)

R-squared	0.885701	Mean dependent var	4.219170
Adjusted R-squared	0.885050	S.D. dependent var	1.721422
S.E. of regression	0.583637	Akaike information criterion	1.767506
Sum squared resid	358.6853	Schwarz criterion	1.800300
Log likelihood	-929.7780	Hannan-Quinn criterion	1.779935
F-statistic	1359.945	Durbin-Watson stat	2.075597
Prob (*F*-statistic)	0.000000		

Table 8.7: Estimating a Fixed Effects model

We could also use the OLS estimates in a model for "stacked" data, imposing the restriction that the intercepts are all equal among units (four restrictions in our example). If this hypothesis is not rejected, OLS and fixed effects become equivalent. We c
an test this hypothesis by means of an *F*-statistic that can be written as:

$$\frac{(SSR_{OLS} - SSR_{FE})/J}{SSR_{FE}/T - K}$$

where, in our example, it is $j = 4$, $T = 1060$, $K = 7$. We then estimate by OLS on pooled data by clicking again on "Estimate" but choosing the "Cross section: none" option. From the resulting output, we select Sum squared resid = 359.0067. Then, we can write
= @fdist (((359.0067-358.6853)/4)/(358.6853/(1060-7)) , 4 , 1053)
which gives us a *p*-value of 0.9181. Hence, the fixed effects are not really necessary as the all the intercepts are the same.

Now let us try to estimate the model with random effect. From the same pool object, we click on "Estimate", choose "Cross section: random" and confirm. We get Table 8.8.

Dependent variable: Y?
Method: Pooled EGLS (cross-section random effects)

Variable	Coefficient	Std. Error	*t*-Statistic	Prob.
(C)	3.974253	0.028601	138.9570	0.0000
2 X?	1.962200	0.023058	85.09732	0.0000
3 X?	0.719249	0.022037	32.63857	0.0000
Random Effects (Cross)				
1-C	0.000000			
2--C	0.000000			
3--C	0.000000			
4-C	0.000000			
5--C	0.000000			
	Effects Specification			
			S.D.	Rho
Random cross-section			0.000000	0.0000
Random Idiosyncratic			0.583637	1.0000
	Weighted Statistics			
R-squared	0.885598	Mean dependent var		4.219170
Adjusted R-squared	0.885382	S.D. dependent var		1.721422
S.E. of regression	0.582792	Sum squared resid		359.0067
F-statistic	4091.192	Durbin-Watson stat		2.072890
Prob (*F*-statistic)	0.000000			
	Unweighted Statistics			
R-squared	0.885598	Mean dependent var		4.219170
Sum squared resid	359.0067	Durbin-Watson stat		2.072890

Table 8.8: Estimating a Random Effects model

The differences in the estimated coefficients with respect to the Fixed Effects model are minor, and the null hypothesis that the personal effects are 0 for all units is not rejected, supporting again the fact that the OLS estimates can be used in this model.

8.7 An empirical analysis with simulated data when N>T.

Let us now consider an example of panel data model in which the longitudinal dimension (number of units) is higher than the temporal dimension. The data are contained in the “example_simul3_chap8” workfile.
We can think of the Lpt series as the logarithm of the number of patents in a given year for a given unit, for example a country, while the State MLA series are the logarithm of R&D spending for the same unit five years earlier (since MLA series are already delayed, in regressions we won’t have to use lagged values). The underlying idea is that five years is the time it takes for the new research, generated by increased spending, to develop a technological innovation to be patented. We then want to determine whether there is a significant relation between research spending and registered patents by comparing different models.
We have 45 units with seven observations each. We first generate the “pool” object in EViews, as we have seen in the previous sections: “New Object”, “Object”, “pool”. The identifiers here range from 01 to 45. When the number of identifiers gets big, as in this case, we can use MS Excel and write 1 in a1, the formula “= a1 +1” in the second cell, copy and paste it up to the 45th cell (in this case). Copying column 1, we can paste it in the initial EViews window of the pool object. There should be a 0 in front of the first nine identifiers in this case (01,...,09), given the variables’ names in the workfile.
We initially estimate by OLS the stacked model, by clicking on “Estimate” from the pool object (Figure 8.5). Among the dependent variables we include “lpt?” and among the regressors a constant and “MLA?”. By clicking on the “OK” button we get Table 8.9. Based on the output in this table, it seems that R&D spending has a positive and significant effect on patents. The Durbin-Watson test leads us to reject the null hypothesis of no autocorrelation of the errors, but this may be due to the fact that we are imposing unsatisfied restrictions on the coefficients, which can greatly reduce the goodness of fit.

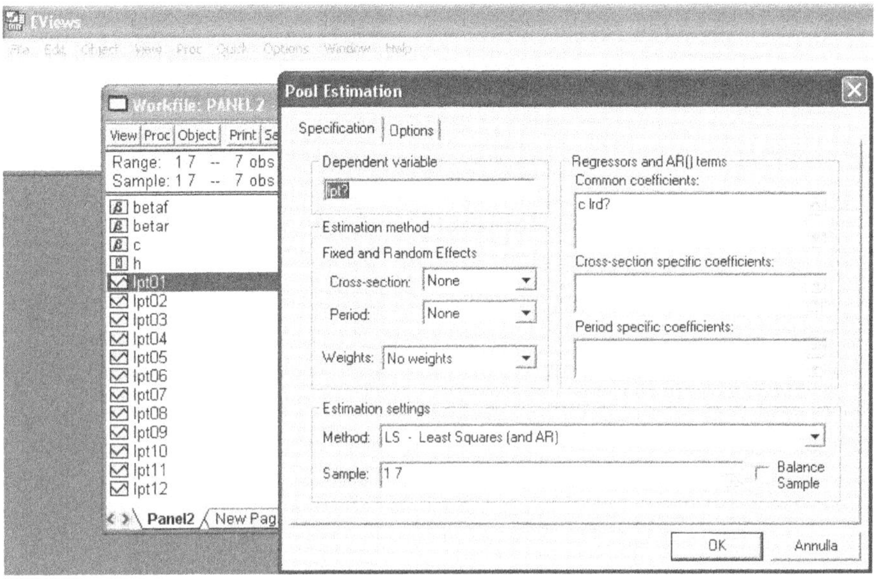

Figure 8.5: OLS estimates of the effects of R&D spending on patents

Dependent variable: LPT?
Method: Pooled Least Squares

Variable	Coefficient	Std. Error	t-Statistic	Prob.
(C)	1.438115	0.102622	14.01366	0.0000
MLA?	0.845443	0.034983	24.16696	0.0000

R-squared	0.651075	Mean dependent var	3.666780
Adjusted R-squared	0.649961	S.D. dependent var	1.350565
S.E. of regression	0.799050	Akaike information criterion	2.395543
Sum squared resid	199.8445	Schwarz criterion	2.419368
Log likelihood	-375.2980	Hannan-Quinn criterion	2.405062
F-statistic	584.0421	Durbin-Watson stat	0.273753

Table 8.9: OLS estimates of the panel model

We can also run a regression on the average of the (log) patents for each unit on a constant and on the average of the (log) R&D expenditures. Each average therefore has seven elements, and we get it from a sample of 45 units. Clicking on "Estimate", we use "@mean (lpt?)" as dependent variable and "c @mean (MLA?)" as explanatory variables (changing the sample size as indicated in Figure 8.6). By doing so we obtain the output in Table 8.10.

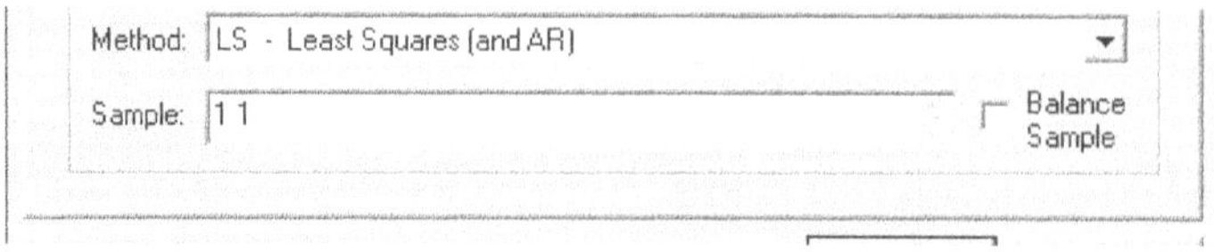

Figure 8.6: Specification of a cross-section regression using sample averages of variables

The coefficients are similar to those obtained previously, as also results from the 95% confidence intervals shown in Figure 8.7. The intervals are given by dashed lines, while the ellipse depicts the OLS coefficients vector, as in the examples in Chapter 3. The two regression confidence intervals with 7 * 45 observations are contained in those that are based on averages, which are narrower and more informative. This is due to the greater variability of the regressors other than the constant in the stacked model, which significantly reduces the standard error of the estimated coefficients.

Dependent variable: @MEAN (LPT?)
Method: Pooled Least Squares
Total pool (balanced) observations: 45

Variable	Coefficient	Std. Error	*t*-Statistic	Prob.
@MEAN (MLA?)	1.400239	0.060987	22.95962	0.0000
R-squared	0.418752	Mean dependent var		3.698719
Adjusted R-squared	0.418752	S.D. dependent var		1.462107
S.E. of regression	1.114705	Akaike information criterion		3.077028
Sum squared resid	54.67296	Schwarz criterion		3.117176
Log likelihood	-68.23314	Hannan-Quinn criterion		3.091995
Durbin-Watson stat	0.000000			

Table 8.10: OLS estimates based on average values

To take advantage of the panel data model approach, we can proceed to estimate by Fixed Effects. Re-using "lpt?" and "MLA?" as variables, as well as a constant, and choosing "cross section: fixed", we get Table 8.11. The FE model fits much better the data: R^2 is higher, the effect of research spending much lower, and the intercept seems to vary widely across units. As mentioned, the unit specific intercept is calculated as C + Fixed Effects; for example, for unit 10 it is 3.151387 + 2.220883. These coefficients can be interpreted as the autonomous capacity to produce scientific and technological innovation (independent on the amount of R&D).

The elasticity, assumed to be constant across units, is given by the coefficient of the MLA. It is interesting to see if it is equal to 1 (1 means that a 1% increase in

R&D spending in levels – not in the logs –, results in a 1% increase in the number of patents five years later). To test this hypothesis, we can run a Wald test for $\beta = 1$, against the alternative hypothesis $\beta \neq 1$, at a 5% confidence level. The null hypothesis is strongly rejected, since the *p*-value for c(2) = 1 is equal to 0.000.

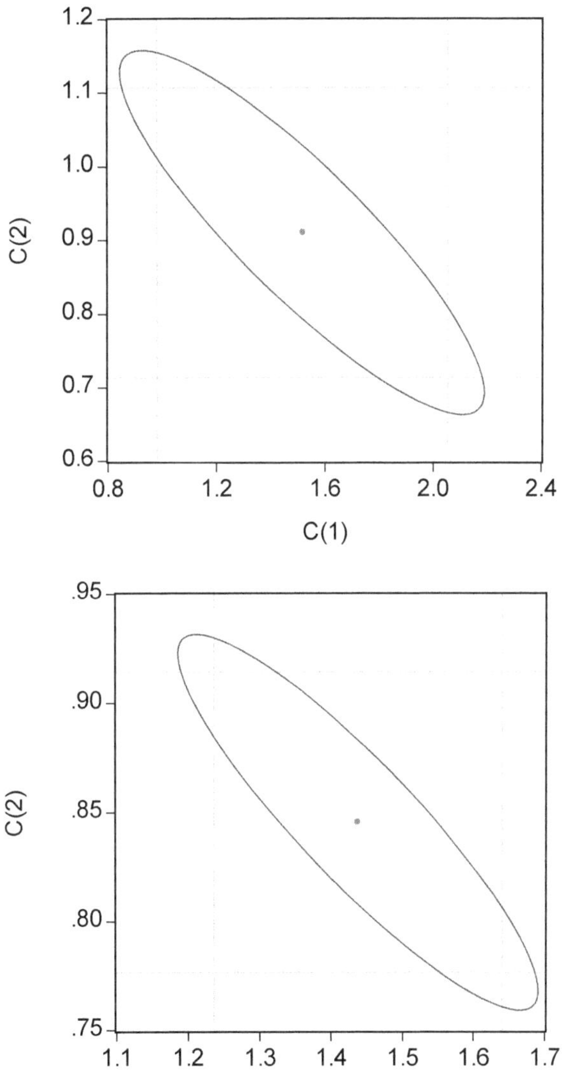

Figure 8.7: 95% confidence intervals for the R&D spending coefficient obtained by OLS on average (left) and stacked OLS (right)

To assess whether the fixed effects are necessary, we can compare the FE and stacked OLS estimates, as seen before. Specifically, we implement an *F*-test based on the statistic:

$$\frac{(SSR_{OLS} - SSR_{FE})/J}{SSR_{FE}/T-K},$$

whose *p*-value is obtained with the following EViews command:
= @fdist (((199.8445-36.14795)/44)/(36.14795/(315-46)) , 269 , 44)
It returns a *p*-value of 0, and therefore we can reject the null hypothesis of equality of all the fixed effects (i.e., FE are needed).

Dependent variable: LPT?
Method: Pooled Least Squares
Sample: 1 7
Included observations: 7
Cross-sections included: 45
Total pool (balanced) observations: 315

Variable	Coefficient	Std. Error	*t*-Statistic	Prob.
(C)	3.151387	0.216762	14.53847	0.0000
THE MLA?	0.195514	0.081854	2.388560	0.0176
Fixed Effects (Cross)				
01 — C	0.129627			
02 — C	-0.032795			
03 — C	-0.127125			
04 — C	-0.348044			
05 — C	1.118766			
06 — C	-1.395677			
07 — C	-0.427812			
08 — C	1.348012			
09 — C	0.611453			
10 — C	2.220883			
11 — C	-0.401384			
(personal effects from 11-45 omitted)				

Effects Specification

Fixed cross-section (dummy variables)

R-squared	0.936886	Mean dependent var	3.666780

Adjusted R-squared	0.926328	S.D. dependent var	1.350565
S.E. of regression	0.366577	Akaike information criterion	0.964988
Sum squared resid	36.14795	Schwarz criterion	1.512983
Log likelihood	-105.9857	*F*-statistic	88.73679
Durbin-Watson stat	1.426859	Prob (*F*-statistic)	0.000000

Table 8.11: Fixed Effects estimates

We save the previous results for later use, by renaming the pool object (e.g., pool01) and writing
vector = betaf pool01. @coef
matrix varf = pool01. @cov
Let us now try to further enrich the model by allowing for random effects. In EViews, we click on "estimate" and choose "Cross-section: random", as in Figure 8.8.

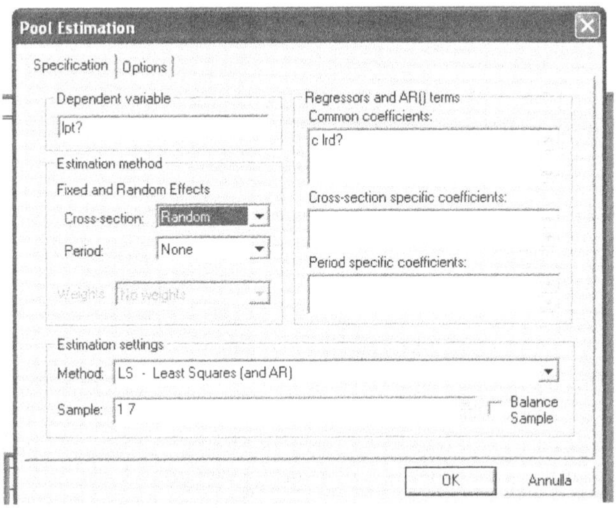

Figure 8.8. Random Effects estimation in EViews

The results, presented in Table 8.12, are quite different from those obtained by Fixed Effects. In particular, the coefficient of R&D spending increases, although the Wald test for the null hypothesis c(2) = 1 leads us to reject again the null hypothesis.

Dependent variable: LPT?
Method: Pooled EGLS (cross-section random effects)
Sample: 1 7, Included observations: 7
Cross-sections included: 45
Total pool (balanced) observations: 315
Swamy and Arora estimator variances of component

Variable	Coefficient	Std. Error	*t*-Statistic	Prob.
(C)	2.291323	0.188467	12.15768	0.0000
THE MLA?	0.521779	0.058873	8.862866	0.0000
Random Effects (Cross)				
01 — C	0.693024			
02 — C	-0.203605			
03 — C	-0.133744			
04 — C	-0.107578			
05 — C	0.750428			
06 — C	-1.390708			
07 — C	0.185554			
08 — C	0.865309			
09 — C	0.530091			
10 — C	1.492412			
(personal effects from 11-45 omitted)				
	Effects Specification			
			S.D.	Rho
Random cross-section			0.703828	0.7866
Random Idiosyncratic			0.366577	0.2134
	Weighted Statistics			
R-squared	0.185495	Mean dependent var		0.708237
Adjusted R-squared	0.182893	S.D. dependent var		0.425705
S.E. of regression	0.384812	Sum squared resid		46.34917
F-statistic	71.28238	Durbin-Watson stat		1.127092
Prob (*F*-statistic)	0.000000			
	Unweighted Statistics			
R-squared	0.555653	Mean dependent var		3.666780
Sum squared resid	254.4973	Durbin-Watson stat		0.205266

Table 8.12: Random Effects estimation

A possible explanation for the differences in results between Fixed and Random Effects is the potential correlation of research spending with Random Effects, which determines the inconsistency of the RE estimators, while we have verified that the FE estimators remain consistent. We then run a Hausman test. First, we save the data of interest for the Hausman test:
vector betar = panel3. @coef
matrixvarr = panel3. @cov
and combine them with those obtained with the FE model to calculate:
vector h = @transpose (betaf-betar) * @inverse (varf-varr) * (betaf-betar)
whose associated *p*-value is obtained with the command:
= @chisq (.1 h)
We get a *p*-value of 0.000, and therefore the most reliable estimators in this example are those based on Fixed Effects.

8.8 An empirical analysis of the effects of public capital in the Italian regions

In this section we study the impact of the stock of (total industrial) capital, labor (units of employees in the industrial sector), and total public capital on the real total industrial value added at 1990 prices. We are particularly interested in assessing the impact of public capital as a production input.
We have the series for the 20 Italian regions with annual data from 1974 to 1991. We can then use the techniques studied in this chapter to analyze the effects of the explanatory variables on the industrial value added, taking into account the relations among units. In this example, the time dimension is higher compared to the previous case, $T = 18$, but remains lower than the longitudinal dimension, $N = 20$.
In the "example_kg_chap8.wf1" workfile, there are series labeled "Y_?" , "UL?", "KK?" and "KG?" for each region, where, following the syntax of the "pool" objects in EViews, each index takes on the name of a region, for a total of 80 series. Y is the industrial value added, UL the units of employees of the industrial sector, KK the total industrial capital stock and KG the total public capital stock. For example, the standardized series of the Lazio region (e.g. series y_st_laz = (y_laz-@mean (y_laz))/@stdev (y_laz)) are shown in Figure 8.9:

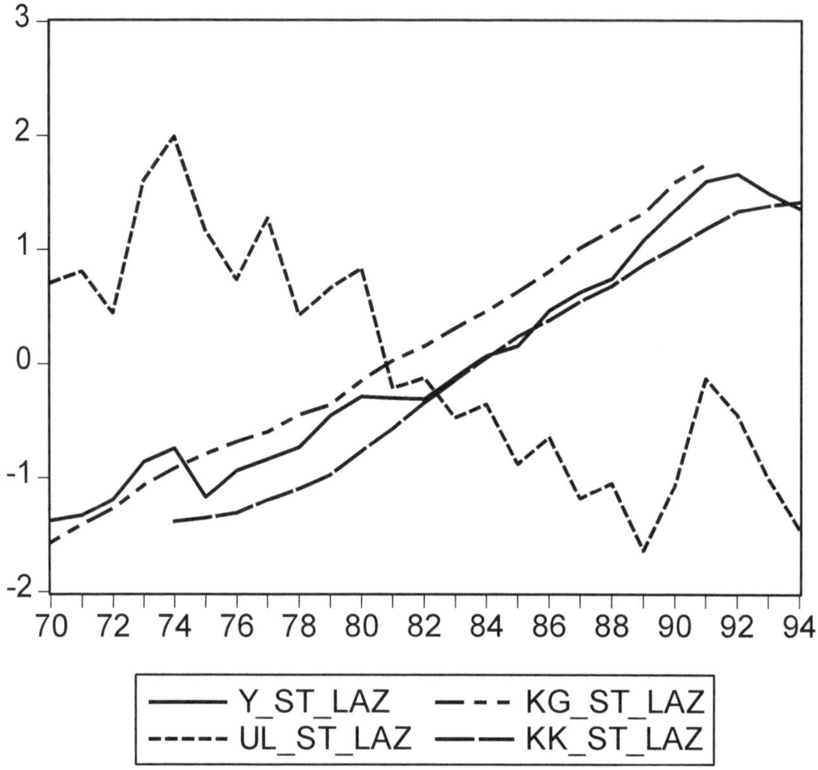

Figure 8.9: Evolution of the dependent variable and regressors in the Lazio region

The number of employees in the industrial sector is very irregular, with a long-term decreasing trend. The stock of industrial capital and that of public capital are in constant growth, and the value added too, although with adjustments and a marked volatility.

From the "pooled object" with the 20 regional identifiers, we start our analysis by estimating by stacked OLS, with results reported in Table 8.13.

Dependent variable: LOG (Y?)
Method: Pooled Least Squares, Sample (adjusted) 1974:
1991
Included observations: 18 after adjusting endpoints
Number of cross-sections used: 20
Total panel (balanced) observations: 360

Variable	Coefficient	Std. Error	*t*-Statistic	Prob.
(C)	4.014009	0.209513	19.15873	0.0000
LOG (UL _?)	0.944488	0.021222	44.50613	0.0000
LOG (KK _?)	0.087527	0.022072	3.965601	0.0001
LOG (KG _?)	-0.082917	0.031236	-2.654507	0.0083
R-squared	0.962259	Mean dependent var		9.119367
Adjusted R-squared	0.961941	S.D. dependent var		1.134052
S.E. of regression	0.221240	Sum squared resid		17.42517
F-statistic	3025.546	Durbin-Watson stat		0.712189
Prob (*F*-statistic)	0.000000			

Table 8.13: Stacked OLS estimation

The stock of public capital is significant, but its coefficient is negative. Since it is the variable of interest, one may wonder if this estimate is reliable or not. The coefficients of employees and capital stock have unexpected values too, which leads us to think that the model is somewhat misspecified. In particular, it could be that Fixed Effects are needed to capture regional heterogeneity.
Using the Fixed Effects model, we obtain the estimation results in Table 8.14.

Dependent variable: LOG (Y?)
Method: Pooled Least Squares
Sample (adjusted) 1974: 1991
Included observations: 18 after adjustments
Cross-sections included: 20
Total pool (balanced) observations: 360

Variable	Coefficient	Std. Error	*t*-Statistic	Prob.
(C)	1.303250	1.154329	1.129011	0.2597
LOG (UL _?)	0.400560	0.116861	3.427665	0.0007
LOG (KK _?)	0.267288	0.053989	4.950799	0.0000
LOG (KG _?)	0.285743	0.102706	2.782148	0.0057
Fixed Effects (Cross)				

PIE--C	0.527209
VDA-C	-0.036563
LOM--	0.507301
TAA-C	0.166616
FRI--C	0.369601
FVG--C	0.292200
LIG--C	-0.178769
ER-C	0.471700
TOS-C	0.292202
UMB--C	0.138299
MAR-C	-0.117192
LAZ-C	0.225064
ABR--C	-0.272113
MOL--C	-0.213132
CAM--C	-0.180410
PUG--C	-0.053506
BAS--C	-0.892324
CAL-C	-0.480805
SAR-C	-0.250572
SIC-C	-0.314807

Effects Specification

Fixed cross-section (dummy variables)

R-squared	0.985718	Mean dependent var	9.119367
Adjusted R-squared	0.984786	S.D. dependent var	1.134052
S.E. of regression	0.139881	Akaike information criterion	-1.034286
Sum squared resid	6.594019	Schwarz criterion	-0.786007
Log likelihood	209.1715	*F*-statistic	1057.231
Durbin-Watson stat	1.830877	Prob (*F*-statistic)	0.000000

Table 8.14: Fixed Effects estimation

The public capital coefficient now becomes positive and significant, and the estimated elasticity is very similar to that of private capital.

The sum of the estimated coefficients is informative about the characteristics of the aggregate production function. In particular, given the Cobb-Douglas function

$$Y = \alpha L^{C_L} K_{PRV}^{C_{PRV}} K_{PBL}^{C_{PBL}}$$

we have:

$$\ln Y = \ln \alpha + C_L \ln L + C_{PRV} \ln K_{PRV} + C_{PBL} \ln K_{PBL}.$$

Therefore, the sum of the coefficients, excluding the intercept, tells us if we have decreasing, constant or increasing returns. The null hypothesis that the sum of the coefficients is 1 is strongly rejected in case of "stacked" OLS estimation while, for example, a sum equal to 0.95 is not rejected, as can be seen in Table 8.15. Thus, simple OLS estimation suggests decreasing returns to scale, but we know that the results from this estimation method are not reliable.

Wald Test:
Pool: POOLOLS

Test Statistic	Value	df	Probability
F-statistic	9.373435	(1, 356)	0.0024
Chi-square	9.373435	1	0.0022

Null Hypothesis Summary:

Normalized Restriction (= 0)	Value	Std. Err.
-1 + C (2) + C (3) + C (4)	-0.050902	0.016626

Wald Test:
Pool:
POOLOLS

Test Statistic	Value	df	Probability
F-statistic	0.002943	(1, 356)	0.956800
Chi-square	0.002943	1.000000	0.956700

Null Hypothesis Summary:

Normalized Restriction (= 0)	Value	Std. Err.
-0.95 + C (2) + C (3) + C (4)	-0.000902	0.016626

Table 8.15: Testing returns to scale in the stacked model

With Fixed Effects, the assumption of constant returns to scale is not rejected, as can be seen in Table 8.16.

Wald Test:
Pool: POOLFE

Test Statistic	Value	df	Probability
F-statistic	0.080743	(1, 337)	0.7765
Chi-square	0.080743	1	0.7763

Null Hypothesis Summary:

Normalized Restriction (= 0)	Value	Std. Err.
-1 + C (2) + C (3) + C (4)	-0.046408	0.163320

Table 8.16: Testing returns to scale in the Fixed Effects model

Do we get different results with the Random Effects model?

Dependent variable: LOG (Y?)
Method: Pooled EGLS (cross-section random effects)
Sample (adjusted) 1974: 1991

Variable	Coefficient	Std. Error	*t*-Statistic	Prob.
(C)	1.335998	0.330814	4.038517	0.0001
LOG (UL _?)	0.639767	0.029905	21.39330	0.0000
LOG (KG _?)	0.167855	0.055208	3.040412	0.0025
LOG (KK _?)	0.257146	0.035337	7.276941	0.0000
Random Effects (Cross)				
PIE--C	0.235424			
VDA-C	0.300333			
LOM--	0.115034			
TAA-C	0.251482			
FRI--C	0.114387			
FVG--C	0.263643			
LIG--C	-0.121403			
ER-C	0.265026			
TOS-C	0.114082			
UMB--C	0.159708			
MAR-C	-0.164970			
LAZ-C	0.134062			

ABR--C	-0.168326		
MOL--C	0.089118		
CAM--C	-0.240631		
PUG--C	-0.101257		
BAS--C	-0.538723		
CAL-C	-0.304286		
SAR-C	-0.107733		
SIC-C	-0.294973		
	Effects Specification		
		S.D.	Rho
Random cross-section		0.116126	0.4080
Random Idiosyncratic		0.139881	0.5920
	Weighted Statistics		
R-squared	0.801908	Mean dependent var	2.490722
Adjusted R-squared	0.800238	S.D. dependent var	0.348465
S.E. of regression	0.155745	Sum squared resid	8.635358
F-statistic	480.3803	Durbin-Watson stat	1.405968
Prob (*F*-statistic)	0.000000		
	Unweighted Statistics		
R-squared	0.936672	Mean dependent var	9.119367
Sum squared resid	29.23860	Durbin-Watson stat	0.415240

Table 8.17: Random Effects estimation

The results shown in Table 8.17 are different in terms of the magnitude of the elasticity of the value added to public capital, which gets slightly smaller, while the labor one increases. In addition, we still do not reject at the 5% level the null hypothesis of constant returns to scale, but we do reject it at 10% significance, as shown in Table 8.18. The bottom of the table shows indeed that the data do not reject the null hypothesis of increasing returns to scale (for example, such that the sum of the coefficients is equal to 1.1).

Wald Test:
Pool: POOLRE

Test Statistic	Value	df	Probability
F-statistic	3.750041	(1, 356)	0.0536
Chi-square	3.750041	1	0.0528

Null Hypothesis Summary:

Normalized Restriction (= 0)	Value	Std. Err.
-1 + C (2) + C (3) + C (4)	0.064767	0.033445

Wald Test:
Pool: POOLRE

Test Statistic	Value	df	Probability
F-statistic	1.109740	(1, 356)	0.292900
Chi-square	1.109740	1.000000	0.292100

Null Hypothesis Summary:

	Value	Std. Err.
Normalized Restriction (= 0)		
-1.1000000000000001 + C (2) + C (3) + C (4)	-0.035233	0.033445

Table 8.18: Testing returns to scale in the Random Effects model

In order to compare the Fixed and Random Effects estimators, we use the Hausman test. We focus on the coefficient of public capital, since it is the main parameter of interest. The test is then computed as
$(0.286\text{-}0.168)^2/(0.103^2\text{-}0.055^2) = 1.838$.
The p-value, calculated on the basis of an $F(1,336)$-distribution is approximately 0.18. Therefore, the results obtained by random effects can be considered more reliable.
Now, we can ask ourselves whether there are differences in the effects of public capital across geographical areas. We know, for example, from growth theory, that the effects could be greater in less developed areas. Hence, we divide the units into three geographical macro-areas:

North	Piemonte, Val d'Aosta, Lombardia, Liguria, Trentino Alto Adige, Veneto, Friuli-Venezia Giulia, Emilia Romagna
Center	Toscana, Umbria, Marche, Lazio
South	Abruzzo, Molise, Campania, Puglia, Basilicata, Calabria, Sicilia, Sardegna

In the North sample (identifiers ER, FVG, LOM, PIE, TAA, VDA, FRI) a random effects estimation produces the results reported in Table 8.19. The corresponding estimates for the Centre (identifiers TOS, UMB, MAR, LAZ) and to the South (identifiers ABR, MOL, CAM, PUG, BAS, CAL, SIC, SAR) are shown, respectively, in Tables 8.20 and 8.21.
It is interesting to note that public capital has a significantly higher elasticity in the Centre and in the South than in the North, as expected, while labor and private capital productivity is higher in the North.

Dependent variable: LOG (Y?)
Method: Pooled EGLS (cross-section random effects)
Sample (adjusted) 1974: 1991
Included observations: 18 after adjustments
Cross-sections included: 7
Total pool (balanced) observations: 126
Swamy and Arora estimator variances of component

Variable	Coefficient	Std. Error	*t*-Statistic	Prob.
(C)	1.224196	0.512781	2.387368	0.0185
LOG (UL _?)	0.560906	0.063270	8.865285	0.0000
LOG (KG _?)	0.258932	0.094299	2.745846	0.0069
LOG (KK _?)	0.244770	0.084303	2.903456	0.0044
Random Effects (Cross)				
ER — C	0.023998			
FVG — C	0.012155			
LOM — C	-0.019932			
PIE — C	0.035264			
TAA — C	-0.013235			
VDA-C	-0.001517			
FRI — C	-0.036734			
	Effects Specification			
			S.D.	Rho
Random cross-section			0.048339	0.0497
Random Idiosyncratic			0.211315	0.9503

Weighted Statistics

R-squared	0.960649	Mean dependent var	6.838349
Adjusted R-squared	0.959681	S.D. dependent var	1.044851
S.E. of regression	0.209801	Sum squared resid	5.370013
F-statistic	992.7663	Durbin-Watson stat	2.121720
Prob (*F*-statistic)	0.000000		

Unweighted Statistics

R-squared	0.978728	Mean dependent var	9.529403
Sum squared resid	5.514421	Durbin-Watson stat	2.066158

Table 8.19: Random Effects estimation, Northern Italy

Dependent variable: LOG (Y?)
Method: Pooled EGLS (cross-section random effects)
Sample (adjusted) 1974: 1991
Included observations: 18 after adjustments
Cross-sections included: 4
Total pool (balanced) observations: 72
Swamy and Arora estimator variances of component

Variable	Coefficient	Std. Error	*t*-Statistic	Prob.
(C)	0.509138	0.111275	4.575482	0.0000
LOG (UL _?)	0.467274	0.022106	21.13825	0.0000
LOG (KK _?)	0.130729	0.008623	15.15994	0.0000
LOG (KG _?)	0.477397	0.017954	26.59056	0.0000
Random Effects (Cross)				
TOS — C	8.45E-12			
UMB — C	4.37 E-12			
MAR — C	-6.85 E-12			
LAZ-C	-5.97 E-12			

Effects Specification

	S.D.	Rho
Random cross-section	8.04 E-08	0.0000
Random Idiosyncratic	0.039404	1.0000

Weighted Statistics

R-squared	0.971572	Mean dependent var	9.376742
Adjusted R-squared	0.970318	S.D. dependent var	0.674873

S.E. of regression	0.116271	Sum squared resid	0.919285
F-statistic	774.6675	Durbin-Watson stat	0.113176
Prob (*F*-statistic)	0.000000		
	Unweighted Statistics		
R-squared	0.971572	Mean dependent var	9.376742
Sum squared resid	0.919285	Durbin-Watson stat	0.113176

Table 8.20: Random Effects estimation, Central Italy

Dependent variable: LOG (Y?)
Method: Pooled EGLS (cross-section random effects)
Sample (adjusted) 1974: 1991

Variable	Coefficient	Std. Error	*t*-Statistic	Prob.
(C)	0.047075	0.469180	0.100334	0.9202
LOG (UL _?)	0.479473	0.045357	10.57106	0.0000
LOG (KG _?)	0.444769	0.069488	6.400696	0.0000
LOG (KK _?)	0.156238	0.034558	4.521001	0.0000
Random Effects (Cross)				
CAM--C	0.087983			
PUG--C	0.216324			
MOL--C	0.191287			
CAL-C	-0.267937			
SAR-C	0.092144			
SIC-C	-0.049301			
BAS--C	-0.401823			
ABR--C	0.131323			

	Effects Specification			
			S.D.	Rho
Random cross-section			0.108168	0.6538
Random Idiosyncratic			0.078714	0.3462

	Weighted Statistics		
R-squared	0.799458	Mean dependent var	1.459634
Adjusted R-squared	0.795160	S.D. dependent var	0.192229
S.E. of regression	0.087001	Sum squared resid	1.059691
F-statistic	186.0355	Durbin-Watson stat	0.423054
Prob (*F*-statistic)	0.000000		

Unweighted Statistics			
R-squared	0.931036	Mean dependent var	8.634301
Sum squared resid	7.592217	Durbin-Watson stat	0.059048

Table 8.21: Random Effects estimation, Southern Italy

The main concepts of this chapter

When we have data with both a temporal and a longitudinal dimension, we must explain y_{it} with x_{it}, where t= 1.2, ..., T and i= 1.2, ..., N, using all available data in order to obtain more efficient estimators than those derived unit-by-unit, taking into account possible differences across units.
In a SUR (Seemingly Unrelated Regression) model, for each of the N units, we have $\underset{T\times 1}{y_i} = \underset{T\times k}{x_i} \underset{k\times 1}{\beta_i} + \underset{T\times 1}{e_i}$, 1, ..., N., namely, we assume that the underlying linear model assumptions are satisfied, and we add the condition $E(e_i e_j') = \sigma_{ij} I_T = \Phi$.
Since the errors in each unit are correlated with the errors in the other units, the information contained in the latter have explanatory power in each correlated unit. It is possible to then apply the GLS estimation method, which yields the *SUR estimator*:

$$\hat{\beta}_{SUR} = (X'\Phi^{-1}X)^{-1}X'\Phi^{-1}Y$$
$$V(\hat{\beta}_{SUR}) = (X'\Phi^{-1}X)^{-1}$$

As the error variance matrix is generally unknown, to get a feasible (FGLS) estimator we must estimate the $\frac{N(N+1)}{2}$ parameters σ_{ij}, $i, j = 1, \ldots, N$. For each of them, we can use the consistent estimators $\hat{\sigma}_{ij} = \frac{1}{T}\hat{e}_i'\hat{e}_j = \frac{1}{T}\sum_{t=1}^{T}\hat{e}_{it}\hat{e}_{jt}$ $, i, j = 1, \ldots, N$, where $\hat{e}_i$ is obtained by unit-by-unit OLS estimation. Replacing σ_{ij} with $\hat{\sigma}_{ij}$ in the Φ matrix, we obtain $\hat{\Phi}$ and we use it in place of Φ, getting $\hat{\beta}_{FGLS} = (X'\hat{\Phi}^{-1}X)^{-1}X'\hat{\Phi}^{-1}Y$.

Asymptotically, we have that $\hat{\beta}_{SUR} \sim N\left[\beta,(X'\Phi^{-1}X)^{-1}\right]$. This result implies that, at least in large samples, we can apply the same inferential tools we have studied in the simple linear regression model.

To test the null hypothesis $\sigma_{ij}= 0,\ i,j=1,\dots,N$, that is, the units are uncorrelated and therefore we can use OLS estimation, against the alternative hypothesis "at least one correlation σ_{ij} is non-0 ", the test statistic is $\lambda = T\sum_{i=2}^{N}\sum_{J=1}^{i-1}\hat{r}_{ij}^{\ 2}$ with $\lambda \underset{H_0}{\overset{a}{\sim}} \chi^2(N(N-1)/2)$ after defining $\hat{r}_{ij}^{\ 2} = \dfrac{\hat{\sigma}_{ij}^{\ 2}}{\hat{\sigma}_{ii}\hat{\sigma}_{jj}}$.

We have then defined the Fixed Effects (FE) model as:

$y_{it} = \beta_{1i} + \sum_{j=2}^{K}\beta_j x_{jit} + e_{it}$, $E(e_i)=0$, $E(e_i e'_i) = \sigma_{ii}I$ and $Cov(e_{it}e_{jt})=0$.

Thus, the intercept is different for each unit, while the coefficients β_j are common. We can then estimate the parameters $\beta = \left[\beta_{11}\beta_{12}\dots\beta_{1N}\beta_s\right]'$ by OLS, obtaining the *Fixed Effects estimator*: $\hat{\beta}_{FE} = (X'X)^{-1}X'y$. In this model, the variance of the error, which is common to all units, is estimated by using the entire sample: $\hat{\sigma}_e^{\ 2} = \dfrac{\hat{e}'\hat{e}}{NT-(N+K-1)}$.

We have then analyzed a generalization of the FE model based on the idea that the unit-specific terms are stochastic rather than deterministic. We have defined the Random Effects (RE) model as: $y_{it} = \beta_{1i} + \sum_{j=2}^{K}\beta_j x_{jit} + e_{it}$,

$E(e_i)=0$ and $E(e_i e'_i) = \sigma_{ii}I, Cov(e_{it}e_{jt})=0$, $\beta_{1i} = \overline{\beta}_1 + \mu_i$,

$E(\mu_i)=0,\ Var(\mu_i)=\sigma_\mu^{\ 2},\ E(\mu_i\mu_j)=0,\ E(\mu_i e_{jt})=0$. The population parameters are therefore $\beta = (\overline{\beta}_1, \beta_2,\dots,\beta_K)$. Expressing the model in vectorial form as

$$\underset{(T\times 1)}{y_i} = \underset{(T\times 1)}{J}\ \underset{(1\times 1)}{\overline{\beta}_1} + \underset{(T\times(K-1))}{x_{si}}\ \underset{((K-1)\times 1)}{\beta_s} + \underset{(1\times 1)}{\mu_i}\ \underset{(T\times 1)}{J} + \underset{(T\times 1)}{e_i}$$

,

the variance-covariance matrix of the error term $\underset{(T\times 1)}{\varepsilon = \mu_i J + e_i}$ is:

$\underset{(T\times T)}{V = \sigma_\mu^{\ 2}JJ'} + \underset{(T\times T)}{\sigma_e^{\ 2}I}$, hence,

$$\underbrace{\Gamma}_{NT\times NT} = E(\varepsilon\varepsilon') = \begin{bmatrix} \underbrace{V}_{T\times T} & 0 & \dots & 0 \\ 0 & \underbrace{V}_{T\times T} & \dots & 0 \\ \dots & & \dots & 0 \\ 0 & 0 & \dots & \underbrace{V}_{T\times T} \end{bmatrix}$$

The GLS estimator for the parameters $\beta = (\overline{\beta}_1, \beta_2, ..., \beta_K)$, known as the *Random Effects estimator*, is $\hat{\beta}_{RE} = \left(X'\Gamma^{-1}X\right)^{-1}(X'\Gamma^{-1}y)$.
Choosing between Fixed and Random Effects: when $T \to \infty$ and N is fixed, $\hat{\beta}_{FE}$ and $\hat{\beta}_{RE}$ become identical. But for small T the differences can be significant. If the hypothesis that the coefficients are constant between units does not hold, both $\hat{\beta}_{FE}$ and $\hat{\beta}_{RE}$ are biased and inconsistent. The SUR model is more general in this respect but it is also over-parameterized, so that its estimation requires panel data with a substantial temporal dimension. Finally, the models presented in this chapter assume that x_j and y_j do not affect y_i, $i \neq j$. Cross-unit dependence occurs only via errors cross-correlation in the SUR model, and via the common coefficients for the regressors in FE and RE models. This assumption is reasonable for panels with a very large longitudinal dimension. With few units, it is likely that there are direct effects of one unit on the others, which requires a proper extension of the specification.

Exercises

1) Derive the SUR estimator for a two-dimensional system and show that if the regressors are the same in each equation it is numerically equal to the OLS estimator applied equation-by-equation.

2) Consider the generic model $y_{it} = \beta_{1it}+\beta_{2it}x_{it}+\beta_{3it}z_{it}+\varepsilon_{it}$ with $\varepsilon_{it} \sim N(0, \sigma_\varepsilon^2)$ and $i=1,...,N$; $t=1,...,T$.
a) Is it possible to estimate this model?
b) Which restrictions would you impose to obtain the SUR model? And to obtain the Fixed and Random Effects models?
c) Apply such restrictions and write the FE and RE models. How do you estimate their parameters?

3) Consider the following two-unit model:

$$y_{1t} = \beta_1+\gamma x_{1t} +\varepsilon_{1t}$$
$$y_{2t} = \beta_2+\gamma x_{2t} +\varepsilon_{2t}$$
$$\varepsilon_{it} \sim N(0,\sigma_\varepsilon^2);\ i=1,2; \forall t$$
$$Cov(\varepsilon_{1t},\varepsilon_{2t}) = 0$$
$$Cov(\varepsilon_{it},\varepsilon_{is}) = 0\ \forall s \neq t$$
$$\beta_i=\beta+\mu_i;$$
$$E(\mu_i)=0;$$
$$VAR(\mu_i)=\sigma_\mu^2$$
$$Cov(\mu_1,\mu_2)=0$$
$$Cov(\varepsilon_{it},\mu_i)=0$$
$$i=1,2;$$

a) Group the equations for the two units into a single model.
b) Obtain the variance-covariance matrix of the errors in the model obtained in (a) and the GLS estimator of the parameters.

4) Employment and wage productivity
In the "example_wage_chap8.wf1" workfile you find the series "?e", "?w", and "?pr" that measure, respectively, the number of employees (employment), real wages and productivity in Italian regions over the period 1995-2007. Using the available data, evaluate the effects of the number of employees and productivity on real wages, using an appropriate panel method for both the whole Italy and the three macro-areas North, Center and South, as defined in section 8.8.

9. Models for Qualitative Data

9.1 The linear regression model with a binary dependent variable

So far we have dealt with modelling continuous variables, such as aggregate consumption, investment or inflation. Many economic choices, however, are qualitative, especially at the microeconomic level. For example, whether a firm produces or not a given good, or a person decides to enter or not in the labor force, or a bank approves or not a mortgage, and so on. In this chapter we will analyze models that can handle this type of variables.
A feature of all the examples given above is that there are two possible choices, which we can associate with the values *0* and *1*. For example, we set the value to *1* if the firm undertakes to produce the good and *0* otherwise, or *1* if the person decides to enter the workforce and *0* otherwise. Then, explaining such choices is equivalent to explaining a *binary dependent variable*,

$$y_i = \{0,1\},\ i=1,\ldots N. \tag{9.1.1}$$

To start with, let us consider what happens if we try to model a binary dependent variable with the standard linear regression model:

$$y_i = x_i\beta + \varepsilon_i, \tag{9.1.2}$$

where x_i groups the explanatory variables, which can be unit-specific or common. For example, the bank's decision to approve or not the mortgage typically depends on the applicant's profile, but also on the overall economic and financial conditions.
From the model in (9.1.2) it is

$$E(y_i) = x_i\beta \tag{9.1.3}$$

As y_i can only be either *0* or *1*, if we define

$$P_i = \text{Prob}(y_i = 1), \tag{9.1.4}$$

we have:

$$E(y_i) = 1 * P_i + 0 * (1 - P_i). \tag{9.1.5}$$

From (9.1.3) and (9.1.5), it follows that

$$P_i = x_i\beta \tag{9.1.6}$$

Since the β parameter is fully unrestricted, it can happen that $P_i \notin [0,1]$, which instead should not happen as P_i is a probability.
A second problem with using the linear model for a binary dependent variable relates to the properties of the error term. From (9.1.1), (9.1.2) and (9.1.4), we get:

$$\varepsilon = \begin{cases} 1 - x_i\beta & \text{\it with probability } P_i = x_i\beta \\ -x_i\beta & \text{\it with probability } 1 - P_i = 1 - x_i\beta \end{cases} \tag{9.1.7}$$

Hence,

$$E(\varepsilon_i) = (1 - x_i\beta)P_i + (-x_i\beta)(1 - P_i) = (1 - x_1\beta)x_i\beta - x_i\beta(1 - x_i\beta) = 0 \tag{9.1.8}$$

But

$$\begin{aligned} E(\varepsilon_i^2) &= (1 - x_i\beta)^2 P_i + (-x_i\beta)^2(1 - P_i) \\ &= (1 - x_i\beta)^2 x_i\beta + (x_i\beta)^2(1 - x_i\beta) \\ &= (1 - x_i\beta)x_i\beta = P_i(1 - P_i) = E(y_i)[1 - E(y_i)] \end{aligned} \tag{9.1.9}$$

Then, the errors of the linear model show a particular form of heteroscedasticity, which would require the use of an appropriate GLS estimator. However, the fact that the linear model may lead to estimated probabilities outside the *[0.1]* range suggests that it is necessary to adopt a different specification when the dependent variable is binary.

9.2 The LOGIT and PROBIT models: specification

We begin the search for a model to explain binary dependent variables trying to identify the determinants of the underlying binary choice, $P_i = Prob\ (y_i = 1)$, which as we have seen in (9.1.6) coincides with $E(y_i)$.

Each choice of unit *i* (*0* or *1*) is associated with a certain utility level, U_{i0} or U_{i1}. Let us suppose that utility depends on characteristics of the choice (z_i) and of the unit making the choice (w_i), and on a random factor:

$$\begin{aligned} U_{i0} &= z_{i0}\delta + w_i\gamma_0 + \varepsilon_{i0} \\ U_{i1} &= z_{i1}\delta + w_i\gamma_1 + \varepsilon_{i1} \end{aligned} \tag{9.2.1}$$

Unit *i* chooses *1* if

$$y_i^* = U_{i1} - U_{i0} > 0\,, \tag{9.2.2}$$

So that

$$y_i = \begin{cases} 1 & if \quad y_i^* > 0 \\ 0 & if \quad y_i^* \leq 0 \end{cases} \tag{9.2.3}$$

We can re-write y_i^* as:

$$\begin{aligned} y_i^* &= (z_{i1} - z_{i0})\delta + w_i(\gamma_1 - \gamma_0) + \varepsilon_{i1} - \varepsilon_{i0} \\ &= \left[(z_{i1} - z_{i0}) \quad w_i\right]\begin{bmatrix} \delta \\ (\gamma_1 - \gamma_0) \end{bmatrix} + \varepsilon_i^* \\ &= x_i\beta + \varepsilon_i^* \end{aligned} \tag{9.2.4}$$

Therefore,

$$P_i = \Pr\{y_i = 1\} = \Pr\{y_i^* > 0\} = \Pr\{\varepsilon_i^* > -x_i\beta\} \qquad (9.2.5)$$

To complete the model, we have to select a distribution for the error ε^* so that we can calculate $\Pr\{\varepsilon_i^* > -x_i\beta\}$. The two most common choices are:

$$\text{Standard Normal: } F(t) = \int_{-\infty}^{t} (2\pi)^{-\frac{1}{2}} \exp\left[-\frac{x^2}{2}\right] dx \rightarrow \textit{PROBIT Model} \qquad (9.2.6)$$

$$\text{Logistic: } F(t) = \frac{1}{[1 + \exp(-t)]} \rightarrow \textit{LOGIT Model} \qquad (9.2.7)$$

where F denotes the cumulative distribution function and f the density function.

Both distributions are symmetric and centered on 0, with a unit variance for the standard Normal, while the variance of the Logistic is $\pi^2/3 > 1$. The two distributions are very similar and the Logistic one is easier to use.

Next we have:

$$\begin{aligned} P_i &= E(y_i) = \Pr\{\varepsilon_i^* > -x_i\beta\} = 1 - \Pr\{\varepsilon_i^* \leq -x_i\beta\} \\ &= 1 - F(-x_i\beta) = F(x_i\beta) \end{aligned} \qquad (9.2.8)$$

as both the Normal and Logistic distribution are symmetric.

For example, for the LOGIT model, from (9.2.7) and (9.2.8) we get:

$$P_i = \frac{1}{[1 + \exp(-x_i\beta)]} \qquad (9.2.9)$$

Note that if we instead assume that ε^* has a uniform distribution, i.e.,

$$F(x_i\beta) = x_i\beta \qquad (9.2.10)$$

then we get the linear regression model, as:

$$P_i = F(x_i\beta) = x_i\beta = E(y_i) \Rightarrow y_i = x_i\beta + \varepsilon_i \qquad (9.2.11)$$

9.3 The LOGIT and PROBIT models: estimation and interpretation of estimated coefficients

Having specified the model for the binary dependent variable, we now consider parameter estimation. It is convenient to distinguish two cases: first, where we have a panel of observations, hence many temporal observations, T_i, for each cross-sectional unit, *i = 1, ...N*. Second, where we have a pure cross-sectional sample, hence, a single observation per unit.

Estimation with panel data

When T_i observations are available for each unit *i*, we can estimate $P_i = \Pr\{y_i = 1\}$ as

$$\hat{P}_i = \frac{r_i}{T_i}, \tag{9.3.1}$$

where r_i is the number of *1* in the T_i observations for the *i*-th unit.

Therefore, it is:

$$\hat{P}_i = P_i + e_i = F(x_i\beta) + e_i, \tag{9.3.2}$$

Where

$$E(e_i) = 0, \text{ since } E(\hat{P}_i) = P_i, \tag{9.3.3}$$

$$Var(e_i) = \frac{P_i(1-P_i)}{T_i} \tag{9.3.4}$$

For the LOGIT model, we then have:

$$\begin{aligned} &P_i = \frac{1}{[1+\exp(-x_i\beta)]} \Rightarrow P_i[1+\exp(-x_i\beta)] = 1 \Rightarrow \exp(-x_i\beta) = \frac{1}{P_i} - 1 \Rightarrow \\ &\exp(x_i\beta) = \frac{P_i}{1-P_i} \Rightarrow \log(\frac{P_i}{1-P_i}) = x_i\beta \end{aligned} \tag{9.3.5}$$

Using a Taylor expansion to linearize the model in (9.3.5), it can be shown that:

$$\log(\frac{\hat{P}_i}{1-\hat{P}_i}) \cong \log(\frac{P_i}{1-P_i}) + \frac{e_i}{P_i(1-P_i)} \cong x_i\beta + u_i \tag{9.3.6}$$

With

$$E(u_i) = 0, \qquad Var(u_i) = \frac{1}{T_i P_i (1-P_i)} \tag{9.3.7}$$

We can now estimate the parameters of (9.3.6) by feasible GLS, using $\hat{P}_i$ instead of P_i to estimate $Var(u_i)$.

For the PROBIT model, we follow similar steps. First, it is:

$$P_i = F(x_i\beta) \Rightarrow x_i\beta = F^{-1}(P_i). \tag{9.3.8}$$

Then, we replace P_i with the estimator in (9.3.1), apply a Taylor expansion to linearize the model, and get:

$$F^{-1}(\hat{P}_i) \cong F^{-1}(P_i) + \frac{e_i}{f\left(F^{-1}(P_i)\right)} = x_i\beta + u_i, \tag{9.3.9}$$

with

$$E(u_i) = 0, \qquad Var(u_i) = \frac{P_i(1-P_i)}{T_i\left[f\left(F^{-1}(P_i)\right)\right]^2}, \tag{9.3.10}$$

where, as mentioned, F denotes the cumulative distribution function and f the density function.

Finally, as for the LOGIT model, we can estimate the model in (9.3.9) by feasible GLS, replacing P_i with $\hat{P}_i$ in $Var(u_i)$.

Interpretation of the estimated parameters

Having obtained estimators for the LOGIT and PROBIT parameters β, $\hat{\beta}$, let us now see how to use them to assess how the probability of outcome y_i changes for a marginal variation in a regressor, assuming all the others remain constant. Thus, the quantity of interest is:

$$\frac{\partial P_i}{\partial x_{ij}} \tag{9.3.11}$$

where x_{ij} denotes the *j*-th regressor for unit *i*.

The estimated coefficients $\hat{\beta}$ measure the effects of Δx_i not on P_i but on $F^{-1}(P_i)$ for the PROBIT model, and on $\log(\frac{P_i}{1-P_i})$ for the LOGIT model. However, we can also assess the effects of Δx_i on P_i by means of the formulae:

$$\text{PROBIT: } \frac{\partial P_i}{\partial x_{ij}} = f(x_i\beta)\beta_j\,, \tag{9.3.12}$$

$$\text{LOGIT: } \frac{\partial P_i}{\partial x_{ij}} = \frac{\beta_j e^{-x_i\beta}}{\left[1+e^{-x_i\beta}\right]^2}\,, \tag{9.3.13}$$

where β can be replaced by $\hat{\beta}$.

Note that for both models, the effects depend not only on the coefficient of x_{ij}, β_j, but also on all the other coefficients, β, and on the value of all regressors, x_i. Note also that the sign of the effect of the *j*-th variable is the same as that of the coefficient β_j. Finally, by convention, results usually are presented for the "average" unit in the sample.

Estimation with cross-sectional data

Maximum Likelihood Estimation of LOGIT and PROBIT models can be used with both panel and cross-sectional data, but it is particularly useful with the latter. We can write the likelihood function as:

$$L = \prod_1^N f(y_i) = \prod_1^N P_i^{y_i}(1-P_i)^{1-y_i} = \prod_1^N F(x_i\beta)^{y_i}\left[1-F(x_i\beta)^{1-y_i}\right] \tag{9.3.14}$$

the log likelihood as:

$$\log(L) = \sum_{i=1}^{N} y_i \log[F(x_i\beta)] + \sum_{i=1}^{N}(1-y_i)\log[1-F(x_i\beta)], \tag{9.3.15}$$

and obtain the Maximum Likelihood Estimators, $\hat{\beta}_{MLE}$, as those maximizing (9.3.15).
The optimization problem has no analytical solution, contrary to the OLS case, but numerical methods can be used to obtain $\hat{\beta}_{MLE}$. For example, the so-called Newton-Raphson algorithm suggests starting with an arbitrary initial value and then iteratively applying the expression

$$\hat{\beta}_{n+1} = \hat{\beta}_n - \left[\frac{\partial^2 \log(L)}{\partial\beta\partial\beta'}\right]_{\hat{\beta}_n}^{-1}\left[\frac{\partial \log(L)}{\partial\beta}\right]_{\hat{\beta}_n} \tag{9.3.16}$$

so long as the difference between $\hat{\beta}_{n+1}$ and $\hat{\beta}_n$ is larger than a pre-specified very small value.
As we saw in Chapter 2, the Maximum Likelihood Estimators are asymptotically Normal, consistent and asymptotically efficient, when the model assumptions are met. And as we saw in Chapter 4, the feasible GLS estimators, to be used in case of repeated observations, are also consistent and asymptotically Normal, again when the model assumptions are met. The asymptotic Normality result for the estimators can also be used to make inference about the parameters β, by means of t- or F-tests.

9.4 Model evaluation

The evaluation of the explanatory power of LOGIT and PROBIT models is complicated by the binary nature of the dependent variable. However, if the model is linearized as in (9.3.6) or (9.3.9), then the coefficient of determination provides useful information on the explanatory power of the model.
When the sample consists of a panel with repeated observations for each unit, we can also compare the nonparametric estimators of P_i, $\hat{P}_i$ in (9.3.1), with

those resulting from LOGIT or PROBIT estimation, P_i^*. The former are always consistent but not necessarily efficient, while the latter are consistent only if the model is correctly specified. It is also possible to build the test statistic:

$$\sum_1^N \frac{T_i(\hat{P}_i - P_i^*)}{P_i^*(1-P_i^*)}, \tag{9.4.1}$$

whose asymptotic distribution, under the null hypothesis of correct model specification, is $\chi^2(N-k)$.
If the model is instead estimated by Maximum Likelihood, a simple measure of its explanatory capability is given by:

$$1-\frac{\log\left[L\left(\hat{\beta}_{2,MLE},\hat{\beta}_{3,MLE},\ldots,\hat{\beta}_{k,MLE}\right)\right]}{\log\left[L\left(\hat{\beta}_2=0,\hat{\beta}_3=0,\ldots,\hat{\beta}_k=0\right)\right]} \tag{9.4.2}$$

which is 0 if all the regressors, except the intercept, are non-significant, and the more negative it is, the better the explanatory power of the model.
Finally, the model can be evaluated according to the percentage of correct predictions ($y_i^P = y_i$), using for example the rule:

$$y_i^P = 1 \text{ if } P^*_i = F(x_i\hat{\beta}) \geq 0.5. \tag{9.4.3}$$

This percentage should be compared to the sample average of y_i, which represents the expected percentage using a simple random rule.

9.5 An empirical analysis with simulated data

In the workfile "example_simul_chap9", the Y variable is dichotomous: it takes values 1 and 0 depending on whether the economy is, respectively, in expansion or recession. The regressors $X1$ and $X2$ are economic indicators, such as investment growth, an interest rate spread etc.
Using a linear model has several disadvantages, as we have seen: the errors are heteroscedastic and, above all, the fitted value $\hat{\beta}\, x_t$ could be smaller than 0 or

larger than 1. For these reasons, it is preferable to carry out the estimation of PROBIT and LOGIT models.
The variables are graphed in Figure 9.1:

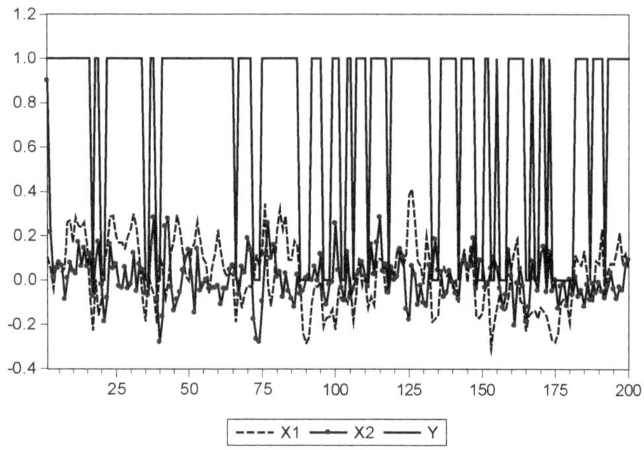

Figure 9.1: Graph of the variables

The variable *X1* shows a positive correlation with *Y*, while *X2*, which is less volatile, seems also less correlated with the target variable. Cross-correlations are shown in Table 9.1 and confirm the lower linear dependence between *Y* and *X2*

	X1	X2	Y
X1	1.000000	-0.022671	0.596316
X2	-0.022671	1.000000	0.324925
Y	0.596316	0.324925	1.000000

Table 9.1: Cross-correlations among the variables

We can estimate LOGIT and PROBIT models using EViews as follows. First, we create a new Equation object in the workfile and we select "BINARY" in the "Estimation settings" window (Figure 9.2).

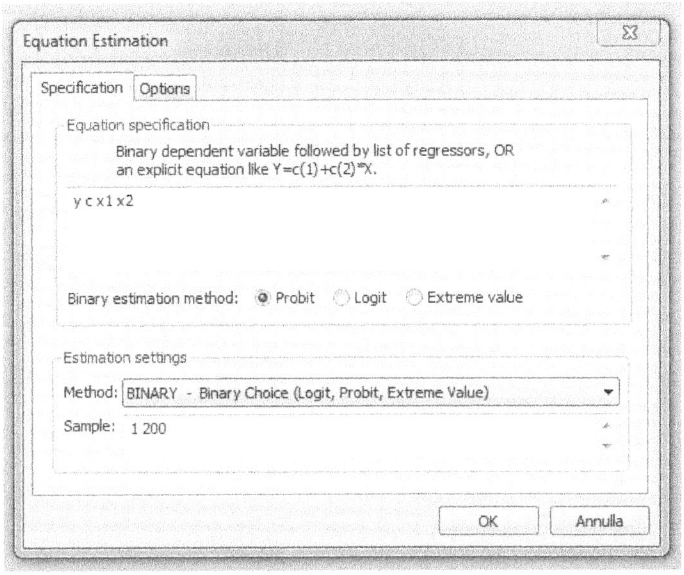

Figure 9.2: EViews windows to estimate binary models

By choosing the PROBIT model, based as we have seen on the standard Normal distribution, we get Table 9.2. It reports the estimated coefficients and some other results that have a similar interpretation as in OLS regressions.
Specifically, in Table 9.2, *z*-statistics (and their *p*-values) test the null hypothesis that the coefficients are equal to 0, against the alternative hypothesis that the coefficients are non-0. In the example, all variables are strongly significant
The LR statistic (and its *p*-value) test the null hypothesis that all coefficients, except the intercept, are all non-significant, against the alternative hypothesis that at least one of them is different from 0. In the example, we strongly reject the null hypothesis. The LR statistic is asymptotically distributed as $\chi^2_{(q)}$, where the number of degrees of freedom is equal to the number of restrictions (i.e. the number of regressors minus one, the intercept).
The so-called McFadden R^2, taking values in the 0-1 interval, grows with the goodness of fit of the model.

Dependent variable: Y
Method: ML - Binary PROBIT (Quadratic hill climbing)

Variable	Coefficient	Std. Error	z-Statistic	Prob.
C	1.940793	0.357342	5.431185	0.0000
X1	20.96295	3.847454	5.448526	0.0000
X2	18.68916	3.492415	5.351357	0.0000

McFadden R-squared	0.732739	Mean dependent var	0.750000
S.D. dependent var	0.434099	S.E. of regression	0.221435
Akaike info criterion	0.330581	Sum squared resid	9.659552
Schwarz criterion	0.380055	Log likelihood	-30.05806
Hannan-Quinn criterion	0.350602	Deviance	60.11612
Restr. Deviance	224.9341	Restr. log likelihood	-112.4670
LR statistic	164.8179	Avg. log likelihood	-0.150290
Prob(LR statistic)	0.000000		
Obs with Dep=0	50	Total obs	200
Obs with Dep=1	150		

Table 9.2: Estimated PROBIT model

Dependent variable: Y
Method: ML - Binary Logit (Quadratic hill climbing)

Variable	Coefficient	Std. Error	z-Statistic	Prob.
C	3.494784	0.701079	4.984863	0.0000
X1	37.28424	7.339369	5.080033	0.0000
X2	33.63476	6.868456	4.896990	0.0000

McFadden R-squared	0.730489	Mean dependent var	0.750000
S.D. dependent var	0.434099	S.E. of regression	0.221443
Akaike info criterion	0.333112	Sum squared resid	9.660255
Schwarz criterion	0.382586	Log likelihood	-30.31115
Hannan-Quinn criterion	0.353133	Deviance	60.62231
Restr. deviance	224.9341	Restr. log likelihood	-112.4670
LR statistic	164.3117	Avg. log likelihood	-0.151556
Prob(LR statistic)	0.000000		
Obs with Dep=0	50	Total obs	200
Obs with Dep=1	150		

Table 9.3: Estimated LOGIT model

By changing the distribution of the success probability and choosing the Logistic, we get Table 9.3.
The results are very similar to PROBIT estimation, since the two models differ strongly only in the tails and in this example there are few extreme values.
As discussed in Section 9.3, the estimated coefficients do not represent the effects of marginal changes in the regressors on the probability that y takes the value 1. For example, the estimated total differential of the conditional expectation of the likelihood of expansion in the 120-th observation, where *X1* and *X2* are both close enough to their sample mean, is, for the PROBIT model:

$$\hat{d}E(y_{120} \mid x'_{120}\,\beta) = \frac{\hat{\partial}}{\partial x_{1,120}} E(y_{120} \mid x'_{120}\,\beta)\ dx_1 + \frac{\hat{\partial}}{\partial x_{2,120}} E(y_{120} \mid x'_{120}\,\beta)\ dx_2 =$$
$$= f\left(-x'_{120}\,\hat{\beta}\right)\hat{\beta}_1 dx_1 + f\left(-x'_{120}\,\hat{\beta}\right)\hat{\beta}_2 dx_2 = 0.366\cdot dx_1 + 0.326\cdot dx_2,$$

where f indicates the standard Normal density. These estimates were obtained with the following EViews commands:

```
=@dnorm(-
(1.940793+0.025895*20.96295+0.000936*18.68916))*20.96295
=@dnorm(-
(1.940793+0.025895*20.96295+0.000936*18.68916))*18.68916
```

or:

```
vector(3) o120
o120(1)=1
o120(2)=x1(120)
o120(3)=x2(120)
vector                                              b1=@dnorm(-
@transpose(o120)*eq01.@coefs)*eq01.@coefs(2)
vector                                              b2=@dnorm(-
@transpose(o120)*eq01.@coefs)*eq01.@coefs(3)
```

Therefore, in t=120 the variation in the conditional expectation is positive if the regressors do not decrease.

9.6 Leading indicators for GDP growth

The workfile "example_growth_chap9" contains the binary quarterly series RGR that takes the value 1 if the annual growth rate of the Italian economy is greater than 1% and 0 otherwise:

$$y_i = \begin{cases} 1 & if \quad y_i^* > 0.01 \\ 0 & if \quad y_i^* \leq 0.01 \end{cases}$$

It also contains the oil price series, OILP. Data is quarterly and ranges from 1980 to 2012.

From Figure 9.3, it is evident that periods with marked increases in oil prices are followed by periods of low growth, such as during the oil crisis of the early 1990s (after the Iraqi invasion of Kuwait) and in the period preceding the global financial crisis. Hence, the oil price seems to be a leading indicator for GDP growth, although in some periods it may send the wrong signal.

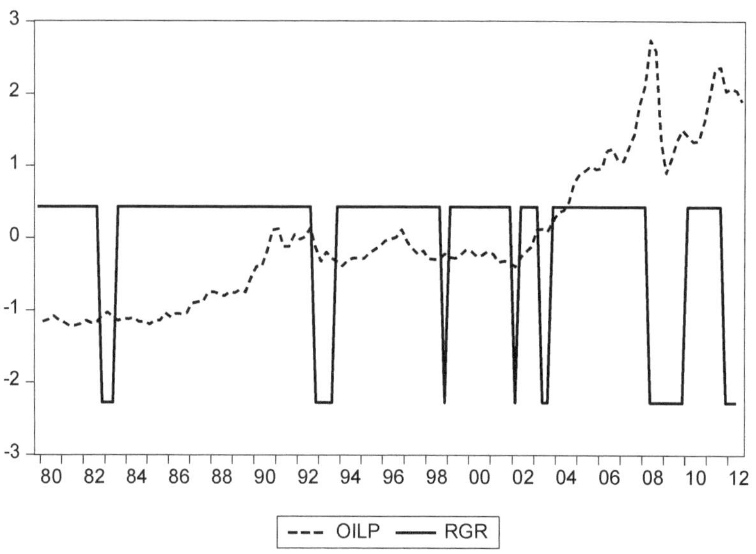

Figure 9.3: Italian GDP growth above 1% and oil prices (standardized variables)

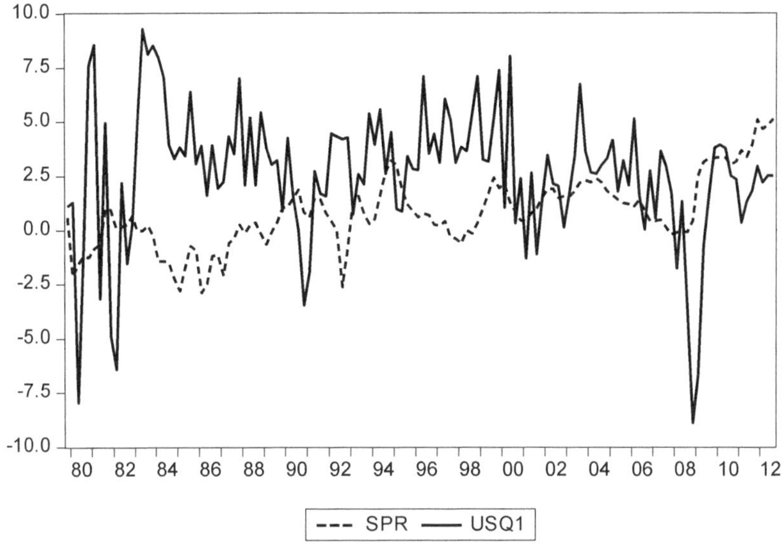

Figure 9.4: Term spread and US GDP growth

We also have time series reporting the difference between the long-term and short-term interest rates, the so-called term spread (SPR), and the quarterly GDP growth in the US (USq1). The spread is a leading indicator, because when it shrinks it means that future interest rates are expected to decline, which in turn is typically associated with a slowdown in growth. Figure 9.4 shows that in fact the spread remained substantially negative during the 1980s, apart from a brief stint in 1982, quite high for the first half of the 1990s, and then decreased until 2000. It also went down sharply at the start of the global financial crisis, and then increased gradually.

The rate of growth in the United States can be also regarded as a leading indicator, given the importance of the US as a trading partner of Italy. Lower US growth is associated with a decline in imports from Italy and thus a reduction in the Italian growth.

To check whether the oil price, term spread and US growth are good leading indicators for Italian growth, let us assess the explanatory power of their lags in a binary model for growth, with RGR as dependent variable.

The proper lag length of the indicators is not theoretically determined, but it has to be selected at the empirical level. A rather well performing LOGIT model is reported in Table 9.4 and specified as in Figure 9.5.

From the Table, we note that the three explanatory variables are highly significant and the coefficients have the expected sign (from 9.3.13 we know that the sign is the same as that of the effect of the regressor on the probability that the dependent variable is equal to 1). In addition, the large lag length of many regressors suggests that the arrival of a period of low growth can be forecast well in advance.
The model residuals, shown in Figure 9.6, indicate, however, that the model forecasts were not accurate in the first part of the sample and in the mid-1990s.

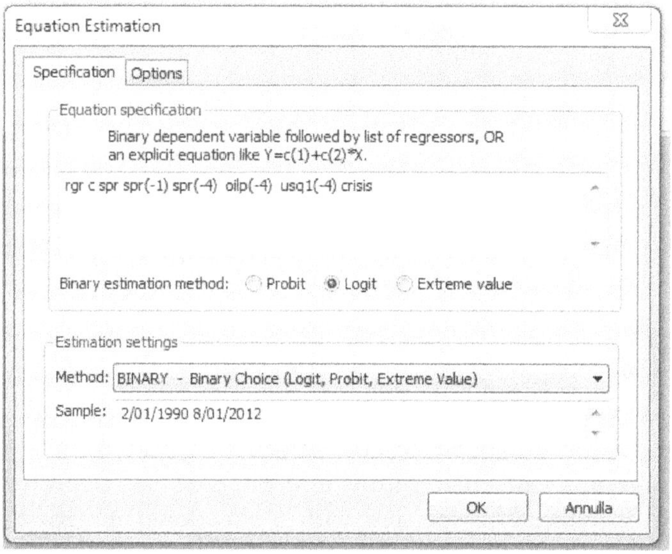

Figure 9.5: Specification of the LOGIT model for the binary variable RGR

Dependent variable: RGR
Method: ML - Binary Logit (Quadratic hill climbing)

Variable	Coefficient	Std. Error	z-Statistic	Prob.
C	4.560470	1.177350	3.873503	0.0001
SPR	-1.428913	0.523179	-2.731215	0.0063
SPR(-1)	1.102897	0.495917	2.223955	0.0262
SPR(-4)	0.727799	0.342689	2.123790	0.0337
OILP(-4)	-0.003759	0.001483	-2.535256	0.0112
USq1(-4)	0.262975	0.108258	2.429161	0.0151
CRISIS	0.891056	1.193749	0.746435	0.4554

McFadden R-squared	0.329419	Mean dependent var	0.833333
S.D. dependent var	0.374166	S.E. of regression	0.308238
Akaike info criterion	0.715387	Sum squared resid	11.30626
Schwarz criterion	0.872958	Log likelihood	-38.06936
Hannan-Quinn criterion	0.779403	Deviance	76.13871
Restr. Deviance	113.5414	Restr. log likelihood	-56.77071
LR statistic	37.40271	Avg. log likelihood	-0.302138
Prob(LR statistic)	0.000001		
Obs with Dep=0	21	Total obs	126
Obs with Dep=1	105		

Table 9.4: Estimated leading indicator, LOGIT model

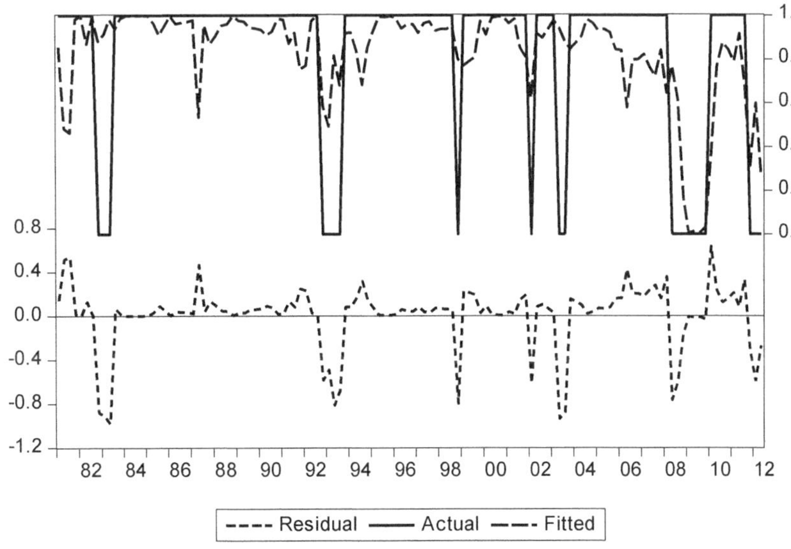

Figure 9.6: Residuals from the LOGIT model

We can calculate the estimated change in the expected value of the dependent variable with respect to a variation of the regressor in a given observation, as seen in the previous Sections. As an illustration, let us consider the estimates obtained in 2000q1 for the oil price of the previous year, entering the following EViews commands:

```
vector(7) o41   '(NB: obs 41 corresponds to 2000:1)
o41(1)=1
o41(2)=spr(41)
o41(3)=spr(41-1)
o41(4)=spr(41-4)
o41(5)=oilp(41-4)
o41(6)=usq1(41-4)
o41(7)=crisis(41)
vector
b5=@dnorm(@transpose(o41)*eq_logit.@coefs)*eq_logit.@coefs(5)
```

The result is very small and equal to -0.03%, whereby the appraisal indicates that at that time the oil price negatively affected growth, but only marginally so.

9.7 An empirical analysis of the sign of stock returns

Let us consider the (daily) monthly returns on the ENEL stock (an Italian energy company) from 2005 to 2012, reported in Figure 9.7 and contained in the workfile "example_enel_chap9_a.wf1".

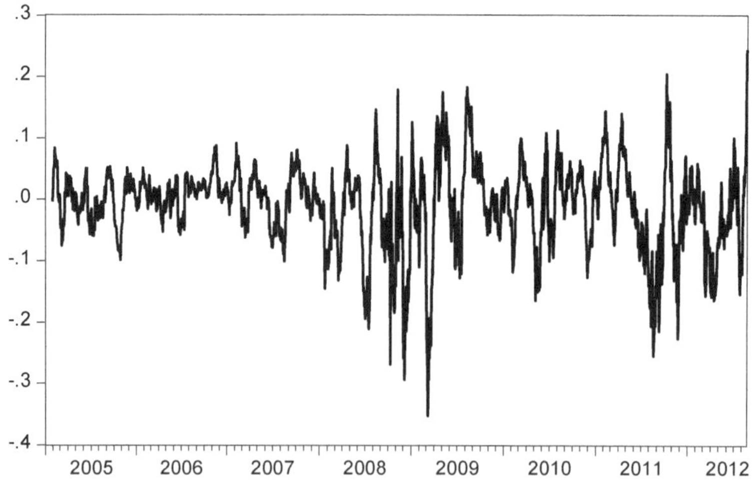

Figure 9.7: ENEL stock returns

We want to evaluate whether it is possible to predict this variable by using the available (lagged) information. In particular, we want to assess whether we can anticipate the increase or reduction in ENEL stock returns. Hence, we build a binary variable (BIN) equal to 1 when the stock price goes up (and then its return is positive) and equal to 0 when the price drops (and therefore the return is negative). As explanatory variables we choose the past stock returns, Earnings per Share (EPS), price-earnings ratio and the FTSE MIB (the stock market index of Borsa Italiana). As usual, we also include a "CRISIS" dummy variable, taking the value 0 until 20 June 2007, and 1 from 21 June 2007 until the end of the sample.
The estimated PROBIT model is reported in Table 9.5. The only significant variable is the past stock returns, whose coefficient changes with the crisis. But this is already an interesting result, because theoretically the sign should not be predictable in an efficient market. We must, however, better evaluate how much past returns help in forecasting future ones. To this aim, we can use the so-called "expectation-prediction table", shown in Table 9.6.

Dependent variable: BIN
Method: ML - Binary PROBIT (Quadratic hill climbing)
Sample (adjusted): 2/01/2005 8/21/2012
Included observations: 1971 after adjustments

Variable	Coefficient	Std. Error	z-Statistic	Prob.
C	0.007523	0.385123	0.019533	0.9844
R_ENEL(-1)	68.30107	5.048410	13.52922	0.0000
FTSE_MIB(-1)	1.01E-05	1.14E-05	0.887082	0.3750
ENEL_EPS(-1)	-0.097210	0.552987	-0.175791	0.8605
ENEL_PE(-1)	-0.013798	0.032793	-0.420767	0.6739
CRISIS*R_ENEL(-1)	-31.10835	5.420608	-5.738905	0.0000

McFadden R-squared	0.631462	Mean dependent var	0.503805
S.D. dependent var	0.500112	S.E. of regression	0.276042
Akaike info criterion	0.516969	Sum squared resid	149.7313
Schwarz criterion	0.533975	Log likelihood	-503.4731
Hannan-Quinn criterion	0.523218	Deviance	1006.946
Restr. deviance	2732.272	Restr. log likelihood	-1366.136
LR statistic	1725.326	Avg. log likelihood	-0.255440
Prob(LR statistic)	0.000000		
Obs with Dep=0	978	Total obs	1971
Obs with Dep=1	993		

Table 9.5: Estimated PROBIT model for the sign of returns of the ENEL stock

Expectation-Prediction Evaluation for Binary Specification
Equation: EQ_PROBIT_CRISI
Success cutoff: C = 0.5

	Estimated Equation			Constant Probability		
	Dep=0	Dep=1	Total	Dep=0	Dep=1	Total
P(Dep=1)<=C	**859**	92	951	0	0	0
P(Dep=1)>C	119	**901**	1020	978	993	1971
Total	978	993	1971	978	993	1971
Correct	859	901	1760	0	993	993
% Correct	87.83	90.74	89.29	0.00	100.00	50.38
% Incorrect	12.17	9.26	10.71	100.00	0.00	49.62
Total Gain*	87.83	-9.26	38.91			
Percent Gain**	87.83	NA	78.43			

Table 9.6: Forecast evaluation of the estimated PROBIT model

The columns called "Estimated Equation" classify the probabilities generated by the PROBIT model, while those entitled "Constant Probability" classify those generated by a model that includes only the intercept.

Let us first focus on the columns that show the PROBIT results (Estimated Equation). The first row counts how many times the probability (as calculated by the model) that the BIN binary variable is equal to 1 is less than the chosen "cutoff" (C = 0.5). In this example, the model predicted a low probability for the event "BIN = 1" (high for "BIN = 0") in 951 cases, and in such cases (in line with the forecast BIN = 0) the event "BIN = 1" occurred 92 times only, while the event "BIN = 0" occurred much more often, 859 times.

The second row counts how many times the probability that the BIN binary variable is equal to 1 is greater than the chosen "cutoff" (C). In this example, the model predicted a high probability for the event "BIN = 1" (low for BIN = 0) in 1020 cases, and the events "BIN = 1" (correct prediction) and BIN = 0 (wrong prediction) occurred 901 and 119 times, respectively.

The fourth row reports the values in bold in the first and second rows, and shows how many times the model worked, i.e., how many times the model returned a high probability for the event "BIN = 0" (and low for BIN = 1) when it actually occurred, and a high probability for the event "BIN = 1" (and low for BIN = 0) when again it actually occurred.

The fifth and sixth rows translate this information in percentage terms. We then see that the model has provided correct predictions in 89.29% of the cases, with a better performance when the returns are positive (BIN = 1) than

when they are negative (BIN = 1), the corresponding percentages are 90.74% and 87.83%.
Before analyzing the last rows of Table 9.6, let us consider the "Constant Probability" columns. As mentioned, they report the results of a model that includes the intercept only, which therefore makes the probability of the "BIN= 1" and "BIN = 0" events equal, in all periods, to their corresponding full-sample frequencies. Therefore, this model assigns to the "BIN = 0" event a probability equal to the number of observations with BIN=0, divided by the total number of observations, i.e. 978/1971. Similarly, the probability assigned to the "BIN = 1" event is equal to the number of observations of BIN equal to 1, divided by the total number of observations, i.e. 993/1971. Therefore, the probability that BIN is 0 is always less than the cutoff (0.5) and the model always predict BIN = 1. As a consequence, it correctly guesses all the 993 cases where in fact BIN = 1, but it misses *all* the cases where BIN = 0.
We are now ready to evaluate the gains we get from the PROBIT model with respect to the naïve model with the intercept only. The results are reported in the seventh row of the "Estimated Equation" column (Total Gain). Let us first consider the case BIN = 0. In this case, as we have seen, the naïve model systematically fails while the PROBIT model correctly anticipates the event in 859 cases. Thus, there is a gain of 859 cases out of 978, which is equal to 87.83% in percentage terms.
Conversely, if BIN = 1, the naïve model with intercept only discovers all 993 cases, while the correctly identified cases using the PROBIT are 901, with a loss of 933-901 = 92 out of 933 cases, equal to 9.26%.
In summary, moving from the naïve model with intercept only to the PROBIT specification yields a gain of 859 cases (those where BIN = 0) and a loss of 92 cases (those where BIN = 1), with a net gain of 859-92 = 767 cases. The overall gain is therefore substantial: 767 cases out of a total of 1971, which is equivalent to 38.91%.
The result we have just obtained is surprising as, according to the efficient markets theory, stock returns should be unpredictable. However, the sample period we have studied includes the global financial crisis, an event of such magnitude that could lead to question the efficient use of information by financial operators. We could then expect that in a quieter period the gains from the PROBIT model are lower. To see whether this is indeed the case, we use the workfile "example_enel_chap9_b.wf1", which contains data for the subperiod 2000 to 2005. With a specification similar to that in Table 9.5, see Table 9.7, indeed we get a slightly worse performance than for the previous sample, see Table 9.8.

Dependent variable: BIN
Method: ML - Binary PROBIT (Quadratic hill climbing)
Sample (adjusted): 5/16/2000 7/12/2005
Included observations: 1346 after adjustments

Variable	Coefficient	Std. Error	z-Statistic	Prob.
C	0.592643	0.488959	1.212050	0.2255
ROENEL(-1)	-1.760065	0.653176	-2.694625	0.0070
ITMIB30(-1)	-7.58E-06	6.21E-06	-1.220814	0.2222
ENEL_EPS(-1)	-0.311921	0.817396	-0.381603	0.7028
ENEL_PE(-1)	-0.018841	0.016745	-1.125169	0.2605

Mean dependent var	0.456166	S.D. dependent var	0.498260
S.E. of regression	0.496526	Akaike info criterion	1.376024
Sum squared resid	330.6078	Schwarz criterion	1.395359
Log likelihood	-921.0644	Hannan-Quinn criterion	1.383266
Restr. log likelihood	-927.7971	Avg. log likelihood	-0.684297
LR statistic (4 df)	13.46543	McFadden R-squared	0.007257
Probability(LR stat)	0.009212		

Obs with Dep=0	732	Total obs	1346
Obs with Dep=1	614		

Table 9.7: Estimated PROBIT model for the sample 2000-05

Dependent variable: BIN
Method: ML - Binary PROBIT (Quadratic hill climbing)
Sample (adjusted): 5/16/2000 7/12/2005,
Prediction Evaluation (success cutoff C = 0.5)

	Estimated Equation			Constant Probability		
	Dep=0	Dep=1	Total	Dep=0	Dep=1	Total
P(Dep=1)<=C	613	488	1101	732	614	1346
P(Dep=1)>C	119	126	245	0	0	0
Total	732	614	1346	732	614	1346
Correct	613	126	739	732	0	732
% Correct	83.74	20.52	54.90	100.00	0.00	54.38
% Incorrect	16.26	79.48	45.10	0.00	100.00	45.62
Total Gain*	-16.26	20.52	0.52			
Percent Gain**	NA	20.52	1.14			

Table 9.8: Forecast evaluation of the PROBIT model over the period 2000-2005

First, the naïve model with intercept only still fails to anticipate any instances when BIN = 1, it always predicts BIN = 0. Conversely, the PROBIT model provides a correct prediction in 54.90% of cases, but is it much better when returns decrease (83.74%) than when they increase (20.52%). In this dataset, therefore, moving from the naïve model to the PROBIT specification, there is a gain of 126 cases (those where BIN = 1) and a loss of 119 cases (those where BIN = 0), with a net gain of 126-119 = 7 cases. The gain becomes therefore minimal, just 0.52%, which confirms our hypothesis that the financial crisis can affect market efficiency.

The main concepts of this chapter

In this chapter we focused on econometric modelling of binary variables, which typically represent the choices of economic agents or alternative states of a given variable: $y_i = \{0,1\}$, $i=1,\ldots N$. As y_i takes only two values, it is $P_i = Prob(y_i = 1) = E(y_i) = x_i\beta$.

In a linear model for a binary variable, the errors are heteroscedastic by construction and, in addition, the GLS estimator does not constrain the fitted values to be in the [0,1] interval, even if they are probabilities. It is therefore preferable to model P_i as $P_i = \Pr\{\varepsilon_i^* > -x_i\beta\}$. Choosing as underlying density a Logistic or a standard Normal we get, respectively, the LOGIT or PROBIT models.

In the case of repeated observations, we can estimate $P_i = \Pr\{y_i = 1\}$ with $\hat{P}_i = \frac{r_i}{T_i}$, where r_i is the number of 1s in the T_i observations for the i-th unit.

In the LOGIT model, using the relationship $\log(\frac{P_i}{1-P_i}) = x_i\beta$ as approximated by a Taylor expansion, we get:

$$\log(\frac{\hat{P}_i}{1-\hat{P}_i}) \cong \log(\frac{P_i}{1-P_i}) + \frac{e_i}{P_i(1-P_i)} \cong x_i\beta + u_i,$$

with $E(u_i)=0$, $Var(u_i)=\dfrac{1}{n_i P_i(1-P_i)}$. We can then apply feasible GLS estimation in the model above, using $\hat{P}_i$ to estimate $Var(u_i)$. For the PROBIT model, we instead get

$$F^{-1}(\hat{P}_i) \cong F^{-1}(P_i) + \frac{e_i}{f\left(F^{-1}(P_i)\right)} = x_i\beta + u_i,$$

with $E(u_i)=0$, $Var(u_i)=\dfrac{P_i(1-P_i)}{n_i\left[f\left(F^{-1}(P_i)\right)\right]^2}$. We can again apply feasible GLS estimation, using $\hat{P}_i$ to estimate $Var(u_i)$.

We have then seen that for the PROBIT model it is:

$$\frac{\partial P_i}{\partial x_{ij}} = f(x_i\beta)\beta_j$$

and for LOGIT:

$$\frac{\partial P_i}{\partial x_{ij}} = \frac{\beta_j e^{-x_i\beta}}{\left[1+e^{-x_i\beta}\right]^2}.$$

When there are no repeated observations, but possibly also in that case, we can use Maximum Likelihood Estimation (MLE). It is:

$$\log(L) = \sum_{i=1}^{N} y_i \log[F(x_i\beta)] + \sum_{i=1}^{N}(1-y_i)\log[1-F(x_i\beta)],$$

the MLE estimators, $\hat{\beta}_{MLE}$, are those that maximize log(L).

In terms of fit, when the model is linearized, the coefficient of determination can be used to assess the explanatory capacity of the original model.

The nonparametric estimators for each unit, $\hat{P}_i$, defined above, are always consistent, but not necessarily efficient. The fitted values from LOGIT or PROBIT models, P_i^*, are instead consistent only if the model is correctly specified. $\hat{P}_i$ and P_i^* can then be used to construct the statistic

$\sum_{1}^{N} \frac{T_i(\hat{P}_i - P_i^*)}{P_i^*(1 - P_i^*)}$ that, under the null hypothesis of correct model specification, has a $\chi^2(N-k)$ asymptotic distribution.

If the model is estimated by MLE, a simple measure of its explanatory power is given by $1 - \frac{\log\left[L\left(\hat{\beta}_{2,MLE}, \hat{\beta}_{3,MLE}, \ldots, \hat{\beta}_{k,MLE}\right)\right]}{\log\left[L\left(\hat{\beta}_2 = 0, \hat{\beta}_3 = 0, \ldots, \hat{\beta}_k = 0\right)\right]}$, which is 0 if all non-constant regressors are not significant, and the more negative the better the explanatory power of the model.

Exercises

1) In the workfile "example_telecom_chap9.wf1", we find the daily series of monthly returns for the Telecom stock (R_TELECOM) from 2005 to 2012, where Telecom is an Italian company. You should check whether it is possible to predict the increases and decreases in the stock returns in period *t* (the BIN variable in the workfile) using information in period *t*-1. The set of possible explanatory variables includes the return (R_TELECOM), the FTSE MIB (FTSE_MIB)), the Earnings per Share (TEL_EPS), the Dividend Yield (TEL_DY) and the price-earnings ratio (TEL_PER). You should estimate a PROBIT model, as seen in this chapter for the ENEL case, evaluating and commenting on the so-called "expectation-prediction table".

2) Consider the model:

$$y_i = x_i\beta + e_i$$

where y_i is binary with values (0,1).

a) Explain what happens if you estimate the model by OLS.

b) Explain how to estimate the parameters using a LOGIT model and having a sample of repeated observations for each unit i.

3) Consider the following model structure for the errors of a model for a binary dependent variable:

$$\varepsilon = \begin{cases} 1 - x_i\beta & \textit{with probability } P_i = x_i\beta \\ -x_i\beta & \textit{with probability } 1 - P_i = 1 - x_i\beta \end{cases}$$

Derive the expected value and variance of the errors.

4) Consider the model:

$$y_i = x_i\beta + e_i$$

where y_i is binary with values (0,1) and a cross-sectional sample is available.

a) Propose a proper estimation method for the LOGIT model.

b) Propose a measure of goodness of fit.

c) Propose an alternative measure of goodness of fit that can be applied in the case of repeated observations.

5) Suppose that an individual *i* chooses her education between the three following possibilities:

- finishing her studies before the end of high school, $D_{1i} = 1$
- completing high school, $D_{2i} = 1$
- going to university, $D_{3i} = 1$

For example, an individual who completes high school but does not enroll for university generates the following observations: $D_{1i} = 0, D_{2i} = 1, D_{3i} = 0$.

Suppose that the individuals' choices are based on the utility function of education:

$$U_i = X_i\beta + \varepsilon_i \qquad \varepsilon_i \sim N(0,1),$$

so that an individual decides to leave high school if $X_i\beta + \varepsilon_i < c_1$; only complete high school if $c_1 \leq X_i\beta + \varepsilon_i < c_2$; enroll for university if $X_i\beta + \varepsilon_i > c_2$.

Write the likelihood function for this model.